Advances in Experimental Medicine and Biology

Volume 1407

Advances in Experimental Medicine and Biology provides a platform for scientific contributions in the main disciplines of the biomedicine and the life sciences. This series publishes thematic volumes on contemporary research in the areas of microbiology, immunology, neurosciences, biochemistry, biomedical engineering, genetics, physiology, and cancer research. Covering emerging topics and techniques in basic and clinical science, it brings together clinicians and researchers from various fields.

Advances in Experimental Medicine and Biology has been publishing exceptional works in the field for over 40 years, and is indexed in SCOPUS, Medline (PubMed), EMBASE, BIOSIS, Reaxys, EMBiology, the Chemical Abstracts Service (CAS), and Pathway Studio.

2021 Impact Factor: 3.650 (no longer indexed in SCIE as of 2022)

Youchun Wang
Editor

Pseudotyped Viruses

Springer

Editor
Youchun Wang
Chinese Academy of Medical Sciences &
Peking Union Medical College
Beijing, China

Institute of Medical Biology, Chinese
Academy of Medical Science & Peking
Union Medical College
Kunming, China

ISSN 0065-2598 ISSN 2214-8019 (electronic)
Advances in Experimental Medicine and Biology
ISBN 978-981-99-0112-8 ISBN 978-981-99-0113-5 (eBook)
https://doi.org/10.1007/978-981-99-0113-5

This Springer imprint is published by the registered company Springer Nature Singapore Pte Ltd.
The registered company address is: 152 Beach Road, #21-01/04 Gateway East, Singapore 189721, Singapore

Preface

The pseudotyped virus has been widely known in biological medical field. In fact, the concept of pseudotyped virus is different. In a broad sense, any recombinant virus which contains foreign gene regardless of having foreign protein on the surface of viral particle or not has been called pseudotyped virus. This kind of pseudotyped virus has wide usage, which can transduct foreign gene into targeted tissue or cell for gene therapy. It can be used as vector for developing vaccine and can also be used as diagnostic control for nucleic acid assay, and others. In narrow sense, the pseudotyped virus has protein of target virus on surface of parent virus with non-complete genome. Some cannot continually replicate after infecting cell, which are called non-replicating pseudotyped virus; others can replicate with hybrid genome of two different viruses at least, which are called replicating pseudotyped virus. Regardless of non-replicating pseudotyped virus or replicating pseudotyped virus, the main protein which mediates entry into cell and induces main immune response must be located on viral particle, and its function must be the same as an authentic virus. This kind of pseudotyped virus can be used as surrogate for authentic virus to study the biological functions of virus, detection of neutralizing antibody, screening viral entry inhibitors, and others. This book is focused on this kind of pseudotyped virus.

Although many viruses especially for emerging viruses have been constructed pseudotyped viruses and they have been widely used as surrogate for authentic viruses, we still face many technological problems. The different constructed strategies or vectors have been used for different viruses. The condition and parameters for construction and detection should be optimized case by case. Some viruses, for example, flavivirus, are still not constructed pseudotyped viruses after optimizing all strategies. Thus, we invited scientists who have worked on pseudotyped virus for many years and have much experience in this field to write this book. Our purpose is to review the progress or problems in pseudotyped virus by sorting out the published papers as much as possible by combining our experience. We hope the book can provide guidance to construct pseudotyped virus and also prospect for further development. Of course, as this field grows rapidly, the book may not cover all the progress.

I thank all contributors from my heart. Most of them are concentrating on research of SARS-COV-2 now. They had hard time reviewing all relevant papers during pandemic period of COVID-19 and contributed their experience. Their selfless dedication has enriched the book. I hope the book can provide help to all the readers.

Beijing, China
Kunming, China

Youchun Wang

Contents

Editor and Contributors

About the Editor

Youchun Wang is professor in the Institute of Medical Biology, Chinese Academy of Medicine Sciences & Peking Union Medical College, Kunming, China. He is former deputy director of the National Institutes for Food and Drug Control (NIFDC), co-director of the WHO Collaborating Center for Standardization and Evaluation of Biologicals, and director of the NMPA Key Laboratory for Quality Research and Evaluation of Biological Products. He holds an MD in Epidemiology from Peking Medical University and PhD in Virology from the University College London (UCL). He has mainly done research works on hepatitis virus, HIV, HPV, and emerging viruses such as Ebola virus, SARS-CoV-2, and others and has also developed new methods for quality control of viral vaccines and diagnostics. Since outbreak of SARS in 2003, he has focused on developing pseudotyped viruses for emerging viruses and other viruses and developed assays with those pseudotyped viruses to detect neutralizing antibody and screen anti-viral compounds. So far, more than 20 kinds of viruses with different serotypes, genotypes, and other variants have been constructed into pseudotyped viruses. The biological characteristics of those viruses had been comprehensively analyzed based on their pseudotyped viruses. He obtained several grants from the Wellcome Trust in the UK, Ministry of Science and Technology, National Foundation of Natural Science, and others in China. He has published more than 300 academic papers, out of which 110 were published in Cell, Nature, Science, Cell Research, Cell Host & Microbes, Nature Protocol, and other SCI journals as the first or corresponding author. He also holds several positions in academic associations, such as former Chairman of Medical Virology and Vice Chairman of Medical Microbiology and Immunology of Chinese Medical Association, former Vice Chairman of Chinese Laboratory Animal Association, former Chairman of Beijing Medical Virology, and Chairman of Vaccine Committee of Chinese Pharmacopoeia.

Contributors

Yu Bai Division of Hepatitis Virus & Enterovirus Vaccines, Institute for Biological Products, National Institutes for Food and Drug Control, Beijing, China
WHO Collaborating Center for Standardization and Evaluation of Biologicals, Beijing, China

Lianlian Bian Division of Hepatitis Virus & Enterovirus Vaccines, Institute for Biological Products, National Institutes for Food and Drug Control, Beijing, China
WHO Collaborating Center for Standardization and Evaluation of Biologicals, Beijing, China

Meina Cai Division of HIV/AIDS and Sex-transmitted Virus Vaccines, National Institutes for Food and Drug Control (NIFDC) and WHO Collaborating Center for Standardization and Evaluation of Biologicals, Beijing, China

Ruifeng Chen Immunotech Applied Science Limited, Beijing, China

Kelly A. S. da Costa Viral Pseudotype Unit, Medway School of Pharmacy, University of Kent and Greenwich at Medway, Chatham, UK

Lisha Cui Minhai biotechnology Co. Ltd, Beijing, China

Qianqian Cui Division of HIV/AIDS and Sex-transmitted Virus Vaccines, National Institutes for Food and Drug Control (NIFDC) and WHO Collaborating Center for Standardization and Evaluation of Biologicals, Beijing, China

Zhimin Cui Division of HIV/AIDS and Sex-transmitted Virus Vaccines, National Institutes for Food and Drug Control (NIFDC) and WHO Collaborating Center for Standardization and Evaluation of Biologicals, Beijing, China

Weijing Huang Division of HIV/AIDS and Sex-transmitted Virus Vaccines, National Institutes for Food and Drug Control (NIFDC) and WHO Collaborating Center for Standardization and Evaluation of Biologicals, Beijing, China

Chen Liang Lady Davis Institute, Jewish General Hospital, McGill Centre for Viral Diseases, Montreal, QC, Canada
Department of Medicine, McGill University, Montreal, QC, Canada

Zhenglun Liang Division of Hepatitis Virus & Enterovirus Vaccines, Institute for Biological Products, National Institutes for Food and Drug Control, Beijing, China
WHO Collaborating Center for Standardization and Evaluation of Biologicals, Beijing, China

Ziteng Liang Division of HIV/AIDS and Sex-transmitted Virus Vaccines, National Institutes for Food and Drug Control (NIFDC) and WHO Collaborating Center for Standardization and Evaluation of Biologicals, Beijing, China

Qianqian Li Jiangsu Recbio Technology Co., Ltd., Taizhou, China

Tao Li Division of HIV/AIDS and Sex-transmitted Virus Vaccines, National Institutes for Food and Drug Control (NIFDC) and WHO Collaborating Center for Standardization and Evaluation of Biologicals, Beijing, China

Shou Liu Division of HIV/AIDS and Sex-transmitted Virus Vaccines, National Institutes for Food and Drug Control (NIFDC) and WHO Collaborating Center for Standardization and Evaluation of Biologicals, Beijing, China

Si Liu Department of Gastroenterology, Beijing Friendship Hospital, Capital Medical University, National Clinical Research Center for Digestive Disease, Beijing Digestive Disease Center, Beijing Key Laboratory for Precancerous Lesion of Digestive Disease, Beijing, China

Caifeng Long Division of Monoclonal Antibody Products, National Institutes for Food and Drug Control (NIFDC) and WHO Collaborating Center for Standardization and Evaluation of Biologicals, Beijing, China

Annan Ming Department of Pathogen Biology, School of Clinical and Basic Medical Sciences, Shandong First Medical University & Shandong Academy of Medical Sciences, Jinan, Shandong, China

Li Min Department of Gastroenterology, Beijing Friendship Hospital, Capital Medical University, National Clinical Research Center for Digestive Disease, Beijing Digestive Disease Center, Beijing Key Laboratory for Precancerous Lesion of Digestive Disease, Beijing, China

Jianhui Nie Division of HIV/AIDS and Sex-transmitted Virus Vaccines, National Institutes for Food and Drug Control (NIFDC) and WHO Collaborating Center for Standardization and Evaluation of Biologicals, Beijing, China

Tingting Ning Department of Gastroenterology, Beijing Friendship Hospital, Capital Medical University, National Clinical Research Center for Digestive Disease, Beijing Digestive Disease Center, Beijing Key Laboratory for Precancerous Lesion of Digestive Disease, Beijing, China

Joanne Marie M. Del Rosario Viral Pseudotype Unit, Medway School of Pharmacy, University of Kent and Greenwich at Medway, Chatham, UK

Magan Solomon Lady Davis Institute, Jewish General Hospital, McGill Centre for Viral Diseases, Montreal, QC, Canada
Department of Medicine, McGill University, Montreal, QC, Canada

Wenjie Tan NHC Key Laboratory of Biosafety, National Institute for Viral Disease Control and Prevention, Chinese Center for Disease Control and Prevention, Beijing, China

Nigel J. Temperton Viral Pseudotype Unit, Medway School of Pharmacy, University of Kent and Greenwich at Medway, Chatham, UK

Lan Wang Division of Monoclonal Antibody Products, National Institutes for Food and Drug Control (NIFDC) and WHO Collaborating Center for Standardization and Evaluation of Biologicals, Beijing, China

Meiyu Wang Division of HIV/AIDS and Sex-transmitted Virus Vaccines, National Institutes for Food and Drug Control (NIFDC) and WHO Collaborating Center for Standardization and Evaluation of Biologicals, Beijing, China
Units Department of Laboratory Medicine, Sichuan Academy of Medical Sciences & Sichuan Provincial People's Hospital, School of Medicine, University of Electronic and Technology of China, Chengdu, China

Wenbo Wang Division of Monoclonal Antibody Products, National Institutes for Food and Drug Control (NIFDC) and WHO Collaborating Center for Standardization and Evaluation of Biologicals, Beijing, China

Wenbo Wang Division of Monoclonal Antibody Products, National Institutes for Food and Drug Control (NIFDC) and WHO Collaborating Center for Standardization and Evaluation of Biologicals, Beijing, China

Xiao Wang Department of Pathogen Biology, School of Clinical and Basic Medical Sciences, Shandong First Medical University & Shandong Academy of Medical Sciences, Jinan, Shandong, China

Youchun Wang Chinese Academy of Medical Science & Peking Union Medical College, Beijing, China
Institute of Medical Biology, Chinese Academy of Medical Science & Peking Union Medical College, Kunming, China

Jiajing Wu Beijing Yunling Biotechnology Co., Ltd, Beijing, China

Xi Wu Division of HIV/AIDS and Sex-transmitted Virus Vaccines, National Institutes for Food and Drug Control (NIFDC) and WHO Collaborating Center for Standardization and Evaluation of Biologicals, Beijing, China

Xing Wu Division of Hepatitis Virus & Enterovirus Vaccines, Institute for Biological Products, National Institutes for Food and Drug Control, Beijing, China
WHO Collaborating Center for Standardization and Evaluation of Biologicals, Beijing, China

Xueling Wu Cell Collection and Research Center, National Institutes for Food and Drug Control (NIFDC), and WHO Collaborating Center for Standardization and Evaluation of Biologicals, Beijing, China

Xueling Wu Cell Collection and Research Center, National Institutes for Food and Drug Control (NIFDC) and WHO Collaborating Center for Standardization and Evaluation of Biologicals, Beijing, China

Si-An Xie Department of Gastroenterology, Beijing Friendship Hospital, Capital Medical University, National Clinical Research Center for Digestive Disease, Beijing Digestive Disease Center, Beijing Key Laboratory for Precancerous Lesion of Digestive Disease, Beijing, China

Junxuan Xu Department of Gastroenterology, Beijing Friendship Hospital, Capital Medical University, National Clinical Research Center for Digestive Disease, Beijing Digestive Disease Center, Beijing Key Laboratory for Precancerous Lesion of Digestive Disease, Beijing, China

Yi Yang Department of Gastroenterology, Beijing Friendship Hospital, Capital Medical University, National Clinical Research Center for Digestive Disease, Beijing Digestive Disease Center, Beijing Key Laboratory for Precancerous Lesion of Digestive Disease, Beijing, China

Fei Yuan State Key Laboratory of Integrated Management of Pest Insects and Rodents, Institute of Zoology, Chinese Academy of Sciences, Beijing, China

Leiliang Zhang Shandong Provincial Hospital Affiliated to Shandong First Medical University, Jinan, Shandong, China
Department of Pathogen Biology, School of Clinical and Basic Medical Sciences, Shandong First Medical University & Shandong Academy of Medical Sciences, Jinan, Shandong, China

Li Zhang Division of HIV/AIDS and Sex-transmitted Virus Vaccines, National Institutes for Food and Drug Control (NIFDC) and WHO Collaborating Center for Standardization and Evaluation of Biologicals, Beijing, China

Nan Zhang Department of Gastroenterology, Beijing Friendship Hospital, Capital Medical University, National Clinical Research Center for Digestive Disease, Beijing Digestive Disease Center, Beijing Key Laboratory for Precancerous Lesion of Digestive Disease, Beijing, China

Aihua Zheng State Key Laboratory of Integrated Management of Pest Insects and Rodents, Institute of Zoology, Chinese Academy of Sciences, Beijing, China

Zehua Zhou Division of HIV/AIDS and Sex-transmitted Virus Vaccines, National Institutes for Food and Drug Control (NIFDC) and WHO Collaborating Center for Standardization and Evaluation of Biologicals, Beijing, China

Shengtao Zhu Department of Gastroenterology, Beijing Friendship Hospital, Capital Medical University, National Clinical Research Center for Digestive Disease, Beijing Digestive Disease Center, Beijing Key Laboratory for Precancerous Lesion of Digestive Disease, Beijing, China

Chapter 1
Pseudotyped Viruses

Youchun Wang, Zehua Zhou, Xi Wu, Tao Li, Jiajing Wu, Meina Cai, Jianhui Nie, Wenbo Wang, and Zhimin Cui

Abstract Pseudotyped viruses have been constructed for many viruses. They can mimic the authentic virus and have many advantages compared to authentic viruses. Thus, they have been widely used as a surrogate of authentic virus for viral function analysis, detection of neutralizing antibodies, screening viral entry inhibitors, and others. This chapter reviewed the progress in the field of pseudotyped viruses in general, including the definition and the advantages of pseudotyped viruses, their potential usage, different strategies or vectors used for the construction of pseudotyped viruses, and factors that affect the construction of pseudotyped viruses.

Keywords Vector · Neutralizing antibody · Virus entry · Inhibitor · Envelope protein · Self-assembled pseudotyped virus

---

Y. Wang (✉)
Chinese Academy of Medical Sciences & Peking Union Medical College, Beijing, China

Institute of Medical Biology, Chinese Academy of Medicine Sciences & Peking Union Medical College, Kunming, China
e-mail: wangyc@nifdc.org.cn; wangyc@imbcams.com.cn

Z. Zhou · X. Wu · T. Li · M. Cai · J. Nie · Z. Cui
Division of HIV/AIDS and Sex-transmitted Virus Vaccines, National Institutes for Food and Drug Control (NIFDC) and WHO Collaborating Center for Standardization and Evaluation of Biologicals, Beijing, China

J. Wu
Beijing Yunling Biotechnology Co., Ltd., Beijing, China

W. Wang
Division of Monoclonal Antibody Products, National Institutes for Food and Drug Control (NIFDC) and WHO Collaborating Center for Standardization and Evaluation of Biologicals, Beijing, China

Y. Wang (ed.), *Pseudotyped Viruses*, Advances in Experimental Medicine and Biology 1407, https://doi.org/10.1007/978-981-99-0113-5_1

Abbreviations

COVID-19	Coronavirus disease-19
Env	Envelope protein
EV 71	Enterovirus 71
FIV	Feline immunodeficiency virus
G	Glycoprotein
hCMV	Human cytomegalovirus
HDV	Hepatitis delta virus
HIV	Human immunodeficiency virus
HPV	Human papillomavirus
IRES	Internal ribosomal entry site
L	RNA-dependent RNA polymerase
L1	Major capsid protein
L2	Minor capsid protein
LTR	Long terminal repeats
M	Matrix protein
MHV	Mouse hepatitis virus
MLV	Murine leukemia virus
N	Nucleoprotein
P	Phosphoprotein
Pol	Polymerase
RCR	Replication-competent retrovirus
S	Spike protein
SARS-CoV-2	Severe acute respiratory syndrome coronavirus 2
SEAP	Secretory alkaline phosphatase
SIV	Simian immunodeficiency virus
VLP	Virus like particle
VSV	Vesicular stomatitis virus

1.1 Introduction

Viral function analysis, detection of viral neutralizing antibodies, screening of antiviral drugs, and others are highly dependent on the culture of authentic viruses. However, some viruses such as HPV cannot be easily cultured. Although some viruses can be cultured, they must be operated in biosafety level 3 or 4 laboratories, especially for emerging viruses. Meanwhile, the emerging viruses may not be easy to obtain before widely spread. Especially mutations for those viruses naturally occurred with high frequency, and their alternation of biological functions needed timely monitoring. All those disadvantages of authentic virus hindered the study of viral function and the development of vaccines and antiviral drugs.

Recently, pseudotyped viruses, an alternative to authentic viruses, have been developed and widely used for studying the viral biological characteristics and evaluating antiviral products. Especially, during the COVID-19 pandemic, the pseudotyped virus for SARS-CoV-2 has been widely used around the world and becomes a powerful tool for studying the SARS-CoV-2. The assays based on pseudotyped viruses for detecting neutralizing antibodies and antiviral drugs for SARS-CoV-2 were also widely used as surrogate assays as authentic virus assays [1–3]. In general, the pseudotyped virus is a virus like particle whose surface protein (the envelope protein of the enveloped virus or the capsid protein of the non-enveloped virus) differs from its core component [4, 5]. The targeted viral protein on the pseudotyped virus has a similar structure as on the live virus and can mediate the pseudotyped virus into susceptible cells. Because the pseudotyped virus has no intact genome of the virus, it only has a one-round infection in the susceptible cells. Compared to the authentic virus, an assay based on the pseudotyped virus has many advantages including (1) being easily constructed based on target protein sequences. If the target protein sequence is available in public, the sequence should be easily synthesized, and the pseudotyped virus should be constructed based on the synthesized sequence, not needing the authentic virus. (2) Some pseudotyped viruses have no complete genome sequence of the virus; it cannot continually replicate in the cell after infection. There is no biosafety concern. Thus, it can be operated in biosafety level 2 laboratories without needing a higher biosafety level laboratory. (3) The pseudotyped viruses normally have reporter genes in their genomes and could be easily quantitated by detecting reporter signal after infecting the cell. (4) The targeting protein including envelope or capsid protein on pseudotyped virus should have a similar structure and show similar function as those on authentic viruses.

Different laboratories around the world constructed a series of pseudotyped viruses by using a different system and also developed assays based on those pseudotyped viruses for research on the virus and evaluating the antiviral products [4, 5]. However, viruses belong to different families and have different biological characteristics. Thus, a variety of constructed strategies or vectors have been developed for those viruses. A suitable system may be selected based on viral biological characteristics before the construction of a pseudotyped virus. In general, for enveloped viruses, backbone vectors such as HIV or VSV vectors are quite often used for constructing the pseudotyped virus. For non-enveloped viruses, the self-assembled pseudotyped virus was constructed without the help of other vectors.

1.2 The Vectors for Enveloped Viruses

1.2.1 Lentiviral Vectors for Enveloped Viruses

Lentiviral vectors are often the favored packaging machine for enveloped pseudotyped viruses due to their excessive efficiency. These vectors had been by and large derived from human immunodeficiency virus (HIV) [4, 6], simian immunodeficiency virus (SIV) [7], feline immunodeficiency virus (FIV) [8], or murine leukemia virus (MLV) [9–11].

The lentivirus genome is mostly composed of gag, pol, and env genes. Additionally, lentiviral vectors consist of two regulatory genes (*tat* and *rev*) and four cogenes (*nef*, *vif*, *vpr*, and *vpu*), which encode proteins that play a necessary function in viral attachment, entry, replication, and release [12]. In addition, the retrovirus genome incorporates two cis-acting sequences, long terminal repeats (LTR), as well as elements for gene expression and transcription. Other key sequence elements include the packaging signal to direct the sequence-specific RNA packaging into newly shaped virions [13], Psi or Y, and the polypurine bundle (Ppt), the beginning point for the synthesis of positive-stranded DNA in the reverse transcription process [14, 15].

1.2.1.1 HIV Vector

The HIV packaging system is the most widely used pseudotyped virus packaging system. There were two-plasmid system [16], three-plasmid system [17–21], and four-plasmid system in the HIV pseudovirus packaging system [6].

For a two-plasmid system, one is expressing plasmid into which the membrane protein for the target virus can be cloned. The other plasmid is the HIV backbone plasmid, a modified pre-viral plasmid that contains the genes necessary for HIV replication. The two plasmids were cotransfected into HEK 293 T cells, and pseudotyped virus based on HIV vector could be obtained by collecting culture medium supernatant after a period of culture. The backbone plasmid of HIV-1 mainly uses pSG3 ΔEnv and pNL4–3 [18–21]. The pSG3 ΔEnv plasmid is modified from the core gene of the HIV SG3 strain and can express all the proteins except Env in the HIV genome [22, 23]. The pNL4–3 plasmid is constructed by the EcoRI segments from λL3 [both orientations], and λL4 was inserted into pN5'. The pNL4–3 plasmid cannot express HIV Env protein and Vpr protein and can express luciferase protein [24–26].

The HIV three-plasmid gadget generally contains a packaging plasmid, a reporter plasmid expressing the reporter gene, and an envelope-expressing plasmid. The packaging construct cannot produce the viral envelope and the protein Vpu. The packaging module [T] and adjoining sequences had been omitted from the 5′ LTR; however, the 5′ splice site was once preserved. The polyadenylation [poly[A]] site derived from the insulin was used instead of the 3′ LTR following the *nef* gene [21]. A CMV promoter had been inserted before the envelope gene in the envelope-expressing plasmid to drive its expression. The reporter plasmid employs the cis-acting sequences in HIV genome required for packaging. The Rev. response factor [RRE] flanked by means of splice signals had been covered in the transfer plasmid [27–32]. This diagram can expand packaging efficiency.

HIV is the cause of AIDS, which may cause some security problems. Some researchers use the SIV vector to develop a similar three-plasmid system.

SIV-derived lentiviral vectors derived from the simian immunodeficiency virus (Siv mac251) can effectively transfer mature human dendritic cells [33]. The Sivmac251 *gag-pol* gene is controlled by human cytomegalovirus (hCMV) direct early promoter and HIV-1 *rev* gene expression unit; the *rev* exons were fused and placed under the control of the 3-hydroxy-3-methylglutaryl-coa reductase (HMG) promoter, HMG intron I, and SV40 polyadenylation sequences.

1.2.1.2 Murine Leukemia Virus Vector

Murine leukemia virus (MLV), a member of gammaretrovirus, has been modified to generate an efficient pseudotyped virus packaging platform. MLV vector includes both the 5′ and 3′ LTR, the initiation signals for plus- and minus-strand DNA synthesis, and the RNA packaging signal. These sequences, by coincidence, are localized on both ends of the viral genome. Other viral sequences, mainly the coding region for the three structure proteins, are removed to avoid the generation of replication-competent retrovirus (RCR) and replaced with the gene of interest. To generate infectious vectors, the viral-encoded proteins will have to be supplied in trans. To facilitate vector production, many so-called packaging cell lines stably expressing MLV Gag, Pol, and Env proteins have been established [34]. However, only dividing cells could be efficiently transduced by MLV vector, which constrains their use in nonproliferating cells such as neurons, hematopoietic stem cells, myofibers, and hepatocytes [35].

1.2.1.3 FIV Vector

FIV vectors containing the CMV promoter enhancer dramatically increase the transcriptional activity of the FIV vectors in mammal cells [36]. In addition, the FIV vectors comprise at least a portion of the functional element of the FIV packaging sequence, located in a vicinity containing 5′ LTR, the first 350 bp of *gag* and *pol*. The FIV genome seems to encode only three regulatory and accessory genes, *vif*, *orf2*, and *rev*. [37, 38]. FIV Vif shows similar function as HIV and is essential for entry in some feline cells [39, 40]. Rev of FIV is additionally comparable to HIV Rev in enabling late-stage gene expression encoded by means of unspliced or single spliced mRNAs with cis-acting RRE [36, 37, 41]. FIV orf2 (open reading frame 2) encodes a trans-activating protein for FIV LTR, even though it is weaker than the trans-activating component for HIV Tat counterpart and does now not seem to have interaction with Tar-like factors [42–45]. However, the FIV accent genes, *vif*, and *orf2* are necessary for transmission for both dividing and nondividing cells, and Rev-RRE is crucial for the production of highly titrated vector particles.

1.2.2 The Vesicular Stomatitis Virus Vector

VSV is a non-segmented negative-sense RNA virus belonging to the order *Mononegavirales*, the family *Rhabdoviridae*. VSV comprises an 11 kb genome that encodes five structure proteins: the nucleoprotein (N), the phosphoprotein (P), the matrix protein (M), the glycoprotein (G), and the RNA-dependent RNA polymerase (L). Each gene is flanked by a conserved transcription start and end signal [46]. VSV has the advantage of broad cell tropism; rapid replication; high virus yields; small, easily manipulated genome; and the ability to accept 4–5 kb gene insertions which can be expressed at high levels [47]. So VSV genome had been engineered for use to study heterologous viral envelope protein, as vaccine vectors, and for oncolytic vectors.

1.2.2.1 VSV Reverse Genetics

Most applications based on VSV require the use of the reverse genetics system, which manipulates the VSV genome based on plasmids and recovered infectious viruses from plasmids. Since the genome of negative-sense RNA viruses must be encapsidated by N to be recognized by the viral RNA polymerase complex, which is composed of P and L, to initialize first round gene transcription which is required for subsequent cycles of transcription and genome replication, the reverse genetics system for VSV requires N, P and L provided in trans.

In 1995, VSV was recovered successfully by two labs independently [48, 49]. The recovery system is composed of a plasmid encoding an anti-genomic of VSV driven by the T7 promoter and terminated by hepatitis delta virus (HDV)-derived ribozyme sequence to produce accurate 3′ ends. N, P, and L were driven by the T7 promoter from the individual plasmids, which are called "helper" plasmids. Recombinant vaccinia virus expressing T7 RNA polymerase was cotransfected with all four plasmids into BHK21 cells. T7 polymerase initiates transcription of the full-length anti-genome of VSV and the viral protein mRNAs, which were translated by the host cell, producing viral proteins required for subsequent transcription and replication. The infectious virus was successfully rescued. Notably, the virus was only recovered from plasmid encoding anti-genome, rather than the genome of VSV. Further study revealed that T7 polymerase could not generate enough full-length transcripts from the genome sequence but generated truncate forms instead. The truncate transcripts could compete with full-length RNAs for N, P, and L proteins expressed from "helper" plasmids but could not be templates for replication, because they lack the 3′ terminus.

The recovery system was further modified by optimizing conditions under which viral RNP proteins are provided. N, P, and L are either expressed transiently from cotransfected plasmids or via stably transfected cell lines. Besides, though vaccinia virus expressing T7 polymerase facilitates the consistence of VSV, the requirement of additional biosafety facilities and the complete removal of the contaminating vaccinia virus during the recovery process make this approach inconvenient. A vaccinia virus-free recovery system had also been established. An internal ribosomal entry site (IRES) element was inserted between the T7 promoter and coding region of N, P, and L proteins, to increase their expression level. T7 polymerase was expressed either from a cotransfected plasmid or through stably transfected cell lines [50].

The development of VSV reverse genetics systems based on plasmid makes it possible to modify the VSV genome, which supplied the basis for the establishment of VSV as a broadly employed tool, for both vaccine and therapeutics.

1.2.2.2 Recombinant VSV Vectors

A study of the VSV genome revealed that each of its genes is flanked by a conserved transcription start sequence and a transcription termination sequence. And further study indicates that when a heterologous gene together with start and terminal sequences is inserted into the genome, it could also be expressed at a high level. These studies provided a theoretical basis for VSV vector manipulations.

Currently, there are two types of VSV-based recombinant vectors: replication-incompetent vectors and replication-competent vectors. In the replication-incompetent vectors, the VSV G gene is deleted. A heterologous glycoprotein from a different virus is provided in trans, resulting in the production of "pseudotyped" virions. The viruses are only capable of replicating for a single round unless glycoprotein is supplied continuously. There are several advantages to use the replication-incompetent vectors. Firstly, the restricted replication ability avoids biosafety concerns. Secondly, the glycoprotein of the pseudotyped virus can be easily changed by cotransfected plasmids expressing different glycoproteins and VSV delta G pseudovirus. In contrast, in replication-competent vectors, the G protein is either kept or replaced by heterologous glycoprotein in the genome. These viruses derived from replication-competent vectors are capable of propagating in cells while solely relying on the glycoprotein for assembly, budding, and subsequent infection. The replication-competent vectors are especially valuable for studying entry pathways and developing vaccines. Both the replication-competent and replication-incompetent vectors could further insert a gene encoding a fluorescent or luminescent reporter for quantitative and high-throughput analysis (Fig. 1.1).

Though VSV doesn't have stringent selectivity for the envelope proteins, pseudotyped virus harboring envelope proteins of flavivirus is hard to generate. This may be due to the different maturation of VSV and flavivirus envelope protein and their assemble process. G protein of VSV is translated in the rough endoplasmic reticulum and then travels through the secretory pathway to the plasma membrane. And Viral RNPs are transported to the site of budding and budding through sites in the host plasma membrane enriched in VSV G protein [51]. In contrast, the flavivirus virion assembly occurs on the ER membrane close to the sites of viral RNA replication [52, 53]. Once assembled, the immature virion traffic to the Golgi apparatus, and the pr peptide is cleaved by host furin. Then, the mature virion is

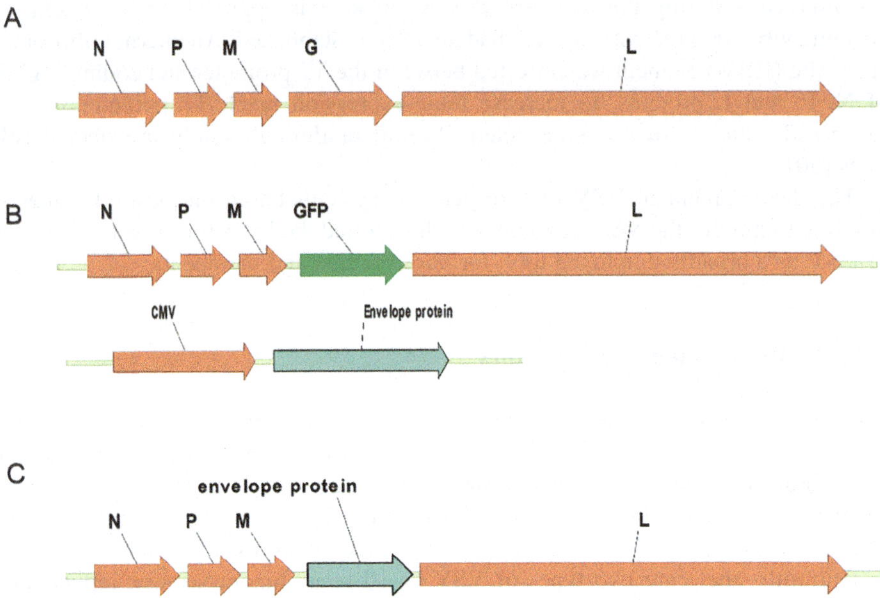

Fig. 1.1 Schematic of genuine VSV and VSV-based packaging system. (**A**) VSV genome expresses five proteins: nucleoprotein (N), phosphoprotein (P), matrix protein (M), glycoprotein (G), and the viral polymerase (L). (**B**) Generation of replication-incompetent VSV with a heterologous viral envelope. A reporter gene replaced the G protein gene of VSV. The incorporated viral envelope protein in the membrane is provided in trans by means of an expression plasmid. (**C**) Generation of a replicating VSV pseudotyped with a heterologous viral envelope. The G protein of VSV is replaced with a heterologous envelope protein to yield a replication-competent pseudotyped virus

released via the secretory pathway. The inefficient cooperation of flavivirus envelope protein to VSV led to the pseudotyped virus growing to low titers.

1.3 Self-Assembled Constructed Pseudotyped Viruses

Viruses can be classified into enveloped viruses and non-enveloped viruses concerning their different external structural features. The pseudotyped viruses for most enveloped viruses have been constructed by using the above vectors. However, some enveloped viruses can also be constructed as pseudotyped viruses by self-assembly, such as SARS-CoV and SARS-CoV-2. Almost all non-enveloped viruses are constructed pseudotyped viruses by self-assembly.

1.3.1 Self-Assembled Pseudotyped Virus for Enveloped Viruses

With the improvement of technical capability and in-depth research, a self-assembled pseudovirus system was gradually developed, in which all structural proteins including viral envelope proteins are co-expressed and co-assembled with replicons carrying reporter genes and viral packaging signals to form detectable pseudotyped viruses, such as the Nipah pseudotyped virus formed by self-assembly of matrix M protein and envelope F/G protein [54] and SARS-CoV-2 pseudotyped viruses carrying all four structural proteins [55]. The advantage is that in addition to the envelope protein, the structural protein of the target virus is also used, which can more realistically reflect the effects of mutations of different proteins at different locations on the physiological changes of the virus.

Severe acute respiratory syndrome coronavirus (SARS-CoV) is an enveloped positive-stranded RNA coronavirus with a full-length viral genome of approximately 31 kb, encoding structural, replication, and regulatory proteins that control viral attachment, entry, gene expression, RNA replication, assembly, and pathogenicity [56, 57]. Coronaviruses typically contain three to four structural proteins [58]. One of the most abundant glycoproteins is the membrane glycoprotein (M), a thrice-spanning membrane protein with a long COOH-terminal cytoplasmic structural domain inside the viral particle and a short NH2-terminal structural domain outside the viral particle. Spike glycoprotein (S) represents a type I membrane glycoprotein that mediates binding and attachment to target cells. In contrast, the nucleocapsid protein (N) is a 50 to 60 kDa internal phosphoprotein that interacts primarily with viral genomic RNA to form the viral core.

It has been shown that encoding SARS coronavirus membrane (M) and nucleocapsid (N) proteins are necessary and sufficient for the formation of virus-like particles [59]. Pseudo-particles can be formed when M and N protein expression plasmids that have been mammalian codon-optimized are cotransfected into human HEK293 cells with or without the S and/or the small envelope (E) proteins. And the absence of viral genomic RNA, protease, or viral polymerase did not affect the formation of SARS coronaviruses, suggesting that this is not necessary for the formation of pseudovirus particles. In contrast, the combined expression of structural proteins does not form pseudovirus particles in the absence of both M and N proteins. However, pseudotyped viral particles formed by M and N proteins alone could not mimic complete virus outgrowth. It was found by transmission electron microscopy that only the expression of the spike (S) glycoprotein was added again to form particles containing a corona-like halo similar to SARS coronavirus outgrowth and to give it a buoyancy density characteristic similar to that of true coronavirus. This suggests that the S glycoprotein is not only essential for viral fusion but is equally important for coronavirus integrity and expulsion outside the cell. Thus, the S, M, and N proteins of SARS coronaviruses are necessary and sufficient for pseudovirus assembly.

In addition, two viral proteins, the M and N proteins, are required for the formation of viral particles in coronaviruses. And surface S proteins are also required for the formation of largely mature coronavirus-like particles and germination. Similarly, three viral gene products, NP, VP35, and VP24, are required for the assembly of the Ebola virus nucleocapsid, of which two gene products, NP and VP35, appear to produce the nucleocapsid, while the third gene product, VP24, may play a catalytic role in the assembly process [59]. The above assembly is different from retroviruses and lentiviruses, which require only a single viral protein, the Gag precursor polyprotein, to line into VLP particles [60, 61]. Previous studies on animal coronaviruses have shown that the E protein is also important for virus assembly and may be related to the presence of structural interactions with the M protein, as in mouse hepatitis virus (MHV) where the M and E proteins are sufficient to form pseudovirus particles [62].

Given that structural protein expression of coronaviruses can self-assemble to form pseudotyped viral particles, there is potential for vaccine development, antibody screening, and development of antiviral drugs. However, specific selection for packaging reporter genes into viral particles is also required to form a detection system. For example, in live viruses, some coronaviruses selectively package positive-stranded genomic RNA (gRNA) into viral particles, rather than other forms of RNA, such as mouse hepatitis virus (MHV) and infectious gastroenteritis virus. Accordingly, further research was advanced to develop strategies for the development of structural proteins to package viral genomes carrying reporter genes into pseudotyped viruses. Lili Kuo et al. showed that the genomic packaging signal (PS) of MHV selectively packages viral genomic RNA into viral particles. MHV PS is a stem-loop RNA structure localized to the replicase gene region, which encodes the nonstructural protein 15 subunits of the viral replicase transcriptional complex [63]. MHV PS is highly conserved in betacoronavirus spectrum A, a subgenus that includes human coronavirus HKU1 (HCoV-HKU1), related viruses in bovine coronavirus (BCoV) and OC43 human coronavirus, all of which have related PS corresponding to MHV PS [64]. However, the structural and functional homology of PS does not extend to other lineages or other genera within betacoronavirus, such as the SARS-CoV of the B lineage [65]. Meanwhile, during coronavirus assembly, the N protein recognizes one or more RNA structures within ORF1ab, allowing the complete viral genome containing this sequence to be packaged into the viral particle, thereby excluding the inclusion of viral subgenomes and host transcripts [63]. Studies reported that different packaging signals were also identified in the nsp15 protein region of SARS-CoV and SARS-CoV-2, with PS580 (nt19785–20,348) predicted to be the packaging signal for SARS-CoV [55, 66]. Based on this, a study has found the key sequence signal PS9 (nt 20,080–21,171) for the correct packaging of the viral genome on the nonstructural protein 15 of SARS-CoV-2 by reporter gene prediction, based on which the four structural proteins can be assembled and the gene sequences carrying the reporter genes can be incorporated to form complete pseudotyped viral particles.

1.3.2 Self-Assembled Pseudotyped Virus for Non-enveloped Viruses

Assembly of non-enveloped virus pseudotyped viruses: the outermost layer of non-enveloped viruses is the protein capsid, which generally consists of several capsid proteins forming polyhedra. Therefore, envelope-free pseudotyped viruses are mainly constructed by two kinds of plasmid expression: one is the capsid protein expression plasmid, i.e., the viral capsid protein sequence is inserted into the eukaryotic expression plasmid; the other is the replicon plasmid, i.e., the reporter gene is used to replace the capsid protein sequence in the viral genome. By cotransfecting eukaryotic cells with the viral capsid protein expression and the RNA-transcribed replicon, a pseudotyped virus with the same capsid protein structure and antigenicity as the original virus and carrying the reporter gene inside can be self-packaged. Based on this principle, Minetaro et al. established three types of pseudotyped poliovirus, PV-1,2,36, which are in good agreement with the cytopathic effect (CPE) assays of live virus [67]. In addition, other researchers have established a method for the preparation of self-assembled pseudoviruses of EV-A71.

There are many types of non-enveloped viruses, and here we focus on the cases of self-assembled pseudotypes formed by enterovirus A71 and human papillomavirus.

Human papillomavirus (HPV) encodes two structural proteins, the viral major capsid protein (L1) and the minor capsid protein (L2), which together form the viral capsid structure [68]. Major capsid protein L1 is the major component of the HPV icosahedral capsid, accounting for 80% to 90% of the total capsid protein and wrapping the genome, while minor protein L2 is located in the long axis of the pentameric center and plays a major role in the packaging of viral genomic DNA and is involved in the process of viral infection of the host [69–71]. In 2004, a pseudotyped virus similar in structure to the natural live virus capsid was prepared using mammalian cells by Buck et al., mainly by co-expressing HPV capsid proteins (L1 and L2) and reporter plasmids, which self-assemble to form particles, and a pseudotyped virus-based neutralizing antibody assay was established based on this method [72]. Since the self-assembled pseudotyped viruses are structurally very similar to live viruses, their detection results are good consistent with the neutralization results of live viruses and can objectively reflect the protective effect of antibodies or vaccine sera, so the WHO has adopted them as the gold standard for HPV neutralizing antibody detection [73, 74]. Moreover, with the increase of research, the optimization of the HPV neutralization assay system is becoming more and more progressive, especially that the inserted reporter gene is gradually evolved from secretory alkaline phosphatase (SEAP) to luciferase reporter plasmid and fluorescent reporter plasmid, realizing the shortening of detection time and the progress of economic feasibility. Specifically, HPV structural genes and Gaussia luciferase (e.g., Gluc) or fluorescent protein (e.g., GFP) reporter plasmids were expressed by cotransfection of mammalian cells, self-assembled to form high-titer pseudotyped HPV viruses, and then the viruses were released by freeze-thawing cells [75].

In addition, HPV self-assembled pseudotyped viruses are also widely used in vaccine production, and the antigen of current commercial prophylactic HPV vaccines is mainly a virus like particle formed by HPV's main capsid protein L1 itself and/or L2 self-assembled VLPs, which has similar antigenic epitopes to the capsid icosahedron of natural live viruses and can induce specific humoral immune response [76].

Enterovirus A71 (EV-A71) is a non-enveloped positive single-stranded RNA virus with a genome packaged in an icosahedral capsid protein consisting of four proteins (VP1-VP4), of which VP1 is located on the outermost side of the capsid and is the immunodominant protein [77]. EV-A71 can cause hand, foot, and mouth disease (HFMD) and serious neurological complications in children, including aseptic meningitis and fatal encephalitis with cardiopulmonary complications [78]. Among them, pEV-A71(FY)-Luc self-assembling pseudotyped virus is an EV-A71 pseudotyped virus containing a firefly luciferase reporter gene that can only be infected in a single round. The expression assembly process was performed by first replacing the capsid region of EV-A71(FY) with the firefly luciferase reporter gene and inserting the T7 promoter at 59 nucleotides from the N-terminal position to construct the EV-A71 replicon in vitro. The EV-A71 replicon was sequentially transfected into cells by enzymatic linearization, RNA transcription, and purification and used to be packaged by the capsid protein and formed EV-A71 pseudotyped enterovirus. The eukaryotic expression plasmid pcDNA6.0 was then used to construct the EV-A71 capsid protein expression with the GFP reporter gene (ensuring transfection efficiency of the structural gene).

In conclusion, the similarity between the pseudotyped virus assembly of enveloped and non-enveloped viruses lies in the presence of one or several structural protein expressions that can self-assemble by folding to form viruslike particles. In addition, to form a pseudotyped virus carrying a reporter gene, the sequence of the reporter gene needs to be substituted or inserted into the sequence information of the target virus so that the virus can successfully package the sequence it recognizes as self into the particle. The difference is that the self-assembly of enveloped viruses is a bit more complex, with an envelope covering the structural capsid, requiring more plasmid expression in the formation of pseudotyped viral particles; it is also because of the presence of the envelope, which needs to complete budding based on the host cell membrane, so most of the pseudotyped enveloped viral particles formed will be discharged outside the cell and free in the supernatant for easier collection, while most of the non-enveloped proteins assembled are present in the cell. Therefore, in the production of self-assembled non-enveloped pseudoviruses, to obtain higher titers of pseudotyped virus, they can be obtained by ultrasonic cleavage or repeated freeze-thawing.

1.4 The Parameters and Conditions for Construction and Package of Pseudotyped Viruses

Efficient construction and high yield of the pseudotyped viruses are important for practical applications. However, many factors including different viruses, vectors, and packaging procedures are quite variable, which decided whether the viruses can be successfully constructed into pseudotyped viruses or produced high yield of pseudotyped viruses if successfully constructed.

1.4.1 Viral Biological Characteristic Affects Pseudotyped Virus Formation and Titer

Some viruses are non-enveloped. The pseudotyped viruses for them are normally constructed by self-assembly methods such as HPV and EV71. However, the packages of HPV and EV71 are also different as mentioned above.

Most enveloped viruses can be constructed into pseudotyped viruses by using different vectors such as HIV and VSV backbone vectors. However, different backbone vectors may be suitable for different viruses. The localization of viral packaging and maturation may largely determine the construction and yield of pseudotyped viruses. Especially, if the envelope proteins are located on the cell membrane surface, it is usually easy to generate high-yield pseudotyped viruses with the usage of either HIV or VSV packaging vectors (filovirus and rhabdovirus). If the glycoprotein is anchored in the membrane of organelles (endoplasmic reticulum or Golgi complex), only the VSV packaging vector can be used to construct pseudotyped viruses. Thus, if the localization of viral packaging from the membrane of organelles into the surface of the cell membrane is artificially changed, the pseudotyped virus could be easy to obtain. In general, three methods could be used for changing their localization. One is to make key mutations on envelop protein. For example, insertion of the N-glycosylation site at N204 induced the transportation of RABV G protein from perinuclear space to cell surface membrane [77], and two-point mutations in the glycoprotein of HTNV: I532K in the Gn and S1094L in the Gc direct HTNV Gn/Gc from the Golgi complex to the cell membrane surface, responsible for the dramatically enhanced Gn/Gc incorporation into VSV pseudotyped viruses [78]. The second is to truncate the cytoplasmic region of the envelope protein, and the intracellular accumulation of envelope protein could be reduced, thereby facilitating the assembly of envelope proteins with the vector core proteins. For example, shortened SARS-CoV S protein with a deletion (Δ) of 8 to 39 amino acids in cytoplasmic tail demonstrated a high titer of SARS-CoV S pseudotyped MLV vector in Δ19 and Δ26 [79], and Δ13, Δ19, and Δ21 in the cytoplasmic tail of SARS-CoV-2 S protein yield high titer of pseudotyped virus due to the enhancement of S surface production and virion incorporation [80–82]. However, truncation of the cytoplasmic region of the envelope protein may have negative consequences, such as altering the intrinsic structure of the surface domain, which may affect pseudotyped virus function and antigenic phenotype. The third is to replace the cytoplasmic tail of targeted envelope proteins with the envelope tail of the corresponding vector HIV-1, VSV, or MLV-A. This method can make the same localization of targeted envelope protein as a backbone vector so that the pseudotyped virus could be easily constructed and produce a high yield. For example, the wild-type HIV-1 Env protein is poorly incorporated into VSV particles, and this could be rescued by replacement of its cytoplasmic tail with that of VSV G [83–85]. The chimeric rabies G protein, which replaces the cytoplasmic region with the VSV G protein, shows increased incorporation onto HIV-1 lentiviral particles and has increased infectivity in vitro in 293 T cells.

Moreover, since the signal peptide of envelope protein is a key determinant for membrane targeting and insertion, optimization or substitution of the signal peptide could also affect the packaging of pseudotyped viruses. For example, the signal peptide of JEV ME protein was replaced with a 54-amino-acid signal peptide from VSV G, enhanced packaging efficiency, and generated a higher number of JEV E pseudotyped lentiviral vectors [86]. The signal peptide of rabies virus G protein was replaced with a 20-amino-acid signal peptide from the heavy chain of human IgG, significantly improving the expression level of the G protein with native conformation by more than 1000-fold [87]. Similarly, the signal peptide of the C-terminal-deleted SARS-CoV-2 S protein increased nine amino acids in the N-terminal or is replaced with the mouse IgGk signal peptide to promote recognition and increase protein levels and finally increase the pseudotyped virus production [77, 83, 86]. However, it should be noted that changes in signal peptides may influence the protein folding and/or glycosylation pattern that impacts antibody-mediated neutralization and receptor binding.

However, some enveloped viruses are still more difficult to obtain pseudotyped viruses such as dengue viruses and yellow fever virus because the structures of their envelop proteins are quite complex.

1.4.2 *The Effects of Envelope Protein Expression*

The increased expression of the envelope protein could substantially improve the pseudovirus titers. Several modifications for envelope protein can increase the expression level. The first is to optimize the codon of the targeted envelope gene, which can drastically increase the expression level, and pseudotyped viruses had been easily constructed. So far, most envelope genes in published papers had been optimized. For example, when constructing pseudotyped virus particles by HIV-1 or VSV packaging systems, SARS-CoV-2, SARS-CoV, MERV-CoV, and influenza virus all utilized codon-optimized target envelop protein gene to increase pseudotyped virus titers [81, 88–92]. And the HPV or RHDV which are self-package forming VLP both increase the pseudotyped virus yield by codon optimization [93, 94]. The second is to make some mutation on envelope protein which can increase expression levels too. For example, compared to the unmodified Ebola virus G protein, lentiviral vectors pseudotyped with the Ebola virus G protein variants, which contain a mutation in the heavily O-glycosylated extracellular domain (amino acids 309 to 489) and the mucin-rich domain, possessed higher titers [95, 96]. Similarly, the single point mutations of D614G, V367F, and N501Y can increase the titer of the SARS-CoV-2 pseudovirus [97–100]. The third is to use a higher efficiency expression vector or modify available plasmid for expressing envelope protein, such as adding Kozak sequence, using different promoters. For example, our lab developed a modification backbone plasmid pSG3ΔEnv.CMVFluc of the HIV packaging system that can improve the pseudoviral yield by 100-fold [101]. In addition, compared with pFIN-EF1α-GFP-2A-mCherH-WPRE, p'HR.cppt.3′ 1.2 kb-UCOE-SFFV-eGFP, and pTYF-EF1α-eGFP vectors, the backbone plasmid pTYF-CMV-eGFP and pTYF-CMV (β-globin intron)-eGFP of the lentiviral packaging system can yield high and prolonged protein expression [102]. Similarly, when using the CMV promoter instead of CAG (a hybrid promoter consisting of a CMV enhancer, beta-actin promoter, splice donor, and a beta-globin splice acceptor), HIV packaging system with dual reporter genes can significantly increase SARS-CoV-2 pseudovirus particles.

1.4.3 The Effects of Packaging System

As mentioned above, pseudotyped viruses for some viruses can be constructed by using different backbone vectors. There are different packaging systems, mainly including the lentiviral vector packaging systems (human immunodeficiency virus (HIV-1), simian immunodeficiency virus (SIV), murine leukemia virus packaging system (MLV), and feline immunodeficiency virus (FIV)) and the vesicular stomatitis virus packaging system (VSV). To date, many pseudotyped virus particles have been obtained by different packaging systems. For example, the SARS-CoV-2 has been successfully constructed by HIV-1, VSV, and MLV packaging systems [99]. The rabies virus and lyssavirus have been successfully constructed using HIV-1 and VSV packaging systems [103]. Pseudotyping of HIV-1-, VSV-, and MLV-based vectors with the GP derived from the LCMV had been successfully obtained [104]. FIV- and VSV-based vectors pseudotyped with the GPs derived from RRV had been obtained [95]. However, the choice of backbone vectors can significantly affect viral yield. For example, compared with HIV-1 packaging system, the rabies virus, lyssavirus, and gammaretrovirus pseudovirus particles constructed by the VSV packaging system have significantly higher pseudotyped virus yield [95, 103]. As for lymphocytic choriomeningitis virus, the titer of the pseudotyped virus as HIV-1 backbone plasmid is lower than MLV and VSV vectors by at least 100 times, while LCMV-pseudotyped lentiviral vectors to construct that pseudovirus revealed vector titers similar to those obtained with MLV and VSV-G packaging system [105]. When comparing MLV and HIV packaging systems, the pseudovirus generated using MLV performed better than HIV as packaging vector in

293 T cells; pseudotyped influenza virus was successfully produced using the MLV packaging system [106]. The similar conclusion was reached by Cosset et al. for the pseudotyped LASV production [107]. In addition, according to our experiences, pseudotyped LASV packaged based on the VSV vector showed a greater titer than that employing HIV system. Furthermore, the hantavirus could only be pseudotyped using the VSV other than HIV system.

1.4.4 Effects of Proteases

The increase of pseudotyped virus titer can be achieved by increasing the expression of the envelope protein. Codon optimization in mammalian cells of targeted envelope proteins, the introduction of enhancers, site mutations, etc. are all to improve the expression of targeted envelope proteins [77, 88]. However, for some viruses, the successful formation of their pseudotyped forms also depends on certain key proteases. For example, the Furin protease cleavage site (682-RRAR-685) is an interesting sequence that exists on both SARS-CoV and SARS-CoV-2 spike proteins [108, 109]. For SARS-CoV-2, not only the variation of the cleavage site (R682Q) was caused in the process of passage on the Vero E6 cell line but also occurred in the process of passage in human patients [110, 111]. All the above changes caused the virus to abandon the 682-RRAR-685 cleavage site in natural selection. Therefore, the researchers mutated 682-RRAR-685 into 682-QRAR-685 on SARS-CoV-2 spike protein to eliminate the cleavage of Furin, which enhanced the production of SARS-CoV-2 pseudotyped virus by 5- or 3.5-fold in cells lacking or containing the serine transmembrane protease 2 (TMPRSS2), respectively [99].

Unlike the packaging of the SARS-CoV-2 pseudotyped virus, there must be enzyme cleavage during the formation of the influenza pseudotyped virus. Cleavage of peptide precursor HA0 into HA1 and HA2 subunits is a necessary step for the influenza virus to obtain fusion activity and infectivity [112]. To simulate the proteolysis characteristics of influenza natural host cells, protease coding plasmids can be transfected together with other necessary plasmids to induce transient expression while producing pseudotyped virus. For example, through cotransfection of TMPRSS2 expression plasmid, HIV backbone vector, and HA expression plasmids, high-titer H3N2 (A/Udorn/307/1972), H4N6 (A/duck/Czechoslovakia/1956), H10N7 (A/chicken/Germany/N49), H14N5 (A/mallard/Astrakhan/263/1982), H15N9 (A/shearwater/West Australia/2576/1979), and H7 LPAI (A/chicken/Italy/1082/1999) pseudotyped particles were obtained, respectively [19]. Similarly, the addition of foreign proteins such as the human airway trypsin (HAT) or TMPRSS4 can also play the same role in the maturation of influenza pseudotyped viruses [113, 114].

1.4.5 Selection of Cell Lines for Pseudovirus Packaging and Detection

The following factors should be considered in the selection of pseudovirus packaging cell lines: [1] the cell lines should show higher transfection efficiency for packaging backbone vectors or viral protein expression vectors. For example, the transfection efficiencies of CHO and HEK 293 T cells for lentiviral vector-based transient transfection of three plasmids were more than 95%, which is the highest compared to HepG2, MCF-7, mef, and Jurkat cells [115]. Vero-E6 produced the highest yield of VSVdG-SARS-CoV-2-S del18 pseudotyped virus compared to BHK21 and 293 T cells [116]. Some engineered cell lines may be more suitable to package some pseudotyped viruses. For example, S Morita [117] established a Plat-E high-efficiency retrovirus packaging cell line based on the 293 T cell line by inducing the expression of the EF1a promoter in cells. Plat-E cells are better in efficiency, stability, and safety, ensuring the high and stable expression of viral structural proteins [3]. According to the characteristics of the pseudotyped virus, select the cell line with high efficiency of the corresponding transfection system. For example, the common cell line of influenza pseudovirus in LV vector and MLV packaging system is HEK293T, which is very sensitive to transfection, while the common cell line in VSV vector packaging system is 293 FT. HEK 293 T/17 is also widely used in the production of high-titer influenza pseudotyped virus [112].

The sensitive cell lines must be selected when the assay based on the pseudotyped virus is established. Some pseudotyped viruses are susceptible to a variety of cell types from human or nonhuman hosts. More suitable cell lines should be screened before the established assay. For example, Huh7 cells showed a high ability to infect the MERS-CoV pseudotyped virus among more than ten cell lines and were selected for assay-based MERS-CoV pseudotyped virus [118]. Some pseudotyped viruses may not be susceptible to available cell lines. The engineering cell lines, which expressed the corresponding receptor, should be constructed. For example, overexpressing α-2,6 sialic acid in MDCK cells for H5N1 [119], and overexpressing the ACE2 receptor in BHK21 cells for SARS-CoV-2 could dramatically enhance the infectivity of the corresponding pseudotyped viruses[116]. Sometimes cell lines with overexpression of proteinase, which increase the virus infection, were also constructed, for example, overexpression of TMPRSS2 enzyme in HEK 293 T cell line. The stably expressing ACE2 and TMPRSS2 [120] were used in the detection of the SARS-CoV-2 neutralizing antibody. Thus, before developing assay based on pseudotyped viruses with different packaging systems, the detection cell lines need to be comprehensively analyzed in combination with the above conditions (Table 1.1).

Table 1.1 Packaging system, packaging cells, and detection cell lines of different pseudotyped viruses

Number	Pseudovirus type	Packaging system [102]	Packaging cell line	Common neutralization detection cell line	References
1	SARS-CoV-2, severe acute respiratory syndrome coronavirus 2	HIV-1, MLV, VSV	HEK 293 T, Vero-E6	Huh 7, 293 T-hACE2, hACE2-TMPRSS2 BHK21-hACE2	[116], [121], [122]
2	SARS-CoV, severe acute respiratory syndrome coronavirus	HIV-1, MLV, VSV	HEK 293 T	Huh 7, 293 T-hACE2	[123]
3	MERS-CoV, Middle East respiratory syndrome coronavirus	HIV-1, MLV, VSV	HEK 293 T	Huh 7, 293 T-hACE2	[118]
4	HIV, indicates human immunodeficiency virus	HIV-1	HEK 293 T	Tzm-bl	[124]
5	RV, rabies virus	HIV-1, VSV	HEK 293 T	HEK 293 T	[125]
6	EBOV, Ebola virus	HIV-1, MLV, VSV	HEK 293 T	HEK 293 T	[126]
7	Influenza virus	HIV-1, MLV, VSV	HEK 293 T/17, 293FT	MDCK, 293 TA, MDCK-London, MDCK-SIAT	[112]
8	Hantavirus	VSV, MLV	HEK 293 T	Vero E6	[127]
9	Nipah virus	HIV-1, VSV	HEK293T	HeLa-USU	[128]
10	hRSV, human respiratory syncytial virus	HIV-1	HEK293T	Hep 2	[129]
11	LASV, Lassa fever virus	HIV-1, MLV	HaCaT	Vero	[130]
12	MARV, Marburg virus	HIV-1	HEK293T	HEK 293 T	[131]
13	CHIKV, Chikungunya virus	HIV-1, VSV	HEK293T	HEK 293 T, HeLa, HBMEC, HepG2	[132]
14	CCHFV, Crimean-Congo hemorrhagic fever virus	VSV	Lenti-X 293 T	BHK-21, HeLa, HepG2, AsPC-1	[133]

1.4.6 The Effect of Packaging Conditions

During the packaging procedure for the pseudotyped virus, several conditions should be optimized, including choice of ratio of different plasmids, transfection reagent, adding sodium butyrate and proteinase, etc. Normally several plasmids should be cotransfected into cell lines for packaging pseudotyped virus. The optimized ratio of the cotransfected plasmids can significantly increase the yield of the pseudotyped virus. For example, when using HIV NL3.4LucR-E- as backbone plasmid to construct influenza H5N1 and H3N2 pseudotyped viruses, the titer of pseudotyped viruses was highest at membrane plasmid: backbone plasmid = 1:3 [88]. Choice of suitable transfected reagents for different plasmids is also important. For example, PEI is selected as the optimal transfection reagent for its less cytotoxicity when packaging HPV pseudotyped virus. PEI can promote the intracellular internalization and subsequent intracellular transport of polymer, improve the transfection efficiency of the binding form (DNA aggregation and protection) and free form, and is easier to be extended to larger production [134]. Adding some proteinase in cell lines can also increase the yield of pseudotyped viruses. For example, when the influenza virus is fused with cells, the full-length HA (HA-0) needs to be cleaved into the globular head (including receptor binding site) HA-1 and elongated stem HA-2 by protease; it is necessary to add exogenous trypsin-like protease to activate the pseudotyped virus in the packaging of influenza pseudotyped particles [19, 135, 136]. Adding sodium butyrate can also increase the yield for some pseudotyped viruses, for example, by optimizing the condition by adding sodium butyrate 21 hours later to induce gene expression [137]. The cell line can also influence pseudotyped virus yield in the packaging system. For example, pseudotyped lentiviruses are produced by transient cotransfection, usually of the 293 T cells [4]. The packaging system can also influence pseudotyped virus, but the packaging system is not invariable and needs to be optimized according to virus type. For example, the rabies pseudotyped virus prepared by the VSV vector was better than the one prepared by the HIV-1 vector, and the pseudotyped virus could achieve a higher titer on the 293 T cell lines [103]. Similarly, using the HIV-1 vector, pseudotyped virus particles are successfully constructed for the Marburg virus and Ebola Zaire (EboZ) viruses [138, 139]. Pseudotyped virus harvest times can also be optimized to increase pseudotyped virus titers. For example, the harvest time of pseudotyped virus in the HIV-1 packaging system is at least 48 h, while that in the VSV packaging system is 24 h. In addition, pseudotyped virus titer can be improved by physical technique, among which the most commonly used methods are ultracentrifugation and ultrafiltration [140, 141]. When there is a large volume of virus-containing cell culture supernatants to be treated, it can be considered the method of membrane-based anion exchange chromatography or precipitation [141].

In general, several strategies for constructing pseudotyped viruses have been developed, and factors that affected the construction of pseudotyped viruses have been optimized. More than ten genera of viruses can be successfully constructed as pseudotyped viruses. However, some of their variants and some viruses from other genera cannot be constructed as pseudotyped viruses. Therefore, their biological characteristics should be deeply understood, and new strategies for developing pseudotyped viruses should be developed for them.

References

1. Bewley, K.R., et al.: Quantification of SARS-CoV-2 neutralizing antibody by wild-type plaque reduction neutralization, microneutralization and pseudotyped virus neutralization assays. Nat. Protoc. **16**, 3114–3140 (2021). https://doi.org/10.1038/s41596-021-00536-y
2. Nie, J., et al.: Establishment and validation of a pseudovirus neutralization assay for SARS-CoV-2. Emerg Microbes Infect. **9**, 680–686 (2020). https://doi.org/10.1080/22221751.2020.1743767
3. Nie, J., et al.: Quantification of SARS-CoV-2 neutralizing antibody by a pseudotyped virus-based assay. Nat. Protoc. **15**, 3699–3715 (2020). https://doi.org/10.1038/s41596-020-0394-5
4. Li, Q., Liu, Q., Huang, W., Li, X., Wang, Y.: Current status on the development of pseudoviruses for enveloped viruses. Rev. Med. Virol. **28** (2018). https://doi.org/10.1002/rmv.1963
5. Sanders, D.A.: No false start for novel pseudotyped vectors. Curr. Opin. Biotechnol. **13**, 437–442 (2002). https://doi.org/10.1016/s0958-1669(02)00374-9
6. Dull, T., et al.: A third-generation lentivirus vector with a conditional packaging system. J. Virol. **72**, 8463–8471 (1998). https://doi.org/10.1128/JVI.72.11.8463-8471.1998
7. Negre, D., et al.: Lentiviral vectors derived from simian immunodeficiency virus. Curr. Top. Microbiol. Immunol. **261**, 53–74 (2002). https://doi.org/10.1007/978-3-642-56114-6_3
8. Poeschla, E., et al.: Identification of a human immunodeficiency virus type 2 (HIV-2) encapsidation determinant and transduction of nondividing human cells by HIV-2-based lentivirus vectors. J. Virol. **72**, 6527–6536 (1998). https://doi.org/10.1128/JVI.72.8.6527-6536.1998
9. Sharma, A., et al.: BET proteins promote efficient murine leukemia virus integration at transcription start sites. Proc. Natl. Acad. Sci. U. S. A. **110**, 12036–12041 (2013). https://doi.org/10.1073/pnas.1307157110
10. De Ravin, S.S., et al.: Enhancers are major targets for murine leukemia virus vector integration. J. Virol. **88**, 4504–4513 (2014). https://doi.org/10.1128/JVI.00011-14
11. El Ashkar, S., et al.: BET-independent MLV-based vectors target away from promoters and regulatory elements. Mol Ther Nucleic Acids. **3**, e179 (2014). https://doi.org/10.1038/mtna.2014.33
12. Federico, M.: From lentiviruses to lentivirus vectors. Methods Mol. Biol. **229**, 3–15 (2003). https://doi.org/10.1385/1-59259-393-3:3
13. Watanabe, S., Temin, H.M.: Encapsidation sequences for spleen necrosis virus, an avian retrovirus, are between the 5′ long terminal repeat and the start of the gag gene. Proc. Natl. Acad. Sci. U. S. A. **79**, 5986–5990 (1982). https://doi.org/10.1073/pnas.79.19.5986
14. Rattray, A.J., Champoux, J.J.: Plus-strand priming by Moloney murine leukemia virus. The sequence features important for cleavage by RNase H. J. Mol. Biol. **208**, 445–456 (1989). https://doi.org/10.1016/0022-2836(89)90508-1
15. Soneoka, Y., et al.: A transient three-plasmid expression system for the production of high titer retroviral vectors. Nucleic Acids Res. **23**, 628–633 (1995). https://doi.org/10.1093/nar/23.4.628
16. Blomer, U., et al.: Highly efficient and sustained gene transfer in adult neurons with a lentivirus vector. J. Virol. **71**, 6641–6649 (1997). https://doi.org/10.1128/JVI.71.9.6641-6649.1997
17. Ao, Z., et al.: Characterization of a trypsin-dependent avian influenza H5N1-pseudotyped HIV vector system for high throughput screening of inhibitory molecules. Antivir. Res. **79**, 12–18 (2008). https://doi.org/10.1016/j.antiviral.2008.02.001
18. Guo, Y., et al.: Analysis of hemagglutinin-mediated entry tropism of H5N1 avian influenza. Virol. J. **6**, 39 (2009). https://doi.org/10.1186/1743-422X-6-39
19. Ferrara, F., et al.: The human transmembrane protease serine 2 is necessary for the production of group 2 influenza a virus pseudotypes. J Mol Genet Med. **7**, 309–314 (2012)

20. Naldini, L., et al.: In vivo gene delivery and stable transduction of nondividing cells by a lentiviral vector. Science. **272**, 263–267 (1996). https://doi.org/10.1126/science.272.5259.263
21. Trono, D., Feinberg, M.B., Baltimore, D.: HIV-1 gag mutants can dominantly interfere with the replication of the wild-type virus. Cell. **59**, 113–120 (1989). https://doi.org/10.1016/0092-8674(89)90874-x
22. Heyndrickx, L., et al.: Antiviral compounds show enhanced activity in HIV-1 single cycle pseudovirus assays as compared to classical PBMC assays. J. Virol. Methods. **148**, 166–173 (2008). https://doi.org/10.1016/j.jviromet.2007.11.009
23. Wei, X., et al.: Emergence of resistant human immunodeficiency virus type 1 in patients receiving fusion inhibitor (T-20) monotherapy. Antimicrob. Agents Chemother. **46**, 1896–1905 (2002). https://doi.org/10.1128/AAC.46.6.1896-1905.2002
24. Wei, X., et al.: Antibody neutralization and escape by HIV-1. Nature. **422**, 307–312 (2003). https://doi.org/10.1038/nature01470
25. Adachi, A., et al.: Production of acquired immunodeficiency syndrome-associated retrovirus in human and nonhuman cells transfected with an infectious molecular clone. J. Virol. **59**, 284–291 (1986). https://doi.org/10.1128/JVI.59.2.284-291.1986
26. He, J., et al.: Human immunodeficiency virus type 1 viral protein R (Vpr) arrests cells in the G2 phase of the cell cycle by inhibiting p34cdc2 activity. J. Virol. **69**, 6705–6711 (1995). https://doi.org/10.1128/JVI.69.11.6705-6711.1995
27. Berkowitz, R.D., Hammarskjold, M.L., Helga-Maria, C., Rekosh, D., Goff, S.P.: 5′ regions of HIV-1 RNAs are not sufficient for encapsidation: implications for the HIV-1 packaging signal. Virology. **212**, 718–723 (1995). https://doi.org/10.1006/viro.1995.1530
28. Buchschacher Jr., G.L., Panganiban, A.T.: Human immunodeficiency virus vectors for inducible expression of foreign genes. J. Virol. **66**, 2731–2739 (1992). https://doi.org/10.1128/JVI.66.5.2731-2739.1992
29. Kaye, J.F., Richardson, J.H., Lever, A.M.: Cis-acting sequences involved in human immunodeficiency virus type 1 RNA packaging. J. Virol. **69**, 6588–6592 (1995). https://doi.org/10.1128/JVI.69.10.6588-6592.1995
30. Parolin, C., Dorfman, T., Palu, G., Gottlinger, H., Sodroski, J.: Analysis in human immunodeficiency virus type 1 vectors of cis-acting sequences that affect gene transfer into human lymphocytes. J. Virol. **68**, 3888–3895 (1994). https://doi.org/10.1128/JVI.68.6.3888-3895.1994
31. Richardson, J.H., Child, L.A., Lever, A.M.: Packaging of human immunodeficiency virus type 1 RNA requires cis-acting sequences outside the 5′ leader region. J. Virol. **67**, 3997–4005 (1993). https://doi.org/10.1128/JVI.67.7.3997-4005.1993
32. Richardson, J.H., Kaye, J.F., Child, L.A., Lever, A.M.: Helper virus-free transfer of human immunodeficiency virus type 1 vectors. J. Gen. Virol. **76**(Pt 3), 691–696 (1995). https://doi.org/10.1099/0022-1317-76-3-691
33. Sandrin, V., et al.: Lentiviral vectors pseudotyped with a modified RD114 envelope glycoprotein show increased stability in sera and augmented transduction of primary lymphocytes and CD34+ cells derived from human and nonhuman primates. Blood. **100**, 823–832 (2002). https://doi.org/10.1182/blood-2001-11-0042
34. Miller, A.D.: Retroviral vectors. Curr. Top. Microbiol. Immunol. **158**, 1–24 (1992). https://doi.org/10.1007/978-3-642-75608-5_1
35. Uchida, N., et al.: HIV, but not murine leukemia virus, vectors mediate high efficiency gene transfer into freshly isolated G0/G1 human hematopoietic stem cells. Proc. Natl. Acad. Sci. U. S. A. **95**, 11939–11944 (1998). https://doi.org/10.1073/pnas.95.20.11939
36. Poeschla, E.M., Wong-Staal, F., Looney, D.J.: Efficient transduction of nondividing human cells by feline immunodeficiency virus lentiviral vectors. Nat. Med. **4**, 354–357 (1998). https://doi.org/10.1038/nm0398-354
37. Elder, J.H., Phillips, T.R.: Feline immunodeficiency virus as a model for development of molecular approaches to intervention strategies against lentivirus infections. Adv. Virus Res. **45**, 225–247 (1995). https://doi.org/10.1016/s0065-3527(08)60062-7

38. Tomonaga, K., Mikami, T.: Molecular biology of the feline immunodeficiency virus auxiliary genes. J. Gen. Virol. **77**(Pt 8), 1611–1621 (1996). https://doi.org/10.1099/0022-1317-77-8-1611
39. Shacklett, B.L., Luciw, P.A.: Analysis of the vif gene of feline immunodeficiency virus. Virology. **204**, 860–867 (1994). https://doi.org/10.1006/viro.1994.1609
40. Tomonaga, K., et al.: Identification of a feline immunodeficiency virus gene which is essential for cell-free virus infectivity. J. Virol. **66**, 6181–6185 (1992). https://doi.org/10.1128/JVI.66.10.6181-6185.1992
41. Phillips, T.R., et al.: Identification of the rev transactivation and rev-responsive elements of feline immunodeficiency virus. J. Virol. **66**, 5464–5471 (1992). https://doi.org/10.1128/JVI.66.9.5464-5471.1992
42. de Parseval, A., Elder, J.H.: Demonstration that orf2 encodes the feline immunodeficiency virus transactivating (tat) protein and characterization of a unique gene product with partial rev activity. J. Virol. **73**, 608–617 (1999). https://doi.org/10.1128/JVI.73.1.608-617.1999
43. Sparger, E.E., et al.: Regulation of gene expression directed by the long terminal repeat of the feline immunodeficiency virus. Virology. **187**, 165–177 (1992). https://doi.org/10.1016/0042-6822(92)90305-9
44. Tomonaga, K., et al.: The feline immunodeficiency virus ORF-A gene facilitates efficient viral replication in established T-cell lines and peripheral blood lymphocytes. J. Virol. **67**, 5889–5895 (1993). https://doi.org/10.1128/JVI.67.10.5889-5895.1993
45. Waters, A.K., et al.: Influence of ORF2 on host cell tropism of feline immunodeficiency virus. Virology. **215**, 10–16 (1996). https://doi.org/10.1006/viro.1996.0002
46. Rodriguez, L.L., Pauszek, S.J., Bunch, T.A., Schumann, K.R.: Full-length genome analysis of natural isolates of vesicular stomatitis virus (Indiana 1 serotype) from north, central and South America. J. Gen. Virol. **83**, 2475–2483 (2002). https://doi.org/10.1099/0022-1317-83-10-2475
47. Haglund, K., Forman, J., Krausslich, H.G., Rose, J.K.: Expression of human immunodeficiency virus type 1 gag protein precursor and envelope proteins from a vesicular stomatitis virus recombinant: high-level production of virus-like particles containing HIV envelope. Virology. **268**, 112–121 (2000). https://doi.org/10.1006/viro.1999.0120
48. Lawson, N.D., Stillman, E.A., Whitt, M.A., Rose, J.K.: Recombinant vesicular stomatitis viruses from DNA. Proc. Natl. Acad. Sci. U. S. A. **92**, 4477–4481 (1995). https://doi.org/10.1073/pnas.92.10.4477
49. Whelan, S.P., Ball, L.A., Barr, J.N., Wertz, G.T.: Efficient recovery of infectious vesicular stomatitis virus entirely from cDNA clones. Proc. Natl. Acad. Sci. U. S. A. **92**, 8388–8392 (1995). https://doi.org/10.1073/pnas.92.18.8388
50. Harty, R.N., Brown, M.E., Hayes, F.P., Wright, N.T., Schnell, M.J.: Vaccinia virus-free recovery of vesicular stomatitis virus. J. Mol. Microbiol. Biotechnol. **3**, 513–517 (2001)
51. Lyles, D.S.: Assembly and budding of negative-strand RNA viruses. Adv. Virus Res. **85**, 57–90 (2013). https://doi.org/10.1016/B978-0-12-408116-1.00003-3
52. Cortese, M., et al.: Ultrastructural characterization of zika virus replication factories. Cell Rep. **18**, 2113–2123 (2017). https://doi.org/10.1016/j.celrep.2017.02.014
53. Welsch, S., et al.: Composition and three-dimensional architecture of the dengue virus replication and assembly sites. Cell Host Microbe. **5**, 365–375 (2009). https://doi.org/10.1016/j.chom.2009.03.007
54. Irie, T., Licata, J.M., McGettigan, J.P., Schnell, M.J., Harty, R.N.: Budding of PPxY-containing rhabdoviruses is not dependent on host proteins TGS101 and VPS4A. J. Virol. **78**, 2657–2665 (2004). https://doi.org/10.1128/jvi.78.6.2657-2665.2004
55. Syed, A.M., et al.: Rapid assessment of SARS-CoV-2-evolved variants using virus-like particles. Science. **374**, 1626–1632 (2021). https://doi.org/10.1126/science.abl6184
56. Drosten, C., et al.: Identification of a novel coronavirus in patients with severe acute respiratory syndrome. N. Engl. J. Med. **348**, 1967–1976 (2003). https://doi.org/10.1056/NEJMoa030747

57. Ksiazek, T.G., et al.: A novel coronavirus associated with severe acute respiratory syndrome. N. Engl. J. Med. **348**, 1953–1966 (2003). https://doi.org/10.1056/NEJMoa030781
58. Narayanan, K., Maeda, A., Maeda, J., Makino, S.: Characterization of the coronavirus M protein and nucleocapsid interaction in infected cells. J. Virol. **74**, 8127–8134 (2000). https://doi.org/10.1128/jvi.74.17.8127-8134.2000
59. Huang, Y., Yang, Z.Y., Kong, W.P., Nabel, G.J.: Generation of synthetic severe acute respiratory syndrome coronavirus pseudoparticles: implications for assembly and vaccine production. J. Virol. **78**, 12557–12565 (2004). https://doi.org/10.1128/JVI.78.22.12557-12565.2004
60. Huang, Y., Kong, W.P., Nabel, G.J.: Human immunodeficiency virus type 1-specific immunity after genetic immunization is enhanced by modification of gag and pol expression. J. Virol. **75**, 4947–4951 (2001). https://doi.org/10.1128/JVI.75.10.4947-4951.2001
61. Wills, J.W., Craven, R.C., Achacoso, J.A.: Creation and expression of myristylated forms of Rous sarcoma virus gag protein in mammalian cells. J. Virol. **63**, 4331–4343 (1989). https://doi.org/10.1128/JVI.63.10.4331-4343.1989
62. Vennema, H., et al.: Nucleocapsid-independent assembly of coronavirus-like particles by co-expression of viral envelope protein genes. EMBO J. **15**, 2020–2028 (1996)
63. Kuo, L., Masters, P.S.: Functional analysis of the murine coronavirus genomic RNA packaging signal. J. Virol. **87**, 5182–5192 (2013). https://doi.org/10.1128/JVI.00100-13
64. Carstens, E.B., Ball, L.A.: Ratification vote on taxonomic proposals to the international committee on taxonomy of viruses (2008). Arch. Virol. **154**, 1181–1188 (2009). https://doi.org/10.1007/s00705-009-0400-2
65. Chen, S.C., Olsthoorn, R.C.: Group-specific structural features of the 5′-proximal sequences of coronavirus genomic RNAs. Virology. **401**, 29–41 (2010). https://doi.org/10.1016/j.virol.2010.02.007
66. Hsieh, P.K., et al.: Assembly of severe acute respiratory syndrome coronavirus RNA packaging signal into virus-like particles is nucleocapsid dependent. J. Virol. **79**, 13848–13855 (2005). https://doi.org/10.1128/JVI.79.22.13848-13855.2005
67. Arita, M., Iwai-Itamochi, M.: Evaluation of antigenic differences between wild and Sabin vaccine strains of poliovirus using the pseudovirus neutralization test. Sci. Rep. **9**, 11970 (2019). https://doi.org/10.1038/s41598-019-48534-1
68. Nguyen, H.P., Ramirez-Fort, M.K., Rady, P.L.: The biology of human papillomaviruses. Curr. Probl. Dermatol. **45**, 19–32 (2014). https://doi.org/10.1159/000355959
69. Buck, C.B., et al.: Arrangement of L2 within the papillomavirus capsid. J. Virol. **82**, 5190–5197 (2008). https://doi.org/10.1128/JVI.02726-07
70. Frazer, I.H., Leggatt, G.R., Mattarollo, S.R.: Prevention and treatment of papillomavirus-related cancers through immunization. Annu. Rev. Immunol. **29**, 111–138 (2011). https://doi.org/10.1146/annurev-immunol-031210-101308
71. Holmgren, S.C., Patterson, N.A., Ozbun, M.A., Lambert, P.F.: The minor capsid protein L2 contributes to two steps in the human papillomavirus type 31 life cycle. J. Virol. **79**, 3938–3948 (2005). https://doi.org/10.1128/JVI.79.7.3938-3948.2005
72. Buck, C.B., Pastrana, D.V., Lowy, D.R., Schiller, J.T.: Efficient intracellular assembly of papillomaviral vectors. J. Virol. **78**, 751–757 (2004). https://doi.org/10.1128/jvi.78.2.751-757.2004
73. Pastrana, D.V., et al.: Reactivity of human sera in a sensitive, high-throughput pseudovirus-based papillomavirus neutralization assay for HPV16 and HPV18. Virology. **321**, 205–216 (2004). https://doi.org/10.1016/j.virol.2003.12.027
74. Sehr, P., et al.: High-throughput pseudovirion-based neutralization assay for analysis of natural and vaccine-induced antibodies against human papillomaviruses. PLoS One. **8**, e75677 (2013). https://doi.org/10.1371/journal.pone.0075677

75. Nie, J., Huang, W., Wu, X., Wang, Y.: Optimization and validation of a high throughput method for detecting neutralizing antibodies against human papillomavirus (HPV) based on pseudovirons. J. Med. Virol. **86**, 1542–1555 (2014). https://doi.org/10.1002/jmv.23995
76. Yousefi, Z., et al.: An update on human papilloma virus vaccines: history, types, protection, and efficacy. Front. Immunol. **12**, 805695 (2021). https://doi.org/10.3389/fimmu.2021.805695
77. Hamamoto, N., et al.: Association between RABV G proteins transported from the perinuclear space to the cell surface membrane and N-glycosylation of the Sequon Asn(204). Jpn. J. Infect. Dis. **68**, 387–393 (2015). https://doi.org/10.7883/yoken.JJID.2014.533
78. Slough, M.M., Chandran, K., Jangra, R.K.: Two point mutations in Old World hantavirus glycoproteins afford the generation of highly infectious recombinant vesicular stomatitis virus vectors. *mBio*. **10** (2019). https://doi.org/10.1128/mBio.02372-18
79. Giroglou, T., et al.: Retroviral vectors pseudotyped with severe acute respiratory syndrome coronavirus S protein. J. Virol. **78**, 9007–9015 (2004). https://doi.org/10.1128/JVI.78.17.9007-9015.2004
80. Fu, X., Tao, L., Zhang, X.: Comprehensive and systemic optimization for improving the yield of SARS-CoV-2 spike pseudotyped virus. Mol Ther Methods Clin Dev. **20**, 350–356 (2021). https://doi.org/10.1016/j.omtm.2020.12.007
81. Havranek, K.E., et al.: SARS-CoV-2 spike alterations enhance Pseudoparticle Titers and replication-competent VSV-SARS-CoV-2 virus. Viruses. **12** (2020). https://doi.org/10.3390/v12121465
82. Yu, J., et al.: Deletion of the SARS-CoV-2 spike cytoplasmic tail increases infectivity in Pseudovirus neutralization assays. J. Virol. (2021). https://doi.org/10.1128/JVI.00044-21
83. Johnson, J.E., Rodgers, W., Rose, J.K.: A plasma membrane localization signal in the HIV-1 envelope cytoplasmic domain prevents localization at sites of vesicular stomatitis virus budding and incorporation into VSV virions. Virology. **251**, 244–252 (1998). https://doi.org/10.1006/viro.1998.9429
84. Johnson, J.E., Schnell, M.J., Buonocore, L., Rose, J.K.: Specific targeting to CD4+ cells of recombinant vesicular stomatitis viruses encoding human immunodeficiency virus envelope proteins. J. Virol. **71**, 5060–5068 (1997). https://doi.org/10.1128/JVI.71.7.5060-5068.1997
85. Owens, R.J., Rose, J.K.: Cytoplasmic domain requirement for incorporation of a foreign envelope protein into vesicular stomatitis virus. J. Virol. **67**, 360–365 (1993). https://doi.org/10.1128/JVI.67.1.360-365.1993
86. Liu, H., et al.: Introducing a cleavable signal peptide enhances the packaging efficiency of lentiviral vectors pseudotyped with Japanese encephalitis virus envelope proteins. Virus Res. **229**, 9–16 (2017). https://doi.org/10.1016/j.virusres.2016.12.007
87. Zhao, R., et al.: Novel strategy for expression and characterization of rabies virus glycoprotein. Protein Expr. Purif. **168**, 105567 (2020). https://doi.org/10.1016/j.pep.2019.105567
88. Garcia, J.M., Lai, J.C.: Production of influenza pseudotyped lentiviral particles and their use in influenza research and diagnosis: an update. Expert Rev. Anti-Infect. Ther. **9**, 443–455 (2011). https://doi.org/10.1586/eri.11.25
89. Grehan, K., Ferrara, F., Temperton, N.: An optimised method for the production of MERS-CoV spike expressing viral pseudotypes. MethodsX. **2**, 379–384 (2015). https://doi.org/10.1016/j.mex.2015.09.003
90. Hu, J., et al.: Development of cell-based pseudovirus entry assay to identify potential viral entry inhibitors and neutralizing antibodies against SARS-CoV-2. Genes Dis. **7**, 551–557 (2020). https://doi.org/10.1016/j.gendis.2020.07.006
91. Nie, Y., et al.: Highly infectious SARS-CoV pseudotyped virus reveals the cell tropism and its correlation with receptor expression. Biochem. Biophys. Res. Commun. **321**, 994–1000 (2004). https://doi.org/10.1016/j.bbrc.2004.07.060
92. Yang, P., et al.: An optimized and robust SARS-CoV-2 pseudovirus system for viral entry research. J. Virol. Methods. **295**, 114221 (2021). https://doi.org/10.1016/j.jviromet.2021.114221

93. Fernandez, E., et al.: Conformational and thermal stability improvements for the large-scale production of yeast-derived rabbit hemorrhagic disease virus-like particles as multipurpose vaccine. PLoS One. **8**, e56417 (2013). https://doi.org/10.1371/journal.pone.0056417
94. Mossadegh, N., et al.: Codon optimization of the human papillomavirus 11 (HPV 11) L1 gene leads to increased gene expression and formation of virus-like particles in mammalian epithelial cells. Virology. **326**, 57–66 (2004). https://doi.org/10.1016/j.virol.2004.04.050
95. Cronin, J., Zhang, X.Y., Reiser, J.: Altering the tropism of lentiviral vectors through pseudotyping. Curr. Gene Ther. **5**, 387–398 (2005). https://doi.org/10.2174/1566523054546224
96. Medina, M.F., et al.: Lentiviral vectors pseudotyped with minimal filovirus envelopes increased gene transfer in murine lung. Mol. Ther. **8**, 777–789 (2003). https://doi.org/10.1016/j.ymthe.2003.07.003
97. Chakraborty, S.: Evolutionary and structural analysis elucidates mutations on SARS-CoV2 spike protein with altered human ACE2 binding affinity. Biochem. Biophys. Res. Commun. **534**, 374–380 (2021). https://doi.org/10.1016/j.bbrc.2020.11.075
98. Gu, H., et al.: Adaptation of SARS-CoV-2 in BALB/c mice for testing vaccine efficacy. Science. **369**, 1603–1607 (2020). https://doi.org/10.1126/science.abc4730
99. Johnson, M.C., et al.: Optimized Pseudotyping conditions for the SARS-COV-2 spike glycoprotein. J. Virol. **94** (2020). https://doi.org/10.1128/JVI.01062-20
100. Korber, B., et al.: Tracking changes in SARS-CoV-2 spike: evidence that D614G increases infectivity of the COVID-19 virus. Cell. **182**, 812-827 e819 (2020). https://doi.org/10.1016/j.cell.2020.06.043
101. Liu, Q., et al.: Antibody-dependent-cellular-cytotoxicity-inducing antibodies significantly affect the post-exposure treatment of Ebola virus infection. Sci. Rep. **7**, 45552 (2017). https://doi.org/10.1038/srep45552
102. Mao, Y., et al.: Lentiviral vectors mediate long-term and high efficiency transgene expression in HEK 293T cells. Int. J. Med. Sci. **12**, 407–415 (2015). https://doi.org/10.7150/ijms.11270
103. Desmaris, N., et al.: Production and neurotropism of lentivirus vectors pseudotyped with lyssavirus envelope glycoproteins. Mol. Ther. **4**, 149–156 (2001). https://doi.org/10.1006/mthe.2001.0431
104. Park, F.: Correction of bleeding diathesis without liver toxicity using arenaviral-pseudotyped HIV-1-based vectors in hemophilia a mice. Hum. Gene Ther. **14**, 1489–1494 (2003). https://doi.org/10.1089/104303403769211691
105. Beyer, W.R., Westphal, M., Ostertag, W., von Laer, D.: Oncoretrovirus and lentivirus vectors pseudotyped with lymphocytic choriomeningitis virus glycoprotein: generation, concentration, and broad host range. J. Virol. **76**, 1488–1495 (2002). https://doi.org/10.1128/jvi.76.3.1488-1495.2002
106. Temperton, N.J., et al.: A sensitive retroviral pseudotype assay for influenza H5N1-neutralizing antibodies. Influenza Other Respir. Viruses. **1**, 105–112 (2007). https://doi.org/10.1111/j.1750-2659.2007.00016.x
107. Cosset, F.L., et al.: Characterization of Lassa virus cell entry and neutralization with Lassa virus pseudoparticles. J. Virol. **83**, 3228–3237 (2009). https://doi.org/10.1128/JVI.01711-08
108. Hoffmann, M., et al.: SARS-CoV-2 cell entry depends on ACE2 and TMPRSS2 and is blocked by a clinically proven protease inhibitor. Cell. **181**, 271-280 e278 (2020). https://doi.org/10.1016/j.cell.2020.02.052
109. Li, F., Li, W., Farzan, M., Harrison, S.C.: Structure of SARS coronavirus spike receptor-binding domain complexed with receptor. Science. **309**, 1864–1868 (2005). https://doi.org/10.1126/science.1116480
110. Kim, J.S., et al.: Genome-wide identification and characterization of point mutations in the SARS-CoV-2 genome. Osong. Public Health Res. Perspect. **11**, 101–111 (2020). https://doi.org/10.24171/j.phrp.2020.11.3.05

111. Ogando, N.S., et al.: SARS-coronavirus-2 replication in Vero E6 cells: replication kinetics, rapid adaptation and cytopathology. J. Gen. Virol. **101**, 925–940 (2020). https://doi.org/10.1099/jgv.0.001453
112. Carnell, G.W., Ferrara, F., Grehan, K., Thompson, C.P., Temperton, N.J.: Pseudotype-based neutralization assays for influenza: a systematic analysis. Front. Immunol. **6**, 161 (2015). https://doi.org/10.3389/fimmu.2015.00161
113. Bertram, S., et al.: TMPRSS2 and TMPRSS4 facilitate trypsin-independent spread of influenza virus in Caco-2 cells. J. Virol. **84**, 10016–10025 (2010). https://doi.org/10.1128/JVI.00239-10
114. Wang, W., et al.: Establishment of retroviral pseudotypes with influenza hemagglutinins from H1, H3, and H5 subtypes for sensitive and specific detection of neutralizing antibodies. J. Virol. Methods. **153**, 111–119 (2008). https://doi.org/10.1016/j.jviromet.2008.07.015
115. Nasri, M., Karimi, A., Allahbakhshian Farsani, M.: Production, purification and titration of a lentivirus-based vector for gene delivery purposes. Cytotechnology. **66**, 1031–1038 (2014). https://doi.org/10.1007/s10616-013-9652-5
116. Xiong, H.L., et al.: Robust neutralization assay based on SARS-CoV-2 S-protein-bearing vesicular stomatitis virus (VSV) pseudovirus and ACE2-overexpressing BHK21 cells. Emerg Microbes Infect. **9**, 2105–2113 (2020). https://doi.org/10.1080/22221751.2020.1815589
117. Morita, S., Kojima, T., Kitamura, T.: Plat-E: an efficient and stable system for transient packaging of retroviruses. Gene Ther. **7**, 1063–1066 (2000). https://doi.org/10.1038/sj.gt.3301206
118. Zhao, G., et al.: A safe and convenient pseudovirus-based inhibition assay to detect neutralizing antibodies and screen for viral entry inhibitors against the novel human coronavirus MERS-CoV. Virol. J. **10**, 266 (2013). https://doi.org/10.1186/1743-422X-10-266
119. Tang, D.J., et al.: A single residue substitution in the receptor-binding domain of H5N1 hemagglutinin is critical for packaging into pseudotyped lentiviral particles. PLoS One. **7**, e43596 (2012). https://doi.org/10.1371/journal.pone.0043596
120. Mekkaoui, L., et al.: Optimised method for the production and titration of lentiviral vectors Pseudotyped with the SARS-CoV-2 spike. Bio Protoc. **11**, e4194 (2021). https://doi.org/10.21769/BioProtoc.4194
121. Chen, M., Zhang, X.E.: Construction and applications of SARS-CoV-2 pseudoviruses: a mini review. Int. J. Biol. Sci. **17**, 1574–1580 (2021). https://doi.org/10.7150/ijbs.59184
122. Sholukh, A.M., et al.: Evaluation of cell-based and surrogate SARS-CoV-2 neutralization assays. J. Clin. Microbiol. **59**, e0052721 (2021). https://doi.org/10.1128/JCM.00527-21
123. He, Y., Lu, H., Siddiqui, P., Zhou, Y., Jiang, S.: Receptor-binding domain of severe acute respiratory syndrome coronavirus spike protein contains multiple conformation-dependent epitopes that induce highly potent neutralizing antibodies. J. Immunol. **174**, 4908–4915 (2005). https://doi.org/10.4049/jimmunol.174.8.4908
124. Mavhandu, L.G., et al.: Development of a pseudovirus assay and evaluation to screen natural products for inhibition of HIV-1 subtype C reverse transcriptase. J. Ethnopharmacol. **258**, 112931 (2020). https://doi.org/10.1016/j.jep.2020.112931
125. Nie, J., et al.: Development of in vitro and in vivo rabies virus neutralization assays based on a high-titer pseudovirus system. Sci. Rep. **7**, 42769 (2017). https://doi.org/10.1038/srep42769
126. Fan, P., et al.: Potent neutralizing monoclonal antibodies against Ebola virus isolated from vaccinated donors. MAbs. **12**, 1742457 (2020). https://doi.org/10.1080/19420862.2020.1742457
127. Ning, T., et al.: Monitoring neutralization property change of evolving Hantaan and Seoul viruses with a novel Pseudovirus-based assay. Virol. Sin. **36**, 104–112 (2021). https://doi.org/10.1007/s12250-020-00237-y
128. Yuan, J., et al.: Mutations in the G-H loop region of ephrin-B2 can enhance Nipah virus binding and infection. J. Gen. Virol. **92**, 2142–2152 (2011). https://doi.org/10.1099/vir.0.033787-0

129. Haid, S., Grethe, C., Bankwitz, D., Grunwald, T., Pietschmann, T.: Identification of a human respiratory syncytial virus cell entry inhibitor by using a novel lentiviral Pseudotype system. J. Virol. **90**, 3065–3073 (2015). https://doi.org/10.1128/JVI.03074-15
130. Cruz, M.A., Parks, G.D.: Enhancement of infectivity of insect cell-derived La Crosse virus by human serum. Virus Res. **292**, 198228 (2021). https://doi.org/10.1016/j.virusres.2020.198228
131. Zhang, L., et al.: A bioluminescent imaging mouse model for Marburg virus based on a pseudovirus system. Hum. Vaccin. Immunother. **13**, 1811–1817 (2017). https://doi.org/10.1080/21645515.2017.1325050
132. Hu, D., et al.: Chikungunya virus glycoproteins pseudotype with lentiviral vectors and reveal a broad spectrum of cellular tropism. PLoS One. **9**, e110893 (2014). https://doi.org/10.1371/journal.pone.0110893
133. Vasmehjani, A.A., et al.: Efficient production of a lentiviral system for displaying Crimean-Congo hemorrhagic fever virus glycoproteins reveals a broad range of cellular susceptibility and neutralization ability. Arch. Virol. **165**, 1109–1120 (2020). https://doi.org/10.1007/s00705-020-04576-9
134. Godbey, W.T., Wu, K.K., Hirasaki, G.J., Mikos, A.G.: Improved packing of poly (ethylenimine)/DNA complexes increases transfection efficiency. Gene Ther. **6**, 1380–1388 (1999). https://doi.org/10.1038/sj.gt.3300976
135. Bertram, S., Glowacka, I., Steffen, I., Kuhl, A., Pohlmann, S.: Novel insights into proteolytic cleavage of influenza virus hemagglutinin. Rev. Med. Virol. **20**, 298–310 (2010). https://doi.org/10.1002/rmv.657
136. Skehel, J.J., Wiley, D.C.: Receptor binding and membrane fusion in virus entry: the influenza hemagglutinin. Annu. Rev. Biochem. **69**, 531–569 (2000). https://doi.org/10.1146/annurev.biochem.69.1.531
137. Olsen, J.C., Sechelski, J.: Use of sodium butyrate to enhance production of retroviral vectors expressing CFTR cDNA. Hum. Gene Ther. **6**, 1195–1202 (1995). https://doi.org/10.1089/hum.1995.6.9-1195
138. Ansarah-Sobrinho, C., Nelson, S., Jost, C.A., Whitehead, S.S., Pierson, T.C.: Temperature-dependent production of pseudoinfectious dengue reporter virus particles by complementation. Virology. **381**, 67–74 (2008). https://doi.org/10.1016/j.virol.2008.08.021
139. Mattia, K., et al.: Dengue reporter virus particles for measuring neutralizing antibodies against each of the four dengue serotypes. PLoS One. **6**, e27252 (2011). https://doi.org/10.1371/journal.pone.0027252
140. Kutner, R.H., Puthli, S., Marino, M.P., Reiser, J.: Simplified production and concentration of HIV-1-based lentiviral vectors using HYPERFlask vessels and anion exchange membrane chromatography. BMC Biotechnol. **9**, 10 (2009). https://doi.org/10.1186/1472-6750-9-10
141. Kutner, R.H., Zhang, X.Y., Reiser, J.: Production, concentration and titration of pseudotyped HIV-1-based lentiviral vectors. Nat. Protoc. **4**, 495–505 (2009). https://doi.org/10.1038/nprot.2009.22

Chapter 2
Assays Based on Pseudotyped Viruses

Jianhui Nie, Xueling Wu, and Youchun Wang

Abstract Pseudotyped viruses are more and more widely used in virus research and the evaluation of antiviral products because of their high safety, simple operation, high accessibility, ease in achieving standardization, and high throughput. The development of measures based on pseudotyped virus is closely related to the characteristics of viruses, and it is also necessary to follow the principles of assay development. Only in the process of method development, where the key parameters that affect the results are systematically optimized and the preliminary established method is fully validated, can the accuracy, reliability, and repeatability of the test results be ensured. Only the method established on this basis can be transferred to different laboratories and make the results of different laboratories comparable. This paper summarizes the specific aspects and general principles in the development of assays based on pseudotyped virus, which is of reference value for the development of similar methods.

Keywords Pseudotyped virus · Optimization · Verification · Specificity · Precision · Linear range · Animal model

J. Nie
Division of HIV/AIDS and Sex-transmitted Virus Vaccines, National Institutes for Food and Drug Control (NIFDC) and WHO Collaborating Center for Standardization and Evaluation of Biologicals, Beijing, China

X. Wu
Cell Collection and Research Center, National Institutes for Food and Drug Control (NIFDC) and WHO Collaborating Center for Standardization and Evaluation of Biologicals, Beijing, China

Y. Wang (✉)
Chinese Academy of Medical Sciences & Peking Union Medical College, Beijing, China

Institute of Medical Biology, Chinese Academy of Medicine Sciences & Peking Union Medical College, Kunming, China
e-mail: wangyc@nifdc.org.cn; wangyc@imbcams.com.cn

Y. Wang (ed.), *Pseudotyped Viruses*, Advances in Experimental Medicine and Biology 1407, https://doi.org/10.1007/978-981-99-0113-5_2

Abbreviations

ACE2	Angiotensin-converting enzyme 2
AID_{50}	50% animal infectious dose
CCR5	C-C chemokine receptor type 5
CD4	Cluster of differentiation 4
CPE	Cytopathic effect
CXCR4	C-X-C chemokine receptor type 4
DEAE	Diethylaminoethyl cellulose
DPP4	Dipeptidyl peptidase 4
HI	Hemagglutination inhibition
HIV	Human immunodeficiency virus
HPV	Human papillomavirus
IC_{50}	50% inhibitory concentration
ID_{50}	50% inhibitory dose
IVIS	In vivo imaging system
MERS-CoV	Middle East respiratory syndrome coronavirus
MN	Micro-neutralization
PBNA	Pseudotyped virus based neutralization assay
PFU	Plaque-forming unit
RFFIT	Rapid fluorescent foci inhibition test
RLU	Relative light unit
SARS-CoV-2	Severe acute respiratory syndrome coronavirus 2
SCARB2	Scavenger receptor class B member 2
SD	Standard deviation
SOP	Standard operating procedures
SV40	Simian virus 40
$TCID_{50}$	50% tissue culture infectious dose
VSV	Vesicular stomatitis virus

2.1 Introduction

Pseudotyped virus refers to the virus particles of which the gene of out-shell protein (envelope protein of enveloped virus or capsid protein of non-enveloped virus) is not included in its defective genome but provided by a co-transfected plasmid [1, 2]. The deletion of the out-shell gene causes the virus particle to have only one-cycle infection but not replicate. Therefore, the pseudotyped virus can be handled in the biosafety level 2 laboratory, which is particularly important for those virus researches which can only be operated in the biosafety level 3 or even level 4 laboratory.

Pseudotyped viruses usually contain reporter genes that are easy to detect, and the amount of virus infection can be reflected by detecting the expression of reporter genes after infection. It has the advantages of having high security, being convenient to operate and ready to obtain, ease in achieving high throughput. The employment of pseudotyped viruses solves the problem that some authentic viruses cannot be cultured or are difficult to obtain and shows a high correlation and replaceability with the live virus methods.

The infection process of the virus usually depends on the interaction between the protein in its out shell and the receptor on the cell surface. Therefore, pseudotyped virus can simulate the infection process of authentic virus and is often used in the study of virus infection mechanism, receptor discovery, and so on. Neutralizing antibodies against the virus usually bind to the coat protein of the virus, thus blocking its binding to the cell surface receptor, thereby inhibiting virus infection. Therefore, pseudotyped viruses are also often used in the study of infection inhibitors such as neutralizing monoclonal antibodies. At present, most successful vaccines can induce specific neutralizing antibodies against viruses [3], and pseudotyped viruses are also an important tool for evaluating vaccine-induced neutralizing antibodies, especially for some viruses that are difficult to culture. Pseudotyped virus based neutralization assay (PBNA) has become the gold standard for some vaccine-elicited antibody evaluation. Pseudotyped viruses can simulate the process of virus infection in vivo and can also be used to establish animal infection models and evaluate corresponding treatment or preventive products.

Regardless of whether the method is applied to the detection of neutralizing antibodies in vitro or the analysis of the protective effect of vaccines in vivo, to ensure that the analysis results are objective, scientific, comparable, and accurate, the characteristics of pseudotyped viruses must be fully considered in the process of assay development. And the method is fully optimized and validated, which is also the premise that a method can be transferred to other laboratories and applied. In this paper, the methods of pseudotyped viruses in vitro and in vivo were systematically reviewed from the aspects of optimization, validation, and standardization. It is expected to provide a reference for the development of this kind of assay in the future.

2.2 Development of In Vitro Assays Based on Pseudotyped Viruses

Most of the in vitro assays based on pseudotyped virus test is expected to stimulate the cellular infection process of the real virus in vivo. The in vitro pseudotyped virus assay is usually used in the evaluation of the neutralizing antibodies, which is referred as the pseudotyped virus based neutralization assay. We use PBNA as a model to demonstrate the process of in vitro assay based on pseudotyped virus. The first step of establishing the PBNA is to optimize the various key parameters. After determining the optimal conditions, the PBNA is systematically validated to ensure the accuracy, repeatability, and comparability of the test results.

2.2.1 *Establishment and Optimization of In Vitro Assays Based on Pseudotyped Viruses*

2.2.1.1 Selection of Target Cells

The target cells in PBNA should carry the receptors needed for pseudotyped virus infection. However, the target cells in PBNA are often different from those infected by the authentic virus, or even cells from different tissue origin. In the PBNA, we should not only consider that the cells are sensitive but also consider the high signal-to-noise ratio to ensure the repeatability of the test. Therefore, the target cells in PBNA are usually cell lines rather than primary cells.

When the optimal cell line has not been identified in a PBNA, a broad range of cell lines will usually be employed to screen the best one. If the live virus culture has been established, you include the same cell line into the screening test for PBNA. For example, in the SARS-CoV-2 PBNA, after examining the cells from different species and tissues, it was found that the Huh7 cells, used in live virus culture, could be effectively infected with pseudotyped virus. So, the first SARS-CoV-2 PBNA was established using the Huh7 cell line [4, 5]. In addition, the selected target cells should be able efficiently express the reporter genes. For example, 293FT [6] or 293TT [7, 8] cells are usually utilized as the target cells in HPV PBNA, both of which contain SV40 T antigen, which can interact with SV40 replication initiation sequence in reporter gene plasmid to improve the expression level of it. The increased signal-to-noise ratio could enhance the sensitivity and repeatability of the method. For some viruses, the distribution of receptors may be limited. The existing cell lines cannot meet the requirements of pseudotyped virus detection. To ensure a high infecting efficiency, it may be necessary to overexpress cell receptors, coreceptors, or enzymes in specific cells to generate a new line. For example, the cellular receptor and coreceptor are necessary for HIV-1 infection, including CD4 [9], CCR5 [10], or CXCR4 [11]. These receptors and coreceptors usually exist in T cells or macrophages. In these kinds of cell lines, after HIV-1 pseudotyped virus infection, the expression level of the reporter gene is relatively low, which cannot achieve the quantification of pseudotyped virus infection. A series of modifications were carried out based on HeLa cells, including the introduction of HIV-1 receptor CD4, coreceptor CXCR4, and CCR5. To facilitate detection, the firefly luciferase gene regulated by HIV LTR was introduced into the cell. When HIV-1 pseudotyped virus infects the cell, the Tat protein it carries can activate the expression of the reporter gene, which reflects the amount of pseudotyped virus infection [12–14]. The determination of target cells indicates that the PBNA has been primarily established.

2.2.1.2 Optimization of Cell Inoculation

It usually takes a certain time (24–72 h) for pseudotyped viruses to generate stable expression of reporter genes from infected cells. At the detection endpoint, cells need to reach an appropriate density to make the expression of reporter genes reach a high plateau, to ensure the stability of the test results. Too low density will yield too low detection value, in the rising stage of reporter gene expression; too high density will lead to poor cell state and decrease the level of reporter gene expression. The amount of cell inoculation is usually optimized from two aspects: its influence on the detection signal-to-noise ratio and the inhibition curve fitting degree. The signal-to-noise ratio is the prerequisite for determining the stability of the test. Only when the signal-to-noise ratio reaches a certain degree can the classical curve be guaranteed. The optimization of the signal-to-noise ratio is usually investigated by titration of the same pseudotyped virus at different cell concentrations. At first, the signal-to-noise ratio will gradually increase with the increase of cell numbers. After reaching the platform, with the further increase of cell dose, the signal-to-noise ratio will decrease. The pseudotyped virus titrations of different cell concentrations were linearly fitted. In the platform with the highest signal-to-noise ratio, the linear correlation coefficient is usually the highest [15, 16].

The effect of cell amount on the fit degree of the inhibition curve should also be considered, which determines the stability of the inhibition test results. Usually, at different cell concentrations, the same neutralizing antibody samples were used to detect their inhibitory effects, the inhibition curves were drawn, respectively, and the correlation coefficients of the inhibition curves at different cell concentrations were compared. The inhibition curve of biological samples is usually a four-parameter curve. The correlation coefficient is expected to be proportional to the detected signal-to-noise ratio. The optimal cell addition was determined by combining the signal-to-noise ratio of the test and the correlation coefficient of the inhibition curve. Usually, the optimal amount of cell addition is a range, and the final amount of cell addition determined for the method is the middle value of this range, which can ensure the robustness of the method [4, 17–24].

2.2.1.3 Optimization of the Amount of Pseudotyped Virus

The determination of pseudotyped virus amount is crucial for the standardization of the PBNA. There is a negative correlation between the amount of pseudotyped viruses and the sensitivity of the method, that is, the more pseudotyped viruses are added, the more inhibitors needed to neutralize these viruses, the lower the sensitivity of the method. However, if too little pseudotyped viruses are added, the signal-to-noise ratio of the assay will be too low, which will reduce the stability and repeatability of the method. Therefore, we should comprehensively consider the above two aspects to determine the amount of pseudotyped viruses. The same samples containing neutralizing antibodies are usually detected with different amounts of pseudotyped viruses. The inhibition curve is drawn according to the test value, and the 50% inhibitory dose or concentration (ID_{50} or IC_{50}) is calculated. First of all, we should ensure that in the case of a certain amount of pseudotyped virus, the inhibition curve can be fitted, and the correlation coefficient should be maintained at a high level (such as $R^2 \geq 0.95$). Secondly, the ID_{50} or IC_{50} values should be comparable in a range of different pseudotyped virus concentrations. The interval with a high linear correlation coefficient and relatively stable detection sensitivity was selected as the allowable range of pseudotyped virus addition. The middle value of the range is taken as the optimal amount of virus addition. To ensure the accuracy of the amount of virus added in each test, the criteria for determining the amount of virus added are usually set up in the conditions under which the data generated is acceptable. It is usually controlled by the ratio of pseudotyped virus control (wells with only pseudotyped viruses and cell) and cell control (wells with only the target cells). For different pseudotyped viruses and detection systems, the criteria are different. For example, in the HIV-1 PBNA based on Tzm-bl cells, the criterion of the ratio is not less than 100 [15]; for the SARS-CoV-2 pseudotyped virus neutralization test based on the VSV system, the criterion of the ratio is not less than 1000 [4, 5]. For the HPV pseudotyped virus neutralization test with a fluorescent reporter gene, the 96-well plate pattern test requires that the number of control holes of the virus should not be less than 200 [25].

2.2.1.4 Optimization of Infection Conditions

Some pseudotyped viruses need different auxiliary conditions to improve their infectivity and the stability of the method. For example, rabies pseudotyped virus using the HIV skeleton system and HIV-1 pseudotyped virus need to be supplemented with DEAE-dextran to achieve adequate level of infection. DEAE-dextran is a kind of polycation reagent, which can counteract the electrostatic force of the interaction between virus and cell surface, thus promoting virus infection, and this effect will not affect the binding and neutralization of antibodies, but it will have certain cytotoxicity at high concentrations [26, 27].

The potency and cytotoxicity of different batches of DEAE-dextran from different sources might be different. Therefore, for each batch of DEAE-dextran, it is necessary to test its potency and cytotoxicity to determine its optimal concentration. A series of diluted DEAE-dextran was incubated with cells and pseudotyped viruses for 48 h to observe the cytotoxic effect and detect the level of reporter expression (relative light unit, RLU). The highest RLU value was obtained at the optimal concentration, at which the cytotoxic effect should not be observed under the light microscope. The optimized concentrations of different batches of DEAE-dextran varied in different laboratories (10–60 μg/ml) [21, 28, 29]. Different pseudotyped viruses have different dependence on DEAE-dextran, so the optimal concentration should be investigated individually for different pseudotyped viruses, comprehensively considering its effects on cell viability and virus infectivity to determine the optimal working concentration [15].

In the PBNA, the process of pseudotyped virus neutralization usually occurs before the cells are infected, that is, the samples to be tested are firstly incubated with the pseudotyped viruses and then added to the target cells to analyze the inhibition ability of the samples to pseudotyped virus infection. According to the different infection characteristics of viruses, the incubation conditions and time are different.

The incubation conditions of most pseudotyped viruses and samples are the same as those of cell culture, that is, incubation at 37 °C, such as HIV-1 [27], rabies [20], Nipah [30], Ebola [31], Marburg [32], MERS-CoV[33], and SARS-CoV-2 [4, 5, 34]. The incubation conditions of HPV pseudovirus is an exception with 0 °C or 4 °C for 1 hour [7, 8, 35]. Later, it was found that due to the relatively slow process of HPV infection, early incubation had almost no effect on the detection results [36]. With the omission of this step, the efficiency of the PBNA could be further improved and more amendable to automatic detection.

2.2.1.5 Optimization of Culture Time

The stable expression of reporter genes in PBNA usually takes a certain amount of time, which varies with the combination of different types of pseudotyped viruses and target cells. Live virus infection test requires virus replication to generate cytopathic effect (CPE), which takes a long time, usually more than 7 days. The pseudotyped virus infection only needs to achieve the effective expression of the report gene, which usually takes a short range of time. For example, the pseudotyped virus system of VSV has a short-expression cycle and can be detected in 24 h [37], while the pseudotyped virus system of HIV usually needs to be cultured for 48 h for detection [14], while the infection process of HPV is slow, and the time from infection to detection usually takes 72 h. A short culture period can improve the detection efficiency of the method, but it needs to reach the platform of high expression of reporter genes to ensure the stability of the method. For the newly established method, it is necessary to study the culture time systematically. When the other conditions of the same sample are fixed, the detection results of different culture time are compared, choosing a steady time range as the optimal.

2.2.1.6 Validation of In Vitro Assays Based on Pseudotyped Viruses

When the key conditions and parameters of the in vitro PBNA are determined, the method needs to be systematically validated to ensure the repeatability and comparability of the test results before it can be widely applied. The verification of the method usually includes the following aspects: specificity, linear range, precision, and accuracy.

2.2.1.7 Specificity

The specificity of the method refers to the ability of the method to accurately detect the target characteristics in the presence of possible confounding factors. The specificity of the method is related to the type of samples to be tested. The most common samples detected in PBNA are serum or plasma samples. Determination of specificity should be carried out for different types of samples. Usually, a certain number of negative samples are selected and detected in PBNA with a low starting dilution. The limit of detection for the method is calculated utilizing mean + 2- or 3-times SD according to the detected value, which is the cutoff value for this measure.

The cutoff value may vary with different sample types, for example, the anticoagulant in the plasma sample may have a high nonspecific effect compared to the serum samples. So, the cutoff value of the plasma sample is usually higher than that of the serum sample, and the sensitivity of the method to the plasma sample is relatively lower than that for the serum sample. The plasma sample with heparin as an anticoagulant has a high nonspecific response in many assays. Especially for the detection of HPV pseudotyped virus neutralization antibody, heparin is a specific inhibitor of HPV entering cells, and heparin anticoagulant plasma cannot be used to detect HPV neutralizing antibody [38].

The sample selection is critical for the determination of nonspecific effects. For some virus infections that cannot occur in a certain area or a certain period, samples can be collected in a population in that area or a certain period. For pathogens that are not prevalent in China, such as Ebola and Nipah, samples from the general population can be collected for nonspecific validation; samples before December 2019 [39] can be selected as nonspecific samples for the SARS-CoV-2 PBNA development. If there is a similar well-accepted validated standard detection method, we can use the classical method to screen the samples for validation of specificity. For example, the HIV-1 antibody diagnostic reagent can be used to screen the samples and then to validate the HIV-1 PBNA using the identified negative ones. However, there is no recognized method for the detection of antibodies to some viruses, such as the HPV antibody test with no gold standard detection reagent, while the correlation between HPV nucleic acid test and antibody test results is relatively poor. According to the transmission characteristics of HPV, HPV neutralization antibody detection is mainly used for vaccine type or high-risk virus neutralization antibody detection. These types of viruses are mainly transmitted through sexual contact, and samples from people without sexual experience could be selected as specificity validation samples.

2.2.1.8 Accuracy

The accuracy of the analytical method indicates the gap between the test results and the true value of the sample. The results of PBNA are usually compared with those of live virus detection methods or other types of well-recognized detection methods. Compared with the results of live virus detection (neutralizing monoclonal antibody or HIV-infected serum), the PBNA based on Tzm-bl cells is more sensitive than the live virus method based on PBMC. Generally speaking, the sensitivity of the former is about three times higher than that of the latter. However, for different samples and different virus strains, the degree of sensitivity is different: the sensitivity of monoclonal antibody samples is different, and for infected serum samples, the sensitivity difference is about 1.1 times [40]. When developing the HIV-1 pseudotyped virus neutralization test based on A3R5 cells, A3R5 was compared with the widely used pseudotyped virus test based on Tzm-bl cells to evaluate the correlation between the two methods and the accuracy of the new method [16, 17].

Although many live virus tests are considered to be classical methods, the operation of live virus tests is relatively tedious; the final result is determined by cytopathic effect or plaque reduction, which relies heavily on human experience; virus strains used in different laboratories continue to mutate in the process of culture, and the repeatability of different laboratory results is not ideal. Therefore, when comparing pseudotyped viruses with live viruses, the correlation between different viruses may be different, so it is impossible to establish a unified standard. The classical rabies neutralization antibody detection RFFIT test, using fluorescence staining to determine the results, usually yields more objective results. When the rabies PBNA was compared with this method, the correlation of the results was very high with a R^2 reaching 0.95 [20], while the correlation between Hantaan and Seoul virus pseudotyped viruses and live viruses was 0.91 and 0.82 [41], respectively. For the correlation between H7N9 PBNA and classical HI and MN methods, R^2 could reach 0.85 and 0.82 [19], respectively. Although the correlation coefficients are different, there is no doubt about the correlation of the methods, and there is a significant correlation in statistics ($p < 0.001$). This also shows that pseudotyped viruses can reflect the process of live virus infecting cells and can be used as an alternative method for the study of live virus infection.

2.2.1.9 Linear Range

The linear range refers to the specific range in which there is a linear relationship between the amount of the test sample and the value of the test signal. The pseudotyped virus inhibition curve is a typical four-parameter curve. The result of the pseudotyped virus detection report is usually ID_{50} or IC_{50}, that is, the titer or concentration of the sample corresponding to the middle point of the four-parameter curve. As long as the detection value is located in the straight-line segment of the inhibition curve, the accuracy of the result can be ensured. In a typical four-parameter curve within the inhibition range of 20% to 80%, there is a linear relationship between the concentration of neutralizing antibodies and the detection. The fitting degree of the linear range is usually represented by analyzing the linear phase relationship of the straight-line section.

2.2.1.10 Precision

The precision of the test refers to the deviation of the test results of the same sample during repeated testing. Precision can be divided into three levels, namely, variation within experiments, variation between experiments, and variation between laboratories. Intra-test variation refers to the difference of repeated test results of the same sample performed by the same operator or in the same test; variation between tests refers to the difference of repeated test results of the same sample among different operators or different test runs; the difference between laboratories refers to the difference in the results of testing the same sample between different laboratories. In the design of precision validation, attention should be paid to include samples with high, medium, and low neutralization antibody levels as far as possible and to examine the precision of the method at different antibody levels as comprehensively as possible.

The pseudotyped virus neutralization test involves cell culture steps, and the method has a higher degree of variation compared with the physicochemical or antibody binding test. Especially in the precision validation between laboratories, the coefficient of variation is usually not used to represent the verification results, for example, in the interlaboratory verification of neutralizing antibodies against HIV-1 pseudotyped viruses. The range of variation within three times the average value of the sample is defined as the acceptance range of the sample. When the test results of different laboratories fall within this range, the test results are considered to be consistent. Unifying the standard operating procedures (SOP) among different laboratories, introducing reference materials, and reporting the relative values of the test results can effectively improve the consistency of test results in different laboratories.

2.3 Development of In Vivo Assays Based on Pseudotyped Viruses

The establishment of animal models of a virus infection is very important for the study of virus infection in vivo and the evaluation of related therapeutic and prophylactic products. However, for some viruses with high-risk levels or that are difficult to obtain, it is difficult to establish an effective infection model because of the limitation of the virus and the facility. The emergence of pseudotyped virus technology has partially solved the accessibility of the virus, and it can be used to build infection models in an environment with low biosafety facilities. Moreover, pseudotyped viruses usually contain reporter genes, which can be expressed in vivo after infection. Through the in vivo imaging system (IVIS) of small animals, the infection of pseudotyped viruses can be observed in real time, and the potency of the antivirals can be evaluated in real time. The establishment of a pseudotyped virus infection model usually includes the following steps: the selection of model animals, the optimization of the infection route, the determination of pseudotyped virus infection dose, and so on.

2.3.1 Selection of Model Animals

For viruses of which the receptors are unknown, the first step for establishment of an infection model is to screen the proper model animals. The infection model of the pseudotyped virus mainly uses the small animals. On one hand, animal infection requires a large amount of pseudotyped virus per body weight, and small animals require relatively low demand for pseudotyped virus and virus titer; on the other hand, the detection of pseudovirus infection model usually uses the IVIS, as large animals cannot carry out real-time monitoring detection. Commonly used animals include mice, rats, guinea pigs, hamsters, and so on. Mouse is the most preferred model animal with low body weight and a variety of strains. First of all, different strains of mice with similar body weight were selected and inoculated with the same dose of pseudotyped virus. The expression of the reporter gene (usually firefly luciferase reporter gene) was observed by the IVIS at different time points, and the animal strain with the highest level of reporter gene expression was selected as the model animal.

For viruses of which infection receptors are identified but do not exist in existing strains of animals, specific animal models could be established. First of all, the model animals containing infection receptors should be firstly constructed. The human SCARB2 knock-in mice [42] infected by enterovirus A71 pseudotyped virus and human DPP4 knock-in mice [33] infected by MERS-CoV pseudotyped virus are constructed by gene knock-in to construct humanized mice with corresponding receptors to make them susceptible to the virus. In addition to gene knock-in, viral vectors carrying receptor genes can also infect animals and obtain viral receptors to construct susceptible animals. For example, animals can be infected with adenovirus vector [43] or adeno-associated virus vector [44] carrying the human ACE2 gene to obtain receptors needed for SARS-CoV-2 pseudotyped virus infection. Using the way of gene knock-in, the model animal construction may be complex, but can achieve the stable expression of the receptor, and can reproduce and passage. Using the way of virus vector infection, the operation is relatively simple and can try different kinds and strains of animals. However, the expression level of the gene transferred in this way is unstable, the expression level of the receptor decreases with the extension of feeding time, and the phenotype of model animals cannot be passed on to offspring.

2.3.2 *Optimization of Infection Pathway*

The choice of infection route is very important for the establishment of a pseudotyped virus infection model. The infection routes of the live virus could act as referred model for the establishment of the pseudotyped virus infection model. For example, the establishment of the model of HPV vaginal infection refers to the infection process of live virus. The natural infection site of HPV occurs in the columnar epithelial cells of the basement of the cervix. The prerequisite for the establishment of infection is the damage of the squamous epithelium and the exposure of columnar epithelial cells. Therefore, when establishing the model of HPV pseudotyped virus infection, the mice were first injected with the long-acting contraceptive Depo-Provera (Pfizer), which changed the hormone level of mice and caused the vaginal squamous epithelium to become thinner [45]. After that, the squamous epithelium was destroyed by physical measures (scrapes the epithelium with a cell brush) or chemical reagent (adding nonoxynol-9 into the vagina), and then the pseudotyped virus was added to infect the mouse [45, 46]. The mouse model of HPV skin infection was established in a similar way [47]. The main target organs of Nipah virus [30] and MERS-CoV[33] infection are the lungs. In the two pseudotyped virus infection tests, the intrathoracic injection was used to make the pseudotyped virus reach the target organ directly, and the infection model was established effectively.

When the natural infection mode is unable to establish the infection model effectively, lots of different challenging ways could be tried, such as intravenous injection, intraperitoneal injection, nasal antivirus, intracerebral injection, and so on. The infecting route with the highest signal value is usually fixed as the optimal challenging pathway. For example, the rabies pseudotyped virus infection model did not completely imitate the infection process of the live virus, in which the signal value could not be detected by intramuscular injection of the pseudotyped virus. Finally, the rabies pseudotyped virus infection model was injected through intravenous injection [20], while pseudotyped viruses of Ebola [31], Marburg [32], and Rift Valley fever [48] were all injected intraperitoneally to establish the animal infection models.

2.3.3 Determination of Infection Dose of Pseudotyped Virus

After the preliminary establishment of the pseudotyped virus infection model, it is necessary to optimize the pseudotyped virus infection dose, to ensure the stability of the method and the comparability of the detection results. Compared with the in vitro cell infection test, the amount of pseudotyped virus needed for animal infection in vivo is much higher. The higher the amount of pseudotyped virus added, the better the stability of the test results. But the addition of more pseudotyped viruses means that more neutralizing antibodies are needed to inhibit pseudotyped virus infection. To make the results of different tests stable and comparable, it is necessary to optimize and determine the injection volume of the pseudotyped virus. Usually, after diluting the pseudotyped virus serially, the model animals are infected, and the dose of pseudotyped virus that causes half of the infection of the animals, namely, AID_{50}, is calculated. To ensure that all models are infected, the usual dose of pseudotyped virus is not less than 10 AID_{50}. For example, the dose of the rabies, Nipah, and Rift Valley fever pseudotyped virus infection model is 40 AID_{50} [20], 50 AID_{50} [30], and 90 AID_{50} [48], respectively.

2.3.4 *Comparison of Pseudotyped and Live Virus Infection Model*

For the established pseudotyped virus infection model, we should try to compare it with the live virus infection model if applicable, especially for the potency evaluation of infection inhibitors. For example, by comparing the protective effects of monoclonal antibodies in the rabies pseudotyped virus infection model and live virus infection model, the comparability of the two methods was established. Rabies monoclonal antibody showed an obvious dose-effect relationship in both models, and the half effective dose for animal protection was also very close to each other [20]. In the comparative study of MERS-CoV pseudotyped virus and live virus infection models, it was found that in the two infection models, the main infected organs of virus and pseudotyped virus were both lungs, and the infected tissue cells were lung bronchial epithelial cells. And the corresponding live virus and the pseudotyped virus can be detected in the brain of both infection models. It is proven that the tissue infection tendency of the pseudotyped virus is the same as that of the live virus. The conversion of infection dose between pseudotyped virus and the live virus can be realized by comparing the infection dose. In the human DPP4 knock-in mouse model, the infection effect of 1 $TCID_{50}$ MERS-CoV pseudotyped virus is equivalent to that of 0.004 6 PFU live virus. This also fully proves that the pseudotyped virus infection model can be used as an alternative measure to the live virus model [33].

2.4 Conclusion

The in vitro and in vivo test methods are critical for the evaluation of the antivirals acting in the infection phase. Usually, the detection methods of the live virus are tedious, long-time-consuming, and requiring high-level biosafety facilities. Moreover, some live viruses are difficult to culture or obtain, which limits the development of evaluation research. The development of pseudotyped virus approaches solves the above problems to some extent and can be used as an alternative to live viruses. And some of them have been widely used in antiviral laboratory research or clinical sample evaluation. For cell-based in vitro tests and animal model based in vivo tests, the degree of variation of the method is relatively large, so it is necessary to fully optimize and validate it in the process of development to ensure the repeatability and comparability of the test results.

Acknowledgments This work was supported by the General Program of National Natural Science Foundation of China [grant number 82172244&82073621] and major project of Study on Pathogenesis and Epidemic Prevention Technology System [2021YFC2302500].

References

1. Sanders, D.A.: No false start for novel pseudotyped vectors. Curr. Opin. Biotechnol. **13**, 437–442 (2002)
2. Li, Q., Liu, Q., Huang, W., Li, X., Wang, Y.: Current status on the development of pseudoviruses for enveloped viruses. Rev. Med. Virol. **28** (2018)
3. Plotkin, S.A.: Vaccines: correlates of vaccine-induced immunity. Clin. Infect. Dis. **47**, 401–409 (2008)
4. Nie, J., et al.: Establishment and validation of a pseudovirus neutralization assay for SARS-CoV-2. Emerg Microbes Infect. **9**, 680–686 (2020)
5. Nie, J., et al.: Quantification of SARS-CoV-2 neutralizing antibody by a pseudotyped virus-based assay. Nat. Protoc. **15**, 3699–3715 (2020)
6. Buck, C.B., Pastrana, D.V., Lowy, D.R., Schiller, J.T.: Efficient intracellular assembly of papillomaviral vectors. J. Virol. **78**, 751–757 (2004)
7. Wu, X., et al.: Detection of HPV types and neutralizing antibodies in Gansu province, China. *J Med Virol.* **81**, 693–702 (2009)
8. Wu, X.L., Zhang, C.T., Zhu, X.K., Wang, Y.C.: Detection of HPV types and neutralizing antibodies in women with genital warts in Tianjin City, China. *Virol Sin.* **25**, 8–17 (2010)
9. Dalgleish, A.G., et al.: The CD4 (T4) antigen is an essential component of the receptor for the AIDS retrovirus. Nature. **312**, 763–767 (1984)
10. Doms, R.W.: Beyond receptor expression: the influence of receptor conformation, density, and affinity in HIV-1 infection. Virology. **276**, 229–237 (2000)
11. Feng, Y., Broder, C.C., Kennedy, P.E., Berger, E.A.: HIV-1 entry cofactor: functional cDNA cloning of a seven-transmembrane, G protein-coupled receptor. *Science.* **272**, 872–877 (1996)
12. Platt, E.J., Wehrly, K., Kuhmann, S.E., Chesebro, B., Kabat, D.: Effects of CCR5 and CD4 cell surface concentrations on infections by macrophagetropic isolates of human immunodeficiency virus type 1. J. Virol. **72**, 2855–2864 (1998)
13. Derdeyn, C.A., et al.: Sensitivity of human immunodeficiency virus type 1 to the fusion inhibitor T-20 is modulated by coreceptor specificity defined by the V3 loop of gp120. J. Virol. **74**, 8358–8367 (2000)
14. Wei, X., et al.: Emergence of resistant human immunodeficiency virus type 1 in patients receiving fusion inhibitor (T-20) monotherapy. Antimicrob. Agents Chemother. **46**, 1896–1905 (2002)
15. Nie, J., et al.: Optimization and proficiency testing of a pseudovirus-based assay for detection of HIV-1 neutralizing antibody in China. J. Virol. Methods. **185**, 267–275 (2012)
16. Sarzotti-Kelsoe, M., et al.: Optimization and validation of a neutralizing antibody assay for HIV-1 in A3R5 cells. J. Immunol. Methods. **409**, 147–160 (2014)
17. Chen, Q., et al.: Development and optimization of a sensitive pseudovirus-based assay for HIV-1 neutralizing antibodies detection using A3R5 cells. Hum. Vaccin. Immunother. **14**, 199–208 (2018)
18. Todd, C.A., et al.: Development and implementation of an international proficiency testing program for a neutralizing antibody assay for HIV-1 in TZM-bl cells. J. Immunol. Methods. **375**, 57–67 (2012)
19. Tian, Y., et al.: Development of in vitro and in vivo neutralization assays based on the pseudotyped H7N9 virus. Sci. Rep. **8**, 8484 (2018)
20. Nie, J., et al.: Development of in vitro and in vivo rabies virus neutralization assays based on a high-titer pseudovirus system. Sci. Rep. **7**, 42769 (2017)
21. Montefiori, D.C.: Measuring HIV neutralization in a luciferase reporter gene assay. Methods Mol. Biol. **485**, 395–405 (2009)
22. Frazer, I.H.: Measuring serum antibody to human papillomavirus following infection or vaccination. Gynecol. Oncol. **118**, S8–S11 (2010)
23. Sarzotti-Kelsoe, M., et al.: Optimization and validation of the HIV-1 neutralizing antibody assay in A3R5 cells. Retrovirology. **9** (2012)

24. Gilson, T.D., Gibson, R.T., Androphy, E.J.: Optimization of human papillomavirus-based pseudovirus techniques for efficient gene transfer. Sci. Rep. **10**, 15517 (2020)
25. Nie, J., Liu, Y., Huang, W., Wang, Y.: Development of a triple-color Pseudovirion-based assay to detect neutralizing antibodies against human papillomavirus. Viruses. **8**, 107 (2016)
26. McLinden, R.J., et al.: Detection of HIV-1 neutralizing antibodies in a human CD4(+)/CXCR4 (+)/CCR5(+) T-lymphoblastoid cell assay system. PLoS One. **8**, e77756 (2013)
27. Sarzotti-Kelsoe, M., et al.: Optimization and validation of the TZM-bl assay for standardized assessments of neutralizing antibodies against HIV-1. J. Immunol. Methods. **409**, 131–146 (2014)
28. Nkolola, J.P., et al.: Breadth of neutralizing antibodies elicited by stable, homogeneous clade a and clade C HIV-1 gp140 envelope trimers in Guinea pigs. J. Virol. **84**, 3270–3279 (2010)
29. Bontjer, I., et al.: Stabilized HIV-1 envelope glycoprotein trimers lacking the V1V2 domain, obtained by virus evolution. J. Biol. Chem. **285**, 36456–36470 (2010)
30. Nie, J., et al.: Nipah pseudovirus system enables evaluation of vaccines in vitro and in vivo using non-BSL-4 facilities. Emerg Microbes Infect. **8**, 272–281 (2019)
31. Liu, Q., et al.: Antibody-dependent-cellular-cytotoxicity-inducing antibodies significantly affect the post-exposure treatment of Ebola virus infection. Sci. Rep. **7**, 45552 (2017)
32. Zhang, L., et al.: A bioluminescent imaging mouse model for Marburg virus based on a pseudovirus system. Hum. Vaccin. Immunother. **13**, 1811–1817 (2017)
33. Fan, C., et al.: A human DPP4-Knockin Mouse's susceptibility to infection by authentic and Pseudotyped MERS-CoV. Viruses. **10** (2018)
34. Huang, Y., et al.: Calibration of two validated SARS-CoV-2 pseudovirus neutralization assays for COVID-19 vaccine evaluation. Sci. Rep. **11**, 23921 (2021)
35. Pastrana, D.V., et al.: Reactivity of human sera in a sensitive, high-throughput pseudovirus-based papillomavirus neutralization assay for HPV16 and HPV18. Virology. **321**, 205–216 (2004)
36. Sehr, P., et al.: High-throughput pseudovirion-based neutralization assay for analysis of natural and vaccine-induced antibodies against human papillomaviruses. PLoS One. **8**, e75677 (2013)
37. Whitt, M.A.: Generation of VSV pseudotypes using recombinant DeltaG-VSV for studies on virus entry, identification of entry inhibitors, and immune responses to vaccines. J. Virol. Methods. **169**, 365–374 (2010)
38. Day, P.M., Baker, C.C., Lowy, D.R., Schiller, J.T.: Establishment of papillomavirus infection is enhanced by promyelocytic leukemia protein (PML) expression. Proc. Natl. Acad. Sci. U. S. A. **101**, 14252–14257 (2004)
39. Zhou, P., et al.: A pneumonia outbreak associated with a new coronavirus of probable bat origin. Nature. **579**, 270–273 (2020)
40. Heyndrickx, L., et al.: International network for comparison of HIV neutralization assays: the NeutNet report II. PLoS One. **7**, e36438 (2012)
41. Ning, T., et al.: Monitoring neutralization property change of evolving Hantaan and Seoul viruses with a novel Pseudovirus-based assay. Virol. Sin. **36**, 104–112 (2021)
42. Zhou, S., et al.: A safe and sensitive enterovirus A71 infection model based on human SCARB2 knock-in mice. Vaccine. **34**, 2729–2736 (2016)
43. Tseng, S.H., et al.: A novel pseudovirus-based mouse model of SARS-CoV-2 infection to test COVID-19 interventions. J. Biomed. Sci. **28**, 34 (2021)
44. Israelow, B., et al.: Mouse model of SARS-CoV-2 reveals inflammatory role of type I interferon signaling. J. Exp. Med. **217** (2020)

45. Cuburu, N., Cerio, R.J., Thompson, C.D., Day, P.M.: Mouse model of cervicovaginal papillomavirus infection. Methods Mol. Biol. **1249**, 365–379 (2015)
46. Roberts, J.N., et al.: Genital transmission of HPV in a mouse model is potentiated by nonoxynol-9 and inhibited by carrageenan. Nat. Med. **13**, 857–861 (2007)
47. Handisurya, A., et al.: Murine skin and vaginal mucosa are similarly susceptible to infection by pseudovirions of different papillomavirus classifications and species. Virology. **433**, 385–394 (2012)
48. Ma, J., et al.: In vitro and in vivo efficacy of a Rift Valley fever virus vaccine based on pseudovirus. Hum. Vaccin. Immunother. **15**, 2286–2294 (2019)

Chapter 3
Application of Pseudotyped Viruses

Qianqian Cui and Weijin Huang

Abstract Highly pathogenic emerging and reemerging viruses have serious public health and socioeconomic implications. Although conventional live virus research methods can more reliably investigate disease pathogenicity and evaluate antiviral products, they usually depend on high-level biosafety laboratories and skilled researchers; these requirements hinder in vitro assessments of efficacy, as well as efforts to test vaccines and antibody drugs. In contrast, pseudotyped viruses (i.e., single-round infectious viruses that mimic the membrane structures of various live viruses) are widely used in studies of highly pathogenic viruses because they can be handled in biosafety level 2 facilities. This chapter provides a concise overview of various aspects of pseudotyped virus technologies, including (1) exploration of the mechanisms of viral infection; (2) evaluation of the efficacies of vaccines and monoclonal antibodies based on pseudovirion-based neutralization assay; (3) assessment of antiviral agents (i.e., antibody-based drugs and inhibitors); (4) establishment of animal models of pseudotyped virus infection in vivo; (5) investigation of the evolution, infectivity, and antigenicity of viral variants and viral glycosylation; and (6) prediction of antibody-dependent cell-mediated cytotoxic activity.

Keywords Pseudotyped virus · Application · Infection mechanism · Neutralizing antibodies · In vitro · In vivo

Abbreviations

ACE2	Angiotensin-converting enzyme 2
ADCC	Antibody-dependent cell-mediated cytotoxicity
BASV	Bas-Congo virus

Q. Cui · W. Huang (✉)
Division of HIV/AIDS and Sex-transmitted Virus Vaccines, National Institutes for Food and Drug Control (NIFDC) and WHO Collaborating Center for Standardization and Evaluation of Biologicals, Beijing, China
e-mail: huangweijin@nifdc.org.cn

Y. Wang (ed.), *Pseudotyped Viruses*, Advances in Experimental Medicine and Biology 1407, https://doi.org/10.1007/978-981-99-0113-5_3

COVID-19	Coronavirus disease 2019
DPP4	Dipeptidyl peptidase 4
EBOV	Ebola virus
EV71	Enterovirus 71
EV71	Enterovirus 71
hACE2	Human angiotensin-converting enzyme 2
HIV	Human immunodeficiency virus
HPV	Human papillomavirus
MARV	Marburg virus
MERS-CoV	Middle East respiratory syndrome coronavirus
NiV	Nipah virus
PBNA	Pseudovirion-based neutralization assay
SARS-CoV	Severe acute respiratory syndrome coronavirus
SARS-CoV-2	Severe acute respiratory syndrome coronavirus 2
VSV	Vesicular stomatitis virus

In recent decades, serious threats to human health and socioeconomic conditions have arisen from major epidemics and pandemics involving pathogenic human viruses such as enteroviruses, dengue virus, influenza viruses, Middle East respiratory syndrome coronavirus (MERS-CoV), Ebola virus (EBOV), Chikungunya virus, Nipah virus (NiV), and (most recently) severe acute respiratory syndrome coronavirus 2 (SARS-CoV-2). To improve control of the major epidemics, more comprehensive studies of these viruses are needed. However, studies of highly pathogenic live viruses are restricted to high-containment biosafety level 3 or 4 laboratories because of their high infectivity and pathogenicity; this restriction limits investigations of the mechanisms of infection by these viruses, as well as the development of corresponding vaccines and treatments. Pseudotyped virus systems are recombinant viral reporter systems that are composed of modified vesicular stomatitis virus (VSV) or lentiviral vectors that include reporter genes as the backbone and the presence of various envelope glycoproteins or capsid surface proteins [1]. Because pseudotyped viruses are single-replication systems, they are safer and can be manipulated in biosafety level 2 laboratories; these changes accelerate the speed of research concerning highly pathogenic viruses. Many studies have used pseudotyped viruses to investigate viral pathogenesis, determine cell or tissue tropism [2], identify viral receptors [3], screen antiviral drugs [4], and detect neutralizing antibodies in convalescent individuals and vaccine recipients [5].

3.1 Analysis of Viral Infectivity

Viral entry is the first essential step in the viral replication cycle. Pseudotyped viruses have been extensively used in studies of the mechanisms of viral entry. They have also been validated as useful tools for in vivo studies, particularly concerning interactions between viral proteins and host cell membranes.

3.1.1 Receptor Usage

Receptor recognition and subsequent interactions are the first steps of viral entry into target cells. Pseudotyped viruses that include capsid or envelope proteins can mimic the processes by which live viruses enter host cells. Specific cell membrane receptors that attach to viral proteins have been identified and characterized using pseudotyped viruses. Pseudotyped viruses packaged with envelope glycoproteins of filoviruses, such as Marburg virus (MARV) and EBOV, have been used to identify target receptors by means of a cDNA encoding folate receptor-α that expresses in cells. Multiple studies have shown that folate receptor-α is a vital cofactor for cellular entry by MARV and EBOV [3, 6]. SARS-CoV-2 spike pseudotyped viruses provided evidence that the host cell factors angiotensin-converting enzyme 2 (ACE2) and transmembrane serine protease 2 play key roles in the early stages of target cell entry by SARS-CoV-2 [7]. Pseudotyped viruses are also used to explore the key residues on receptor-binding domains [8, 9]. Similarly, pseudotyped virus infection assays and binding analysis (surface plasmon resonance) have confirmed that specific residues in human dipeptidyl peptidase 4 (hDPP4) are critical for hDPP4-receptor-binding domain and viral entry of MERS-CoV [9]. Receptor-binding assays demonstrated that EphrinB3 and EphrinB2 are both functional receptors for NiV. EphrinB3 was validated as an alternative receptor using both pseudotyped and live NiV. Furthermore, two key residues in EphrinB3 were shown to affect receptor-ligand binding during NiV entry and infection [8]. Twenty-six animal receptors were screened to determine their binding affinities to the SARS-CoV-2 receptor-binding domain; the results showed that ACE2 receptors from various species could bind to the SARS-CoV-2 receptor-binding domain. These binding results are consistent with the findings from studies of SARS-CoV-2 pseudotyped virus transduction [10].

3.1.2 Cellular Tropism

Virus-receptor interactions play key regulatory roles in viral host range, tissue tropism, and viral pathogenesis. Pseudotyped virus systems can also allow preliminary assessment of viral infection characteristics, such as cell tropism. Many studies have used pseudotyped virus systems to determine the types of cells that can be infected by highly pathogenic viruses [11, 12]. The lentiviral Chikungunya virus glycoprotein system exhibited broad cellular and tissue tropism in vitro, but it could not infect most hematopoietic cells. Notably, cells within joints and muscles showed stronger cell tropism [13, 14]. The cell tropism characteristics of SARS-CoV-2 and SARS-CoV were studied using spike protein-based pseudotyped viruses that could infect various cell lines and primary cells. The results showed that SARS-CoV-2 cell entry could be mediated by ACE2-dependent and ACE2-independent mechanisms, thus providing a putative molecular basis for its broad cell tropism [12]. Li et al. established vesicular stomatitis virus (VSV)-based recombinant SARS-CoV-2 pseudotyped virus particles; they confirmed that the pseudotyped virus displayed cell tropism similar to the cell tropism of the wild-type virus [2]. A mixed-cell pseudotyped virus infection assay was established to determine the degrees of ACE2 dependence during infection among various SARS-like coronaviruses [8]. Steffen et al. added Bas-Congo virus (BASV) glycoprotein to glycoprotein-deficient VSV core particles; they explored the ability of BASV glycoprotein to mediate viral entry into target cells. Diverse tissues and species were identified as targets of BASV-glycoprotein (G) infection in vitro [15]. Additionally, studies of tropism have been conducted using various cell lines and types targeted by emerging pseudotyped viruses, such as Ross River virus [16], MERS-CoV [17], EBOV [18], and influenza virus [19].

3.2 In Vitro Pseudovirion-Based Neutralization Assay (PBNA)

Because the neutralizing antibody level is a valid indicator of immune protection from virus infection, it is used to assess the efficacy of relevant vaccines. The scientific principle of the PBNA is that pseudotyped viruses with various reporter genes can be inhibited from infecting host cells in the presence of serum with neutralizing antibodies. The level of decline in the reporter signal after infection is positively correlated with antibody titer [20]. The PBNA has been used to determine serum immunogenicity after vaccine immunization, as well as the presence of neutralizing antibodies in serum samples from convalescent patients. Importantly, the World Health Organization (WHO) guidelines for vaccine production and control recommend use of the PBNA to assess the efficacies of vaccines against human papillomavirus (HPV) [21], enterovirus 71 (EV71) [22], EBOV [23], human immunodeficiency virus (HIV), and others [24].

3.2.1 Development and Evaluation of Vaccines

Pseudotyped virus technology has been widely used for in vitro/in vivo vaccine evaluation, particularly for vaccines against HPV [25, 26], influenza [27, 28], EV71

[29, 30], and EBOV [23]. Additionally, pseudotyped virus systems have accelerated the process of vaccine development against viruses including HIV [31–33], MERS-CoV [34], and SARS-CoV-2 [35].

John Schiller initially established the PBNA method for HPV [36]. On the basis of the Schiller method, Nie improved the detection flux and detection type [37], thus producing new methods that involved different fluorescent markers for three serotypes of detection; this approach has been used by HPV vaccine research groups for in vivo assessments of the efficacy of HPV vaccine, as well as evaluation of vaccine immunogenicity in clinical trials [38]. In guidelines established by the WHO, neutralizing antibodies are regarded as the gold standard for evaluating the immunoprotective effects of HPV vaccines. The applicability and scientific validity of the PBNA have been proven in the evaluation of HPV vaccines produced by Merck. The PBNA can be used instead of competitive enzyme-linked immunosorbent assays to evaluate the efficacy of vaccines after immunization [25, 26].

Concerning influenza virus, both the hemagglutination inhibition assay and viral microneutralization test are used to evaluate the immune response to influenza vaccines. Systematic analyses of the methods, standard operating procedures, and quality control of the influenza PBNA have shown that the PBNA is strongly correlated with the results of hemagglutination inhibition assay, viral microneutralization test, and individual diffusion methods [27]. Stability assays have shown that pseudotyped viruses are very stable after multiple rounds of freezing and thawing under basic cold chain conditions. Compared with conventional methods, the PBNA has advantages in terms of better sensitivity, improved biological safety, high throughput, and simple operation. In addition to assessing the immunogenicity of influenza vaccines, the PBNA has been used to evaluate cross-protective effects; this application is particularly useful for determining the cross-protective effects of universal influenza vaccine candidates and for serological investigations of seasonal influenza [27, 39].

Neutralization assays based on both pseudotyped and live EBOV have been used to detect the neutralizing activities of equine immunoglobulin fragments against EBOV. Similar results were obtained for pseudotyped and live EBOV; the pseudotyped EBOV-determined 50% neutralization titer (NT50) value (1:20,480) is very close to the live EBOV-determined NT50 value (1:21,333) [40]. In 2020, Janssen received conditional approval of the Ad26.ZEBOV vaccine from the European Commission. The WHO Global Advisory Committee on Vaccine Safety accepted the results concerning the immunogenicity and protective efficacy of Ad26.ZEBOV, which were determined by PBNA during phase 1 and 2 clinical trials [23].

Pseudovirion-based neutralization assays are useful for evaluating the neutralization abilities of existing vaccines; they are also important for assessing vaccine candidates and for early stages of vaccine development. Pseudotyped virus technology has been widely used in the immunogenicity assessment of HIV vaccines in clinical trials worldwide [31, 33, 41]. The method of detecting neutralizing antibodies to pseudotyped viruses has been optimized and verified in the world. To simulate viral diversity, pseudotyped virus libraries have been used in combination with vaccine virus strains to evaluate broad-spectrum vaccines.

Because of the dramatic increase in the number of coronavirus disease 2019 (COVID-19) cases worldwide, many companies and universities have been working to develop vaccines against the highly pathogenic SARS-CoV-2. Researchers have established VSV-based pseudotyped viruses to measure neutralizing antibodies against SARS-CoV-2 [20]. Furthermore, Nie et al. verified the reliability of the pseudotyped virus assay system using serum samples from healthy volunteers and patients with COVID-19 [35]. To screen for potential broad-spectrum and long-lasting vaccines, neutralizing antibodies against SARS-CoV-2 variants have been detected in serum from immunized animals during early phases of research. SARS-CoV-2 pseudotyped viruses for variants of concern will be useful to measure the neutralizing capacities of vaccine candidates in preclinical models and clinical trials; the pseudotyped viruses will also provide insights concerning strategies to prevent the emergence of phenotypes that can escape neutralization.

3.2.2 Screening and Validation of Monoclonal Neutralizing Antibodies

The in vitro efficacies of monoclonal antibodies have been compared between live virus assays and the PBNA; the results of the two methods are closely correlated. Therefore, the PBNA can serve as an alternative in vitro method to assess the efficacies of therapeutic monoclonal antibodies; it has particular advantages over binding activity for in vitro assessment of potency [35]. The PBNA has been used to evaluate the in vitro efficacies of prophylactic and therapeutic antibodies against HIV [32, 42], EBOV [43], MERS-CoV[44], SARS-CoV-2 [45], and other viruses [46].

An HIV-1 pseudotyped virus library including 462 viral strains from China was developed to evaluate broad-spectrum monoclonal neutralizing antibodies against HIV[32]. Monoclonal antibodies isolated from infected survivors exhibited neutralizing activity against recent and previous epidemic variants of EBOV [43]. Lentiviral pseudotyped viruses expressing the MERS-CoV spike protein were used to characterize novel neutralizing monoclonal antibodies against MERS-CoV spike protein [44].

The rapid evolution of SARS-CoV-2 in humans has led to the loss of vaccine efficacy and the urgent need for development of effective COVID-19 treatments [45]. Research efforts have been focused on isolating potential neutralizing monoclonal antibodies from patients recovering from COVID-19, as well as screening for broadly neutralizing antibodies against SARS-CoV-2 variants [5, 47]. Candidate antibodies with neutralizing activity against SARS-CoV-2 variants can be rapidly studied using the PBNA. Multiple monoclonal antibodies have been identified and then evaluated in clinical trials; some have received emergency authorization for use worldwide [48, 49]. A single type of antibody cannot effectively control viruses with rapid mutagenesis. Multiple studies have shown that antibody cocktails provide broad and synergistic neutralization against variants of concern; this approach may prevent SARS-CoV-2 escape. Compared with wild-type SARS-CoV-2, Omicron variants have >30 mutations in the spike protein, which has caused many candidate antibodies to demonstrate less or no efficacy [50]. The PBNA offers a tool for screening a wide range of neutralizing monoclonal antibodies against Omicron variants.

3.3 Screening and Validation of Viral Entry Inhibitors

There is a class of antiviral drugs that target distinct stages of viral infection. The efficacy of antiviral drugs that block the process of viral entry into cells can be evaluated by pseudotyped virus-based inhibition assays. Because pseudotyped viruses are easy to produce, are amenable to automated assays, have low costs, and support efficient screening, numerous high-throughput screening platforms based on pseudotyped viruses have been developed and applied to the identification of entry inhibitors for Lassa virus [51], EBOV [52], MARV [53], influenza virus H5N1 [54], NiV[4], HIV [55], and (most recently) SARS-CoV-2 [54, 56].

Pseudotyped virus assays can be used to screen for potential broad-spectrum viral entry inhibitors, including Food and Drug Administration-approved drugs and traditional Chinese herbal medicines that block the interactions between viral proteins and their receptors. HIV-based pseudotyping platforms have been used for the high-throughput screening of 1200 Food and Drug Administration-approved drugs for MARV, influenza virus H5N1, and Lassa virus. The results showed that putative entry inhibitor small molecules produced broad-spectrum inhibition of MARV, avian influenza virus, and Lassa virus [54]. Similarly, Elshabrawy et al. used a high-throughput screening protocol to characterize a broad-spectrum antiviral small molecule from a chemical library of 5000 compounds with activities against SARS-CoV, EBOV, Hendra virus, and NiV[4]. The principle of the broad-spectrum antiviral small molecule is to prevent cathepsin L-mediated cleavage and thus block the entry of pseudotypes bearing the glycoprotein from these viruses [4]. Additionally, the HIV membrane fusion inhibitors screened by Yuxian He and colleagues have entered clinical trials [57]. Zhao et al. screened compound- or peptide-based MERS-CoV entry inhibitors using MERS-CoV pseudovirus inhibition assay [58]; they demonstrated consistent results with live virus and pseudovirus MERS-CoV inhibition assays.

Pseudotyped virus systems have been used to identify SARS-CoV-2 entry inhibitors. SARS-CoV-2 pseudotyped viruses based on VSV-G or HIV-1 packing systems have been used to identify potential active drugs from diverse arrays of known histamine H1 receptor antagonists and numerous natural products, respectively [56, 59]. EK1 and its lipopeptide, EK1C4, are pan-CoV fusion inhibitors that exhibit potent antiviral activity against pseudotyped SARS-CoV-2 and its variants; they have been validated in various animal models, supporting further clinical development [60].

3.4 Animal Model of Pseudotyped Virus Infection In Vivo

Authentic in vivo models of viral infection are important for gaining an understanding of the mechanisms of infection. Conventional assessments of host-pathogen interactions require lab animals to be euthanized at multiple time points; this enables analysis of the stages of infection and quantification of the amounts of each pathogen in various anatomical tissues. A visual dynamic monitoring mouse model has also been established using pseudotyped viruses to observe dynamic distributions in animals after viral infection; the activities of antibodies, vaccines, and viral inhibitors have been evaluated using the model. Animals are first infected with a high-titer pseudotyped virus; if the vaccine or monoclonal antibody treatment induces a protective effect, the pseudotyped virus is presumed to be neutralized by neutralizing antibodies that can prevent the pseudotyped virus from infecting tissue. Additionally, animal models can be used to conduct in vivo evaluations in an animal biosafety level 2 environment.

Human DPP4-knock-in mice have been infected with live and pseudotyped MERS-CoV. The infection patterns of live and pseudotyped MERS-CoV were similar; both could infect the lungs. The protective effects of neutralizing antibodies were demonstrated by using MERS-CoV live virus and pseudotyped virus infection models; the results of the two methods were closely correlated [61]. After high titers of rabies virus infect Kunming mice, an in vivo imaging mouse model of pseudotyped virus infection enabled observation of the tissue tropism and dynamics of the virus, thus allowing greater visualization of the rabies virus infection process [62]. Although an hACE2 gene knock-in mouse model of live SARS-CoV-2 virus infection caused histopathological changes consistent with pneumonia in experimental mice and played an important role in the promotion of therapeutic agents and vaccines for COVID-19 [63], there is considerable safety risk involved in studies with live virus; moreover, different viral strains are needed to evaluate the cross-protective effects against other coronaviruses. The use of SARS-CoV-2 pseudotyped virus in a mouse model enables noninvasive investigation of the mechanisms of infection and tissue distribution. Yamada et al. [64] utilized VSV-based SARS-CoV-2 pseudotyped virus to construct an optical imaging model in hamsters based on the presence of a luminescence reporter gene; they found that lung luminescence reached a peak at 48 hours after infection.

The bioluminescence model of pseudotyped virus infection facilitates in vivo imaging for preliminary evaluation of the immunogenicity and efficacy of vaccine candidates and monoclonal antibodies. Liu et al. constructed an in vivo bioluminescent imaging model of MARV infection in mice, using the MARV-pseudotyped virus. The bioluminescence model of pseudotyped virus infection was used to evaluate the in vivo efficacies of three novel anti-EBOV monoclonal antibodies; one anti-EBOV monoclonal antibody (M318) exhibited unprecedented potency. Three in vivo models of pseudo-filovirus-infection in mice were successfully established using EBOV, MARV, and Lloviu virus as representative viruses of filovirus genera. The mouse models of pseudo-filovirus-infection were used to evaluate the in vivo activities of filovirus entry inhibitors [65]. Additionally, imaging models have been widely used for quantitative analyses of monoclonal antibodies and vaccines against pathogens such as Rift Valley fever virus [66], EV71 [67], Chikungunya virus [68], and influenza virus H7N9 [69].

3.5 Analysis of Variations in Viral Infectivity and Antigenicity

Variations in amino acid sequences and posttranslational modifications, particularly regarding glycosylation on glycoproteins that are responsible for membrane recognition and fusion, may affect viral structure and function. Pseudotyped virus systems permit rapid investigation of whether a mutation affects viral infectivity and immunogenicity by changing the corresponding genes in the vector derived from the pathogenic virus surface proteins.

3.5.1 Viral Variants

Point mutations (sources of genetic variation) and recombination lead to viral evolution and continuous emergence of new viral variants. For viruses with extensive genetic diversity and rapid viral mutagenesis, pseudotyped virus technology enables rapid construction of different mutant strains and epidemic strains in a manner that is easier to manage than the acquisition of live viruses; this approach can also reduce safety concerns. Knowledge of whether high-frequency amino acid mutations affect viral infectivity and antigenicity can help to predict antibody potency against emerging variants. Pseudotyped viruses have been widely used for in vitro analyses of the antigenicity of viral mutants [70, 71].

A library of representative 83 pseudotyped rabies viruses was constructed using the VSV vector for analyses of evolution, infectivity, and antigenicity. Phylogenetic analyses showed no obvious correlations in terms of phylogeny, infectivity, or antigenicity. Amino acid mutations at several residues in the G gene led to changes in infection outcomes and immune escape [70]. The rapid and continuous evolution of HIV poses a considerable challenge to human immune systems and vaccine development. The genetic subtypes of HIV-1 identified in China have been characterized in terms of neutralization activity by analysis of serum from infected patients; broad monoclonal antibody neutralization has been demonstrated using pseudotyped viruses [72].

Influenza A is the most widespread form of influenza virus. The high genetic complexity of influenza A virus increases the possibility of antigenic drift in the surface protein hemagglutinin (HA); this causes difficulties for infection prevention and control programs. Amino acid mutations in the HA gene alter viral antigenicity in most subtypes of influenza A virus [71, 73]. Additionally, various subtypes or strains of viruses have been used to evaluate the effects of vaccine-induced cross-protection in vivo [74]. The results of cross-reactive neutralizing activity analyses might explain some characteristics of the influenza pandemic. Emerging EV71 variants are driven by gene mutations and recombination, which result in EV71 epidemics. Zhang et al. established a neutralization assay based on pseudotyped viruses from different EV71 strains for the detection of cross-reactive neutralizing antibody responses to EV71 vaccine strains. Furthermore, an inactivated EV71 vaccine derived from EV71-C4a demonstrated broad cross-neutralization of EV71 strains with various genotypes [75]. In 2020, the WHO established recommendations to assure the quality, safety, and efficacy of EV71 vaccines as international standards; they suggested using different EV71 pseudotyped viruses to evaluate the cross-protective effects of candidate vaccines. Specifically, pseudotyped viruses derived from EV71 strains of different subgenogroups should be used in nonclinical and clinical trials to detect neutralizing antibody titers in serum samples from animals and humans, respectively [22].

Currently, SARS-CoV-2 variants constitute global health concerns, and their mutations may affect viral infectivity. Through their ability to synthesize spike-membrane proteins of different circulating coronavirus variants, pseudotyped viruses can be used to quickly monitor changes in viral infectivity caused by mutations involving the spike protein; the application of pseudotyped viruses may be helpful for preventing the spread of the pandemic. Since the COVID-19 pandemic began, our group has constructed pseudotyped viruses related to SARS-CoV-2 variants (particularly variants of concern) to estimate their infectivity. At the beginning of the epidemic, Li et al. observed that mutations containing D614G could strengthen infectivity [76], which is consistent with recent findings [77]. Subsequently, Zhang et al. [78] found that the infectivity of 10 SARS-CoV-2 variants was slightly increased in human cell lines. Additionally, high-frequency mutations have been the focus of recent studies to determine whether enhanced infectivity can be predicted in variants of concern [79]. Pseudotyped viruses can be used to track mutations and provide timely insights for strategies to develop vaccines against SARS-CoV-2. Overall, antibody testing based on the PBNA can be used to understand the prevalence and importance of viral variants.

3.5.2 Variations in Viral Glycosylation

Glycosylation is one of the most important and well-studied posttranslational modifications of various proteins. Envelope glycoproteins on viruses are heavily glycosylated. Glycosylation modifications and their functions in viruses must be characterized to more fully understand viral infection and entry into the host, as well as receptor recognition and immune escape. Envelope-based pseudotyped viruses allow for functional studies of glycosylation on viral glycoproteins.

Multiple studies have examined glycosylation-related mutations in various emerging viruses under natural conditions by using pseudotyping assays [80, 81]. Furthermore, the effects of site-specific glycosylation loss through the deletion of potential N-glycosylation sites have been studied to explore the functions of glycosylation in several human pathogenic viruses. Three mutagenesis strategies (asparagine to glutamine, asparagine to alanine, and serine/tyrosine to alanine) were performed for the deletion of glycans on glycoproteins from the Lassa virus; the results showed a decrease on the glycoprotein complex cleavage when the N89, N119, and N365 glycosylation sites were deleted. Additionally, pseudotyped viruses were constructed with N-glycosylation deletions (N to Q mutagenesis) in the glycoprotein complex to evaluate the effects of glycosylation deletions on viral infectivity. Glycosylation deletions of N89Q and N365Q resulted in the absence of pseudotyped virus infectivity, whereas glycosylation deletions of N109Q and N119Q resulted in the reduction of pseudotyped virus infectivity [80]. In a study by Li et al., most natural or putative glycosylation deletions reduced the infectivity of SARS-CoV-2, indicating the importance of glycosylation for viral infectivity [76].

Compared with N-linked glycosylation, the functions of O-linked viral protein glycosylation have not been fully evaluated. The functions of O-glycans in the paramyxovirus family have been studied using mutagenesis strategies. The loss of O-glycosylation sites in Hendra virus glycoprotein led to changes in membrane fusion, which altered viral infectivity via modified glycoprotein or receptor-induced conformation [81]. Overall, the glycosylation of viral proteins is a key regulatory aspect with respect to many cellular virulence processes, virus-host interactions, and immune evasion for multiple pathogenic viruses. Thus, there is a need for pseudotyped virus studies to explore the glycosylation of viral proteins.

3.6 Antibody-Dependent Cell-Mediated Cytotoxicity (ADCC)

In addition to neutralizing activity, therapeutic antibodies modulate the immune response through ADCC. Because pseudotyped virus surface proteins have a similar conformation with live virus, pseudoviruses can better simulate the entry of live virus into cells; this is superior to the conventional method for evaluating ADCC through plasmid transfection. When the ADCC is activated by monoclonal antibodies, the Fab counterpart of an antibody binds to a specific antigenic determinant or epitope on the target cell; its Fc domain concurrently binds to the Fc gamma receptor of the engineered ADCC effector cells (i.e., Jurkat cells), after the addition of engineered ADCC effector cells and monoclonal antibodies. The expression of the luciferase reporter gene was assessed to determine the ADCC characteristics in vivo. Thus, a method was established to evaluate the ADCC characteristics of antibodies to EBOV; the protective mechanisms of various monoclonal antibodies were studied using an in vivo model of EBOV pseudovirus infection [82].

3.7 Conclusion

Pseudotyped virus systems have some notable limitations: they only contain membrane proteins of the pathogenic virus, they are not replicative or pathogenic, and they only mimic the process of viral entry. Despite these limitations, the continuous improvements in pseudotyped virus technology indicate that there is great potential for further in vivo applications; the findings in studies of pseudotyped viruses are often consistent with the findings in studies of live viruses and the results of other conventional methods. A wider range of applications of pseudotyped virus should be explored in future studies. Pseudotyped virus technology, as a proxy for studies of live virus, is expected to gain wide use in evaluations of product quality, as well as mechanistic research. In conclusion, the pseudotyped virus technology described in this chapter offers a practical tool for determining the seroprevalence and clinical relevance of infection; it will play a key role in the treatment or prevention of infections with highly pathogenic viruses.

Acknowledgments We thank Ryan Chastain-Gross, Ph.D., from Liwen Bianji (Edanz) (www.liwenbianji.cn/) for editing the English text of a draft of this manuscript.

References

1. Li, Q., Liu, Q., Huang, W., Li, X., Wang, Y.: Current status on the development of pseudoviruses for enveloped viruses. Rev. Med. Virol. **28** (2018). https://doi.org/10.1002/rmv.1963
2. Li, H., et al.: Establishment of replication-competent vesicular stomatitis virus-based recombinant viruses suitable for SARS-CoV-2 entry and neutralization assays. Emerg Microbes Infect. **9**, 2269–2277 (2020). https://doi.org/10.1080/22221751.2020.1830715
3. Sinn, P.L., et al.: Lentivirus vectors pseudotyped with filoviral envelope glycoproteins transduce airway epithelia from the apical surface independently of folate receptor alpha. J. Virol. **77**, 5902–5910 (2003). https://doi.org/10.1128/jvi.77.10.5902-5910.2003
4. Elshabrawy, H.A., et al.: Identification of a broad-spectrum antiviral small molecule against severe acute respiratory syndrome coronavirus and Ebola, Hendra, and Nipah viruses by using a novel high-throughput screening assay. J. Virol. **88**, 4353–4365 (2014). https://doi.org/10.1128/JVI.03050-13
5. Zost, S.J., et al.: Potently neutralizing and protective human antibodies against SARS-CoV-2. Nature. **584**, 443–449 (2020). https://doi.org/10.1038/s41586-020-2548-6
6. Chan, S.Y., et al.: Folate receptor-alpha is a cofactor for cellular entry by Marburg and Ebola viruses. Cell. **106**, 117–126 (2001). https://doi.org/10.1016/s0092-8674(01)00418-4
7. Hoffmann, M., et al.: SARS-CoV-2 cell entry depends on ACE2 and TMPRSS2 and is blocked by a clinically proven protease inhibitor. Cell. **181**, 271-280 e278 (2020). https://doi.org/10.1016/j.cell.2020.02.052
8. Roelle, S.M., Shukla, N., Pham, A.T., Bruchez, A.M., Matreyek, K.A.: Expanded ACE2 dependencies of diverse SARS-like coronavirus receptor binding domains. PLoS Biol. **20**, e3001738 (2022). https://doi.org/10.1371/journal.pbio.3001738
9. Song, W., et al.: Identification of residues on human receptor DPP4 critical for MERS-CoV binding and entry. Virology. **471-473**, 49–53 (2014). https://doi.org/10.1016/j.virol.2014.10.006

10. Wu, L., et al.: Broad host range of SARS-CoV-2 and the molecular basis for SARS-CoV-2 binding to cat ACE2. Cell Discov. **6**, 68 (2020). https://doi.org/10.1038/s41421-020-00210-9
11. Ou, X., et al.: Characterization of spike glycoprotein of SARS-CoV-2 on virus entry and its immune cross-reactivity with SARS-CoV. Nat. Commun. **11**, 1620 (2020). https://doi.org/10.1038/s41467-020-15562-9
12. Zhang, F., et al.: SARS-CoV-2 pseudovirus infectivity and expression of viral entry-related factors ACE2, TMPRSS2, Kim-1, and NRP-1 in human cells from the respiratory, urinary, digestive, reproductive, and immune systems. J. Med. Virol. **93**, 6671–6685 (2021). https://doi.org/10.1002/jmv.27244
13. Hu, D., et al.: Chikungunya virus glycoproteins pseudotype with lentiviral vectors and reveal a broad spectrum of cellular tropism. PLoS One. **9**, e110893 (2014). https://doi.org/10.1371/journal.pone.0110893
14. Salvador, B., Zhou, Y., Michault, A., Muench, M.O., Simmons, G.: Characterization of chikungunya pseudotyped viruses: identification of refractory cell lines and demonstration of cellular tropism differences mediated by mutations in E1 glycoprotein. Virology. **393**, 33–41 (2009). https://doi.org/10.1016/j.virol.2009.07.013
15. Steffen, I., et al.: Characterization of the Bas-Congo virus glycoprotein and its function in pseudotyped viruses. J. Virol. **87**, 9558–9568 (2013). https://doi.org/10.1128/JVI.01183-13
16. Sharkey, C.M., North, C.L., Kuhn, R.J., Sanders, D.A.: Ross River virus glycoprotein-pseudotyped retroviruses and stable cell lines for their production. J. Virol. **75**, 2653–2659 (2001). https://doi.org/10.1128/JVI.75.6.2653-2659.2001
17. Muller, M.A., et al.: Human coronavirus EMC does not require the SARS-coronavirus receptor and maintains broad replicative capability in mammalian cell lines. *mBio*. **3** (2012). https://doi.org/10.1128/mBio.00515-12
18. MacKenzie, T.C., et al.: Efficient transduction of liver and muscle after in utero injection of lentiviral vectors with different pseudotypes. Mol. Ther. **6**, 349–358 (2002). https://doi.org/10.1006/mthe.2002.0681
19. Wallerstrom, S., et al.: Detection of antibodies against H5 and H7 strains in birds: evaluation of influenza pseudovirus particle neutralization tests. Infect Ecol Epidemiol. **4** (2014). https://doi.org/10.3402/iee.v4.23011
20. Nie, J., et al.: Quantification of SARS-CoV-2 neutralizing antibody by a pseudotyped virus-based assay. Nat. Protoc. **15**, 3699–3715 (2020). https://doi.org/10.1038/s41596-020-0394-5
21. Huang, W., Wang, Y.: The application of Pseudotyped virus Technology in the Evaluation of prevention and control products for emerging viral infectious diseases *Chinese*. J. Virol. **6**, 1177–1186 (2020). https://doi.org/10.13242/j.cnki.bingduxuebao.003680-en
22. Lei, D., Griffiths, E., Martin, J.: WHO working group meeting to develop WHO recommendations to assure the quality, safety and efficacy of enterovirus 71 vaccines. Vaccine. **38**, 4917–4923 (2020). https://doi.org/10.1016/j.vaccine.2020.05.001
23. Roozendaal, R., et al.: Nonhuman primate to human immunobridging to infer the protective effect of an Ebola virus vaccine candidate. NPJ Vaccines. **5**, 112 (2020). https://doi.org/10.1038/s41541-020-00261-9
24. Organization, W.H. *Consolidated guidelines on HIV, viral hepatitis and STI prevention, diagnosis, treatment and care for key populations*. https://apps.who.int/iris/rest/bitstreams/1453332/retrieve. (2022)
25. Krajden, M., et al.: Assessment of HPV 16 and HPV 18 antibody responses by pseudovirus neutralization, Merck cLIA and Merck total IgG LIA immunoassays in a reduced dosage quadrivalent HPV vaccine trial. Vaccine. **32**, 624–630 (2014). https://doi.org/10.1016/j.vaccine.2013.09.007
26. Krajden, M., et al.: Human papillomavirus 16 (HPV 16) and HPV 18 antibody responses measured by pseudovirus neutralization and competitive Luminex assays in a two- versus three-dose HPV vaccine trial. Clin. Vaccine Immunol. **18**, 418–423 (2011). https://doi.org/10.1128/CVI.00489-10

27. Carnell, G.W., Ferrara, F., Grehan, K., Thompson, C.P., Temperton, N.J.: Pseudotype-based neutralization assays for influenza: a systematic analysis. Front. Immunol. **6**, 161 (2015). https://doi.org/10.3389/fimmu.2015.00161
28. Trombetta, C.M., Perini, D., Mather, S., Temperton, N., Montomoli, E.: Overview of serological techniques for influenza vaccine evaluation: past, present and future. *Vaccines (Basel)*. **2**, 707–734 (2014). https://doi.org/10.3390/vaccines2040707
29. Wu, X., et al.: Development and evaluation of a pseudovirus-luciferase assay for rapid and quantitative detection of neutralizing antibodies against enterovirus 71. PLoS One. **8**, e64116 (2013). https://doi.org/10.1371/journal.pone.0064116
30. Chen, P., et al.: Molecular determinants of enterovirus 71 viral entry: cleft around GLN-172 on VP1 protein interacts with variable region on scavenge receptor B 2. J. Biol. Chem. **287**, 6406–6420 (2012). https://doi.org/10.1074/jbc.M111.301622
31. Sarzotti-Kelsoe, M., et al.: Optimization and validation of the TZM-bl assay for standardized assessments of neutralizing antibodies against HIV-1. J. Immunol. Methods. **409**, 131–146 (2014). https://doi.org/10.1016/j.jim.2013.11.022
32. Nie, J., Huang, W., Liu, Q., Wang, Y.: HIV-1 pseudoviruses constructed in China regulatory laboratory. Emerg Microbes Infect. **9**, 32–41 (2020). https://doi.org/10.1080/22221751.2019.1702479
33. Laher, F., et al.: Safety and immune responses after a 12-month booster in healthy HIV-uninfected adults in HVTN 100 in South Africa: a randomized double-blind placebo-controlled trial of ALVAC-HIV (vCP2438) and bivalent subtype C gp120/MF59 vaccines. PLoS Med. **17**, e1003038 (2020). https://doi.org/10.1371/journal.pmed.1003038
34. Almasaud, A., Alharbi, N.K., Hashem, A.M.: Generation of MERS-CoV Pseudotyped viral particles for the evaluation of neutralizing antibodies in mammalian sera. Methods Mol. Biol. **2099**, 117–126 (2020). https://doi.org/10.1007/978-1-0716-0211-9_10
35. Kalkeri, R., et al.: SARS-CoV-2 spike Pseudoviruses: a useful tool to study virus entry and address emerging neutralization escape phenotypes. Microorganisms. **9** (2021). https://doi.org/10.3390/microorganisms9081744
36. Pastrana, D.V., et al.: Reactivity of human sera in a sensitive, high-throughput pseudovirus-based papillomavirus neutralization assay for HPV16 and HPV18. Virology. **321**, 205–216 (2004). https://doi.org/10.1016/j.virol.2003.12.027
37. Nie, J., Huang, W., Wu, X., Wang, Y.: Optimization and validation of a high throughput method for detecting neutralizing antibodies against human papillomavirus (HPV) based on pseudovirons. J. Med. Virol. **86**, 1542–1555 (2014). https://doi.org/10.1002/jmv.23995
38. Nie, J., Liu, Y., Huang, W., Wang, Y.: Development of a triple-color Pseudovirion-based assay to detect neutralizing antibodies against human papillomavirus. Viruses. **8**, 107 (2016). https://doi.org/10.3390/v8040107
39. Zhou, H., et al.: Sequential immunization with consensus influenza hemagglutinins raises cross-reactive neutralizing antibodies against various heterologous HA strains. Vaccine. **35**, 305–312 (2017). https://doi.org/10.1016/j.vaccine.2016.11.051
40. Cao, Z., et al.: The application of a safe neutralization assay for Ebola virus using lentivirus-based Pseudotyped virus. Virol. Sin. **36**, 1648–1651 (2021). https://doi.org/10.1007/s12250-021-00405-8
41. Montefiori, D.C.: Measuring HIV neutralization in a luciferase reporter gene assay. Methods Mol. Biol. **485**, 395–405 (2009). https://doi.org/10.1007/978-1-59745-170-3_26
42. Barouch, D.H., et al.: Therapeutic efficacy of potent neutralizing HIV-1-specific monoclonal antibodies in SHIV-infected rhesus monkeys. Nature. **503**, 224–228 (2013). https://doi.org/10.1038/nature12744
43. Corti, D., et al.: Protective monotherapy against lethal Ebola virus infection by a potently neutralizing antibody. Science. **351**, 1339–1342 (2016). https://doi.org/10.1126/science.aad5224
44. Goo, J., et al.: Characterization of novel monoclonal antibodies against MERS-coronavirus spike protein. Virus Res. **278**, 197863 (2020). https://doi.org/10.1016/j.virusres.2020.197863

45. Shrestha, L.B., Foster, C., Rawlinson, W., Tedla, N., Bull, R.A.: Evolution of the SARS-CoV-2 omicron variants BA.1 to BA.5: implications for immune escape and transmission. Rev. Med. Virol. **e2381** (2022). https://doi.org/10.1002/rmv.2381
46. Zhang, C., et al.: Epitope clustering analysis for vaccine-induced human antibodies in relationship to a panel of murine monoclonal antibodies against HPV16 viral capsid. Vaccine. **36**, 6761–6771 (2018). https://doi.org/10.1016/j.vaccine.2018.09.035
47. Pinto, D., et al.: Cross-neutralization of SARS-CoV-2 by a human monoclonal SARS-CoV antibody. Nature. **583**, 290–295 (2020). https://doi.org/10.1038/s41586-020-2349-y
48. Jones, B.E., et al.: LY-CoV555, a rapidly isolated potent neutralizing antibody, provides protection in a non-human primate model of SARS-CoV-2 infection. bioRxiv. (2020). https://doi.org/10.1101/2020.09.30.318972
49. Boucau, J., et al.: Monoclonal antibody treatment drives rapid culture conversion in SARS-CoV-2 infection. Cell Rep Med. **3**, 100678 (2022). https://doi.org/10.1016/j.xcrm.2022.100678
50. Shafiq, A., et al.: Investigation of the binding and dynamic features of a.30 variant revealed higher binding of RBD for hACE2 and escapes the neutralizing antibody: a molecular simulation approach. Comput. Biol. Med. **146**, 105574 (2022). https://doi.org/10.1016/j.compbiomed.2022.105574
51. Zhang, X., Tang, K., Guo, Y.: The antifungal isavuconazole inhibits the entry of Lassa virus by targeting the stable signal peptide-GP2 subunit interface of Lassa virus glycoprotein. Antivir. Res. **174**, 104701 (2020). https://doi.org/10.1016/j.antiviral.2019.104701
52. Cote, M., et al.: Small molecule inhibitors reveal Niemann-pick C1 is essential for Ebola virus infection. Nature. **477**, 344–348 (2011). https://doi.org/10.1038/nature10380
53. Zhang, X., et al.: Synthesis and biological evaluation of novel tricyclic matrinic derivatives as potential anti-filovirus agents. Acta Pharm. Sin. B. **8**, 629–638 (2018). https://doi.org/10.1016/j.apsb.2018.01.006
54. Wang, J., et al.: A comparative high-throughput screening protocol to identify entry inhibitors of enveloped viruses. J. Biomol. Screen. **19**, 100–107 (2014). https://doi.org/10.1177/1087057113494405
55. Chong, H., Zhu, Y., Yu, D., He, Y.: Structural and functional characterization of membrane fusion inhibitors with extremely potent activity against human immunodeficiency virus type 1 (HIV-1), HIV-2, and simian immunodeficiency virus. J. Virol. **92** (2018). https://doi.org/10.1128/JVI.01088-18
56. Gonzalez-Maldonado, P., et al.: Screening of natural products inhibitors of SARS-CoV-2 entry. Molecules. **27** (2022). https://doi.org/10.3390/molecules27051743
57. Xu, L., et al.: DNA triplex-based complexes display anti-HIV-1-cell fusion activity. Nucleic Acid Ther. **25**, 219–225 (2015). https://doi.org/10.1089/nat.2015.0535
58. Zhao, G., et al.: A safe and convenient pseudovirus-based inhibition assay to detect neutralizing antibodies and screen for viral entry inhibitors against the novel human coronavirus MERS-CoV. Virol. J. **10**, 266 (2013). https://doi.org/10.1186/1743-422X-10-266
59. Ge, S., et al.: Repositioning of histamine H1 receptor antagonist: doxepin inhibits viropexis of SARS-CoV-2 spike pseudovirus by blocking ACE2. Eur. J. Pharmacol. **896**, 173897 (2021). https://doi.org/10.1016/j.ejphar.2021.173897
60. Xia, S., et al.: Structural and functional basis for pan-CoV fusion inhibitors against SARS-CoV-2 and its variants with preclinical evaluation. Signal Transduct. Target. Ther. **6**, 288 (2021). https://doi.org/10.1038/s41392-021-00712-2
61. Fan, C., et al.: A human DPP4-Knockin Mouse's susceptibility to infection by authentic and Pseudotyped MERS-CoV. Viruses. **10** (2018). https://doi.org/10.3390/v10090448
62. Nie, J., et al.: Development of in vitro and in vivo rabies virus neutralization assays based on a high-titer pseudovirus system. Sci. Rep. **7**, 42769 (2017). https://doi.org/10.1038/srep42769
63. Bao, L., et al.: The pathogenicity of SARS-CoV-2 in hACE2 transgenic mice. Nature. **583**, 830–833 (2020). https://doi.org/10.1038/s41586-020-2312-y
64. Yamada, H., et al.: A novel hamster model of SARS-CoV-2 respiratory infection using a pseudotyped virus. Sci. Rep. **12**, 11125 (2022). https://doi.org/10.1038/s41598-022-15258-8

65. Chen, Q., Tang, K., Zhang, X., Chen, P., Guo, Y.: Establishment of pseudovirus infection mouse models for in vivo pharmacodynamics evaluation of filovirus entry inhibitors. Acta Pharm. Sin. B. **8**, 200–208 (2018). https://doi.org/10.1016/j.apsb.2017.08.003
66. Ma, J., et al.: In vitro and in vivo efficacy of a Rift Valley fever virus vaccine based on pseudovirus. Hum. Vaccin. Immunother. **15**, 2286–2294 (2019). https://doi.org/10.1080/21645515.2019.1627820
67. Zhou, S., et al.: A safe and sensitive enterovirus A71 infection model based on human SCARB2 knock-in mice. Vaccine. **34**, 2729–2736 (2016). https://doi.org/10.1016/j.vaccine.2016.04.029
68. Wu, J., Zhao, C., Liu, Q., Huang, W., Wang, Y.: Development and application of a bioluminescent imaging mouse model for chikungunya virus based on pseudovirus system. Vaccine. **35**, 6387–6394 (2017). https://doi.org/10.1016/j.vaccine.2017.10.007
69. Tian, Y., et al.: Development of in vitro and in vivo neutralization assays based on the pseudotyped H7N9 virus. Sci. Rep. **8**, 8484 (2018). https://doi.org/10.1038/s41598-018-26822-6
70. Cai, M., et al.: Analysis of the evolution, infectivity and antigenicity of circulating rabies virus strains. Emerg Microbes Infect. **11**, 1474–1487 (2022). https://doi.org/10.1080/22221751.2022.2078742
71. Ning, T., et al.: Antigenic drift of influenza a(H7N9) virus hemagglutinin. J. Infect. Dis. **219**, 19–25 (2019). https://doi.org/10.1093/infdis/jiy408
72. Shang, H., et al.: Genetic and neutralization sensitivity of diverse HIV-1 env clones from chronically infected patients in China. J. Biol. Chem. **286**, 14531–14541 (2011). https://doi.org/10.1074/jbc.M111.224527
73. Zhu, R., et al.: HA gene amino acid mutations contribute to antigenic variation and immune escape of H9N2 influenza virus. Vet. Res. **53**, 43 (2022). https://doi.org/10.1186/s13567-022-01058-5
74. Labrosse, B., et al.: Detection of extensive cross-neutralization between pandemic and seasonal a/H1N1 influenza viruses using a pseudotype neutralization assay. PLoS One. **5**, e11036 (2010). https://doi.org/10.1371/journal.pone.0011036
75. Zhang, H., et al.: Analysis of cross-reactive neutralizing antibodies in human HFMD serum with an EV71 pseudovirus-based assay. PLoS One. **9**, e100545 (2014). https://doi.org/10.1371/journal.pone.0100545
76. Li, Q., et al.: The impact of mutations in SARS-CoV-2 spike on viral infectivity and antigenicity. Cell. **182**, 1284-1294 e1289 (2020). https://doi.org/10.1016/j.cell.2020.07.012
77. Zhang, L., et al.: SARS-CoV-2 spike-protein D614G mutation increases virion spike density and infectivity. Nat. Commun. **11**, 6013 (2020). https://doi.org/10.1038/s41467-020-19808-4
78. Zhang, L., et al.: Ten emerging SARS-CoV-2 spike variants exhibit variable infectivity, animal tropism, and antibody neutralization. Commun Biol. **4**, 1196 (2021). https://doi.org/10.1038/s42003-021-02728-4
79. Li, T., et al.: Aggregation of high-frequency RBD mutations of SARS-CoV-2 with three VOCs did not cause significant antigenic drift. J. Med. Virol. **94**, 2108–2125 (2022). https://doi.org/10.1002/jmv.27596
80. Zhu, X., et al.: Effects of N-linked glycan on Lassa virus envelope glycoprotein cleavage, infectivity, and immune response. Virol. Sin. **36**, 774–783 (2021). https://doi.org/10.1007/s12250-021-00358-y
81. Stone, J.A., Nicola, A.V., Baum, L.G., Aguilar, H.C.: Multiple novel functions of Henipavirus O-glycans: the first O-glycan functions identified in the paramyxovirus family. PLoS Pathog. **12**, e1005445 (2016). https://doi.org/10.1371/journal.ppat.1005445
82. Liu, Q., et al.: Antibody-dependent-cellular-cytotoxicity-inducing antibodies significantly affect the post-exposure treatment of Ebola virus infection. Sci. Rep. **7**, 45552 (2017). https://doi.org/10.1038/srep45552

Chapter 4
Pseudotyped Viruses for Retroviruses

Magan Solomon and Chen Liang

Abstract Since the discovery of retroviruses, their genome and replication strategies have been extensively studied, leading to the discovery of several unique features that make them invaluable vectors for virus pseudotyping, gene delivery, and gene therapy. Notably, retroviral vectors enable the integration of a gene of interest into the host genome, they can be used to stably transduce both dividing and nondividing cells, and they can deliver relatively large genes. Today, retroviral vectors are commonly used for many research applications and have become an active tool in gene therapy and clinical trials. This chapter will discuss the important features of the retroviral genome and replication cycle that are crucial for the development of retroviral vectors, the different retrovirus-based vector systems that are commonly used, and finally the research and clinical applications of retroviral vectors.

Keywords Retrovirus · Genome integration · Retroviral vector · Gene therapy · Pseudotyped virus

Abbreviations

ADA-SCID	Adenosine deaminase deficiency SCID
AIDS	Acquired immunodeficiency syndrome
ALSV	Avian leucosis/sarcoma viruses
bNAbs	Broadly neutralizing antibodies
CA	Capsid protein
CAS	*cis*-acting sequences
Cas	CRISPR-associated proteins

M. Solomon · C. Liang (✉)
Lady Davis Institute, Jewish General Hospital, McGill Centre for Viral Diseases, Montreal, QC, Canada

Department of Medicine, McGill University, Montreal, QC, Canada
e-mail: chen.liang@mcgill.ca

Y. Wang (ed.), *Pseudotyped Viruses*, Advances in Experimental Medicine and Biology 1407, https://doi.org/10.1007/978-981-99-0113-5_4

CMV	Cytomegalovirus
COVID-19	Coronavirus disease 2019
CRISPR	Clustered regularly interspaced short palindromic repeats
Env	Envelope glycoprotein
gRNA	Guide RNA
HA	Hemagglutinin
HIV	Human immunodeficiency virus
HSPC	Hematopoietic stem progenitor cells
IAV	Influenza A virus
IFITM	Interferon-induced transmembrane protein
IN	Integrase
LTR	Long terminal repeat
MA	Matrix protein
MLV	Murine leukemia virus
NC	Nucleocapsid protein
ORF	Open reading frame
PBS	Primer binding site
PFV	Prototype foamy virus
PPT	Poly-purine tract
PR	Viral protease
PRR	Pattern recognition receptor
RSV	Rous sarcoma virus
RT	Reverse transcriptase
SIN	Self-inactivating
SIV	Simian immunodeficiency virus
VLP	Viruslike particle
X-SCID	X-linked severe combined immunodeficiency

4.1 Introduction to Retroviruses

Like many other viruses, retroviruses were discovered while researchers sought to identify the causative agent of various diseases, particularly in animal species. The first retrovirus isolated, Rous sarcoma virus (RSV), causes sarcomas in chickens and belongs to avian leucosis/sarcoma viruses (ALSVs) [1]. The identification of several other retroviruses eventually followed while investigating the causative agents of diseases in various mammalian species, including mice (murine leukemia viruses; MLV) [2], cats (feline leukemia virus) [3], cattle (bovine leukemia virus) [4], horses (equine anemia infectious virus) [5], sheep (visna virus) [6], and monkeys (simian immunodeficiency virus; SIV) [7].

Until the late 1960s, replication of these viruses was poorly understood. In 1970, the landmark discovery of reverse transcriptase (RT) by Temin [8] and Baltimore [9] led to the revision of the central dogma of genetic information flow, and the

Retroviridae virus family was established. In 1981, the first case of what later became known as acquired immunodeficiency syndrome (AIDS) was reported in the United States. About 2 years later, Montagnier, Barré-Sinoussi, and colleagues successfully isolated the causative agent of AIDS, which was found to have reverse transcriptase activity suggesting that they had isolated a retrovirus, later named human immunodeficiency virus (HIV) [10]. Since their discovery, distinct features of the retroviral genome structure and replication strategies, including that of HIV, have been extensively characterized, many of which render them an important tool in virus pseudotyping that is particularly useful for both research and therapeutic applications.

4.2 The *Retrovirus* Genome and Replication Cycle: Important Regions for Retroviral Vectors

4.2.1 Genome Structure

Retroviruses have a single-stranded positive-sense RNA genome containing three open reading frames (ORFs) that encode *gag*, *pol*, and *env* genes [11]. The 5′ end of the viral genome contains the R (repeat) region followed by the U5 (unique 5′ region) region, while the 3′ end of the genome contains the U3 region (unique 3′ region) and another R region, which flank either ends of the single-stranded RNA genome (Fig. 4.1). Immediately downstream of the R and U5 regions at the 5′ end of the retroviral RNA is the primer binding site (PBS), an 18-nucleotide sequence, which binds a cellular tRNA that is used as a primer to initiate reverse transcription of viral RNA and the synthesis of the negative-strand DNA. The R regions play a key role in reverse transcription of the viral RNA by mediating the first template switch, which is required to complete DNA synthesis. Between the PBS and the start codon of *gag* exist several secondary RNA structures that function as viral RNA

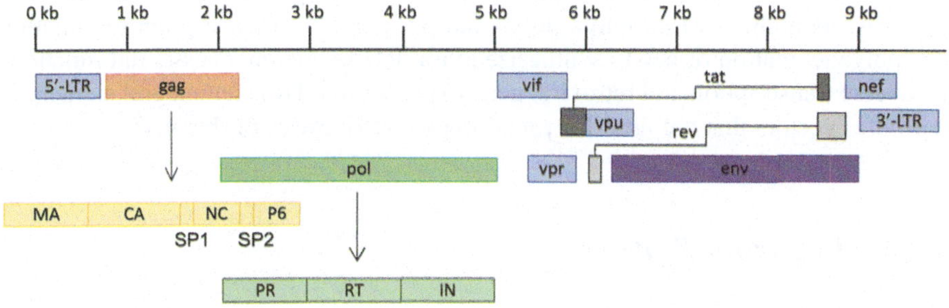

Fig. 4.1 Schematic of HIV-1 genome. The HIV-1 genome contains three open reading frames encoding Gag, Pol, and Env. Gag is transcribed as a polyprotein of six different structural proteins, which are processed by viral protease. Complex retroviruses like HIV-1 also encode regulatory proteins Tat and Rev. as well as accessory proteins Vif, Vpr, Vpu, and Nef

Table 4.1 Important genes and genetic sequences in lentiviral vectors

Gene/genetic sequences	Function	Necessity for HIV-1-based lentiviral vectors
PBS	Binds tRNA primer to initiate reverse transcription of VRNA	Essential
Ψ	Packaging signal required for packaging of transfer RNA	Essential
PPT	Involved in initiating plus-strand DNA synthesis	Essential
pol	Encodes proteins required for viral replication	Essential
gag	Encodes structural proteins	Essential
env	Encodes envelope glycoprotein	Essential
tat	Enhances viral gene expression	Optional
rev	Required for nuclear export of vRNA	Optional
vif/vpr/vpu/nef	Target host restriction factors and enhance transduction efficiency of retroviral vectors	Optional

dimerization and packaging signals (ψ). These signals have a strong affinity for the viral Gag protein and enable the selective encapsulation of two copies of viral genomic RNA into each virus particle. For some retroviruses including HIV-1, this RNA region also functions as the main splicing donor site, which is involved in the production of all spliced viral RNAs that encode viral envelope protein, as well as regulatory and accessory proteins. Adjacent to the 3′ end is a purine-rich sequence known as the poly-purine tract (PPT), which is required to initiate the synthesis of the plus-strand DNA. The PBS, ψ signal, and PPT are essential for a functional retroviral vector.

Once the retroviral genome is reverse transcribed, the double-stranded DNA contains long terminal repeats (LTRs) flanking both the 5′ and 3′ ends (Fig. 4.1). Each LTR contains three regions known as U3, R, and U5. The U3 regions of the viral DNA act as a promoter for cellular RNA polymerase II and as a binding site for other cellular transcription factors that regulate viral RNA synthesis, such as SP1 and NF-kB. In addition to its important role in reverse transcription, the R region also plays an important role in the transcription of the integrated viral DNA called a provirus. Its ability to fold into a polyA hairpin regulates transcription termination and polyadenylation of newly synthesized viral RNAs. Given its essential functions in reverse transcription and retroviral gene expression, LTR is an integral element in retroviral vectors that are used for pseudotyped retroviruses (Table 4.1).

4.2.2 *Retroviral Proteins*

All retroviruses carry the *gag* gene, which encodes the Gag polyprotein [12]. The Gag protein self-assembles to form a viruslike particle (VLP) and thus serves as the structural protein of pseudotyped retrovirus particles. Once the virus particle forms, Gag is processed into mature proteins by viral protease and gives rise to a total of six proteins: the matrix protein (MA), capsid protein (CA), nucleocapsid protein (NC), as well as p6 and two spacer peptides (SP1 and SP2) (Fig. 4.1) [13]. Each of these proteins has important and distinct functions throughout viral replication cycle. While MA has several functions both in the early and later stages of viral replication, it plays a central role in the attachment of the viral Gag protein to cellular membranes. More specifically, the glycine residue at the second position at the N terminus of MA is myristoylated, which anchors Gag to the lipid bilayer of cellular membranes, further strengthened by electrostatic interactions between the positively charged lysines of MA and the negatively charged phosphate groups of the phospholipids of the host cell membrane [14, 15]. CA forms the viral capsid structure that encapsulates viral genomic RNA. In addition to protecting viral RNA from being degraded upon viral entry, CA also shields viral RNA and DNA from being recognized by cellular pattern recognition receptors (PRRs), thereby preventing the activation of the antiviral interferon response. The NC domain of Gag specifically recognizes viral RNA packaging signals. While the p6 peptide does not have a structural role, a short peptide sequence, known as the late domain, is involved in recruiting host factors such as TSG101 and Alix that are involved in the closure of the viral membrane and the release of virus particles from the infected cell at late stages of viral assembly [16, 17]. Finally, SP1 links the CA to NC and has been shown to play a crucial role in the proper multimerization of Gag, which is essential at several stages of the viral replication cycle including viral budding and particle formation [18, 19].

The *pol* gene of retroviruses encodes three enzymes, the viral protease (PR), RT, and integrase (IN). Pol is synthesized as part of the Gag-Pol polyprotein, which occurs upon a ribosomal frameshift at the end of the Gag protein reading frame enabling the ribosomes to bypass the stop codon. When Gag and Gag-Pol assemble to form virus progenies, PR undergoes homodimerization, which activates its protease activity. Polyprotein processing gives rise to RT and IN and processes Gag into its mature proteins enabling the previously noninfectious virions to become infectious. As the signature enzyme of retroviruses, RT is packaged within virus progeny where it catalyzes the reverse transcription of viral RNA into DNA in newly infected cells. In addition to its RNA-dependent DNA polymerase activity, RT also contains RNase H activity, which digests the viral RNA after it has been reverse transcribed into viral DNA. Once viral DNA is synthesized, viral IN inserts viral DNA into cellular DNA to form a provirus.

Like all enveloped viruses, retroviruses encode the Envelope (Env) glycoprotein, which forms a trimer on the surface of virus particles [20]. Env mediates virus entry into the target cell through interacting with its respective cell surface receptor protein. The retroviral Env protein can be replaced by glycoproteins from other viruses to achieve pseudotyping, enabling the pseudovirus to infect various cells of interest. For certain retroviruses like HIV-1, co-receptors are required for virus entry. Upon entry into its host cell, the viral capsid is delivered into the cytoplasm where viral RNA undergoes reverse transcription to synthesize double-stranded DNA while it remains protected by CA. Once the viral DNA is integrated into the cellular DNA, known as the provirus, viral genes are transcribed and translated to give rise to viral proteins for virus progeny production. This unique feature enables pseudotyped retroviruses to deliver a packaged transfer plasmid that contains the gene of interest, also known as a transgene, into the target cell of interest, where it is integrated into the cellular DNA and eventually results in stable expression of the transgene.

In addition to structural proteins that are found in all retroviruses, more complex retroviruses encode two additional groups of genes that confer advantages for viral replication: regulatory and accessory genes. For example, HIV-1 encodes Tat and Rev. proteins, which dramatically enhance viral gene expression during transcription and are required for viral RNA nuclear export, respectively [21, 22]. While accessory genes are often not essential for viral replication in cultured cells, they often contribute to viral pathogenesis by interacting with host cell factors and counteracting host antiviral defense. This concept is elegantly exemplified by HIV-1 accessory proteins Vif, Vpr, Vpu, and Nef [23]. For example, Vif antagonizes cellular APOBEC3G, which otherwise causes hypermutation of viral DNA resulting in a nonfunctional viral genome [24–28]. Meanwhile, Vpu depletes tetherin protein from the cell surface, thereby preventing HIV-1 particles from being sequestered by this host restriction factor [29, 30]. Nef downregulates SERINC5 from the cell surface which otherwise gets incorporated into virus particles and inhibits viral infectivity [31, 32]. Similarly, SIV, a nonhuman primate lentivirus often used as a lentiviral vector, encodes viral protein Vpx, which downregulates SAMHD1, a dNTP hydrolase, which enhances the cells' susceptibility to retroviral infection [33, 34]. These retroviral antagonist proteins can be exploited to enhance the transduction efficiency of retroviral vectors. For example, SIV Vpx has been used to increase the lentiviral transduction of macrophages and dendritic cells which express high levels of SAMHD1 [35].

4.2.3 Viral RNA Replication

Using cellular tRNA that binds to the PBS, RT initiates reverse transcription and synthesizes DNA that is complimentary to the U5 and R regions of the viral genomic RNA [36]. Next, the RNase H domain of RT degrades the U5 and R regions at the 5′ end of the viral RNA. The newly synthesized minus strand DNA along with the tRNA subsequently translocate to the 3′ end via hybridization to the complementary R region at the 3′ end of the viral RNA. This process is known as the first template switch. Next, RT continues to synthesize the minus strand DNA. Once the entire minus strand DNA is completely synthesized, most of the viral RNA is degraded by RNase H. However, the PPT RNA sequence resists degradation by RT and serves as the primer to initiate the synthesis of plus-strand viral DNA. A second PPT, called central PPT (cPPT), resides within the *integrase* gene and generates a DNA flap, which facilitates the nuclear import of the viral DNA. Once the PBS in the positive-strand DNA is synthesized, it mediates the second template switch via hybridization to the complementary PBS sequence within the negative-strand DNA. Finally, RT extends both DNA strands to form a complete copy of double-stranded DNA. The various RNA elements previously described including the LTRs, PBS, PPT, and cPPT are essential for the function of retroviral vectors. Once double-stranded DNA synthesis is complete, IN catalyzes the integration of the newly synthesized viral DNA into the host genome to form the provirus. Unlike other viral vectors that result in the encoded transgene existing as episomal DNA that is lost over time with cell division, the integration of retroviruses enables the stable expression of the inserted gene (Fig. 4.2A). Genomic integration is a key feature that makes retroviral pseudotyped vectors a widely used tool in both research and clinical applications.

4.3 Retroviral Pseudotyping Systems

4.3.1 Overview of Retroviral Pseudotyping Plasmid Systems

Pseudotyped retrovirus production generally requires three key components, each of which is expressed using three separate plasmid vectors [37]. First, the packaging plasmid encodes Gag-Pol, which contains the structural and replication proteins including Gag, PR, RT, and IN. An envelope plasmid encodes the viral glycoprotein of interest that is expressed on the surface of the pseudovirus particle. Finally, a transfer plasmid consists of retroviral vector RNA encoding the transgene of interest to be expressed in the transduced target cells (Fig. 4.2B). These three plasmids are co-transfected into a producer cell type, for example, HEK293 cells, to produce pseudotyped retroviruses that can be harvested and subsequently used to infect the target cell type of interest. While first and second generation retroviral pseudotyping systems involve only these three key components, third-generation systems typically include a fourth plasmid encoding a retroviral regulatory gene. Further separation of the lentiviral genome into an additional plasmid improves safety by preventing the possibility of producing replication-competent recombinant virus. Like previous generation systems, all four plasmids are co-transfected in a producer cell type to produce pseudotyped lentiviral particles [37].

Much consideration has been made to improve the incorporation efficiency of the transgene vector into the pseudotyped retrovirus particles, reverse transcription of the transgene, expression of the gene of interest, and, equally important, safety. Various features that are now commonly found in retroviral vectors have been a part of the efforts to improve these properties. For instance, HIV-1 Tat is essential for HIV-1 replication as it acts as a transactivator for viral transcription. Today, Tat is deleted in commonly used lentiviral vector systems, and instead, the U3 region in the 5′ LTR in the transgene vector is often replaced with a strong viral promoter such as the immediate early gene promoter from cytomegalovirus (CMV) (Fig. 4.2C). The transgene is instead expressed from a heterologous promoter. Another major modification that is widely used to enhance safety is the deletion of the U3 region of the 3′ LTR, which otherwise acts as a strong viral enhancer and promoter resulting in the expression of nearby cellular genes. This safety feature, known as self-inactivating (SIN) retroviral vector, not only prevents the expression of viral RNA from the

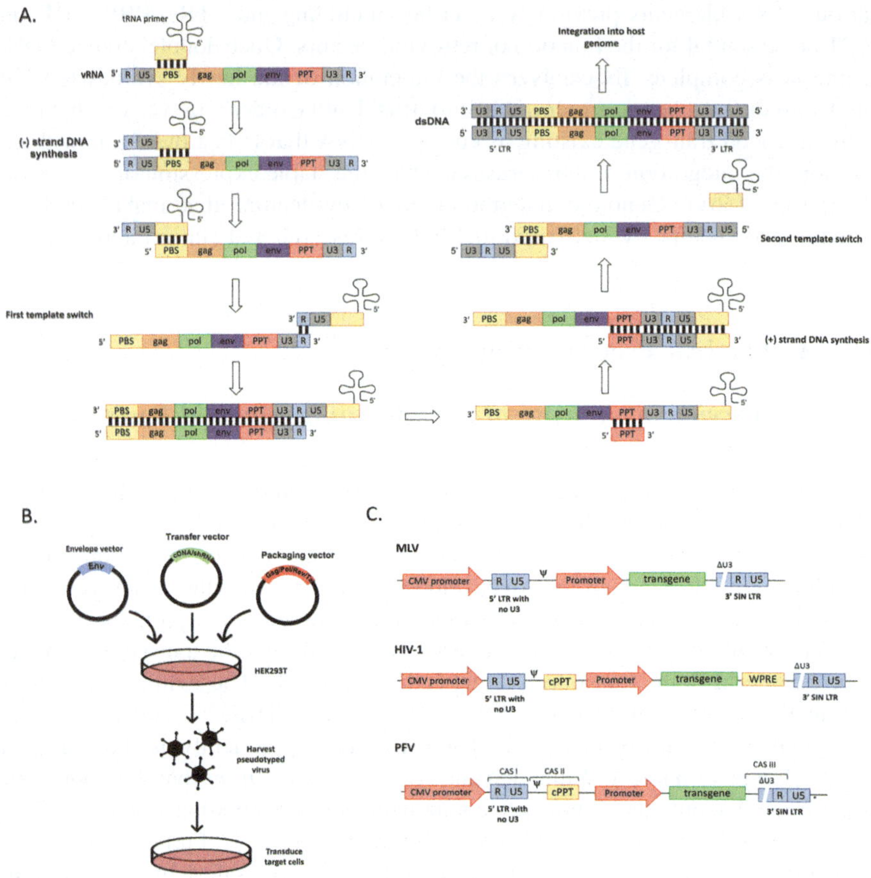

Fig. 4.2 The multistep process of reverse transcription and dsDNA synthesis and schematic of third-generation retroviral plasmid systems. (**A**) A cellular tRNA molecule hybridizes to the PBS on the vRNA where it acts as a primer for RT to initiate cDNA synthesis. Once the R and U5 regions are reverse transcribed by RT, the RNase H domain of RT degrades the R and U5 region at the 5′ end of the vRNA. Next, the R region of the newly synthesized cDNA hybridizes to the R region at the 3′ end of the vRNA. The DNA strand is then extended from the 3′ end of the cDNA by RT. Once cDNA synthesis is completed, the RNase H domain of RT degrades the majority of the vRNA except for the PPT region. The second DNA strand is then extended from the remaining PPT region, which is used as a primer. The cellular tRNA and the remaining vRNA are degraded by RNase H. The PBS region of the second DNA strand subsequently hybridizes with the PBS region of the first DNA strand. Finally, the second DNA strand is extended. The final dsRNA contains a 5′ and 3′ LTR and is integrated into the host genome. (**B**) Retroviral pseudotyping plasmid systems require three key components that are encoded on three separate plasmids: an envelope vector, a transfer vector encoding the transgene, and a packaging vector. The three plasmids are co-transfected into a producer cell type such as the commonly used HEK293T cell line, and pseudotyped viruses are subsequently harvested. Pseudotyped viruses are then used to transduce a target cell line to deliver the transgene of interest. (**C**) Schematic of commonly used transfer vectors

integrated viral DNA but also minimizes the risk of insertional activation of cellular proto-oncogenes neighboring the provirus integration site.

4.3.2 Commonly Used Retrovirus-Derived Transgene Vectors

4.3.2.1 MLV

One of the most widely used retroviral transgene vectors is derived from MLV [38]. MLV is a simple retrovirus that does not encode regulatory genes nor accessory genes. The most commonly used MLV retroviral vector is illustrated in Fig. 4.2C. A CMV promoter is typically used to replace the U3 region in the 5′ LTR to achieve high levels of retroviral vector RNA expression, which serves to enhance the production of pseudotyped viruses. This vector also contains a PBS and an RNA packaging signal, while the U3 region in the 3′ LTR is deleted. It is important to note, however, that MLV-based retroviral vectors can only infect dividing cells. Consequently, their use is restricted to rapidly proliferating cell types.

4.3.2.2 HIV-1 and SIV

Like MLV, transgene vectors derived from HIV-1 are one of the most commonly used retroviral vectors in pseudotyping (Fig. 4.2C) [39]. As opposed to MLV-based retroviral vectors, lentiviral vectors, including HIV-1-based pseudotyped viruses, can transduce both dividing and nondividing cells [40]. Lentiviral vectors were developed after the discovery of HIV-1 in 1983. The organization of lentiviral vectors is similar to that of MLV-based retroviral vectors. However, they contain extra elements to achieve higher transduction efficiency including a cPPT and posttranscriptional response elements such as the woodchuck hepatitis posttranscriptional response element (WPRE). While MLV and other retroviral vector systems use a three-plasmid system, third-generation lentiviral vector systems that are generally used require a fourth regulatory plasmid that encodes Rev., which regulates retroviral RNA nuclear export.

Considering that HIV-1-based lentiviral vectors are derived from the causative agent of acquired immunodeficiency syndrome (AIDS), its use in clinical applications remains controversial. As a result, researchers sought to design vectors derived from nonpathogenic lentiviruses with comparative transduction efficiency in both dividing and nondividing cells and biosafety properties to that of HIV-1-based lentiviral vectors. This led to the development and use of SIV-derived vectors, which are derived from a nonhuman lentivirus that is nonpathogenic in its natural host, nor in humans [41]. Like other lentiviral vector systems, SIV-based vector systems use a four-plasmid system.

4.3.2.3 PFV

In the last 15 years, transfer vectors derived from prototype foamy viruses (PFV), another retrovirus that is nonpathogenic in its natural host nor in humans, have also been developed [42]. Similar to lentiviral vectors, PFV vector systems also use four separate plasmids. Like other retrovirus-derived vectors, PFV-derived vectors contain a CMV promoter that replaces the U3 region and are self-inactivating to enhance safety (Fig. 4.2C). The transfer plasmid contains three *cis*-acting sequences (CAS), which are required for efficient gene transfer. CAS region I (CAS I) is located in the 5′ LTR and includes the PBS where cellular tRNA binds as a primer for reverse transcription. CAS II is located downstream of CAS I and includes RNA packaging, Pol encapsulation, and cPPT sequences. Finally, CAS III is located upstream of the 3′ LTR and contains the 3' PPT. Generally, the transgene is inserted between CAS II and CAS III. PFV has several features that render them important tools in pseudotyping. Like lentiviruses, PFV vectors can be used to transduce both dividing and nondividing cells. PFV also has the largest genome among retroviruses, which allows them to accommodate genes of interest that are up to about 9 kb into the transfer vector [43].

4.4 Applications of Pseudotyped Retroviruses

4.4.1 Functions of Viral Glycoproteins and Cell Entry

As pseudotyped retroviruses can carry heterologous Env proteins, a common application of pseudotyped retroviruses is to investigate the function of viral Env protein from different viruses. More specifically, heterologous Env proteins can be incorporated into MLV- and HIV-based pseudovirus particles, which is especially useful to study the entry step of those enveloped viruses that are highly pathogenic and would otherwise require high-level containment facilities. For example, the study of Ebola virus, a highly pathogenic virus that causes hemorrhagic fever with a high mortality rate in humans, is restricted to containment level 4 facilities. Accordingly, many research groups have used HIV-1 and MLV pseudotyped retroviruses to investigate different steps involved in cell entry of this virus [44, 45]. Similarly, pseudotyped retroviruses are commonly used to study HIV-1 cell entry in a biosafety level 2 containment facility, which is generally more accessible than biosafety level 3 facilities that are otherwise required for live HIV-1.

This platform is equally useful to study cell entry of viruses that rapidly accumulate mutations as a result of adaptions in culture due to the absence of selection pressures that exist in vivo, such as the host immune responses. For example, studies using live HIV-1 to investigate cell entry led to the virus acquiring mutations in the Env glycoprotein due to the error-prone nature of HIV-1 RT and conferred adaptation to tissue culture that resulted in phenotypic changes. As a result, cell entry of culture-adapted HIV-1 strains was inconsistent with what was observed in natural infections in human hosts [46, 47] and thus did not provide a clinically relevant model. Retroviruses pseudotyped with HIV-1 Env were used to overcome this problem that hindered research focusing on HIV-1 entry. In fact, it was the use of pseudotyped HIV-1 that led to a landmark study in which Deng and colleagues identified CC chemokine receptor 5 (CCR5) as a major co-receptor that is required for HIV-1 cell entry [48].

Moreover, due to the error-prone nature of HIV-1 RT, HIV-1 infection in individuals generally results in substantial genetic diversity. As a result, polymorphisms arise in viral proteins including the HIV-1 Env protein that may impact phenotypic properties including viral entry. HIV-1 pseudoviruses provide a practical platform to study the effect of different mutations within HIV-1 Env on cell entry, in particular, those of interest identified in HIV-1 infected individuals. More specifically, a panel of HIV-1 pseudoviruses containing unique mutations of interest in Env can be developed to investigate whether they impact viral entry and elucidate the mechanisms that may be involved.

Pseudotyped retroviruses are also pivotal in studying cell entry of infectious viruses that are either challenging or unable to be propagated in tissue culture. For example, the development of a cell culture system to grow hepatitis C virus (HCV) has been a major challenge since its discovery in the late 1980s. Up until the mid-2000s when the one and only chimeric HCV strain that replicates readily in culture was developed [49], retroviruses pseudotyped with HCV E1 and E2 proteins were pivotal in the discovery of important factors that are essential to mediate viral entry. For example, CD81 and SR-BI were identified as key cellular receptors required for HCV entry using pseudotyped retroviruses [50]. This platform was also used to elucidate the unique functions of E1 and E2, respectively, and how they interact with host receptors at the amino acid levels using a panel of pseudotyped retroviruses expressing E1 and E2 point mutation variants [51–53]. Pseudotyped retroviruses can equally be used to study cell entry of viruses with molecular clones that are unavailable or difficult to acquire.

Furthermore, many surface proteins of enveloped viruses undergo posttranslational modifications including protein glycosylation. The addition of glycans to surface proteins is known to have several important roles throughout viral replication, as well as in stability, antigenicity, and infectivity of enveloped viral particles [54]. Pseudotyped lentiviral particles are a useful tool to study the function and importance of glycosylation of surface proteins of various enveloped viruses. A panel of pseudotyped retroviral glycosylation mutants can be achieved by introducing mutations into potential glycosylation sites within the glycoprotein of interest. The panel of glycosylation mutants can subsequently be used to study whether protein glycosylation impacts various processes of enveloped viruses. For example, Wang et al. used pseudotyped lentiviral particles to create a panel of glycosylation mutants to successfully perform the first systematic study of the biological role of potential N-glycosylation sites on HIV-1 Env in viral infectivity and antibody-mediated neutralization to gain insight into the various functions of Env throughout the viral replication cycle [55]. Pseudotyped lentiviruses have equally played an important role in studying protein glycosylation of the Spike protein expressed on the surface of severe acute respiratory syndrome coronavirus 2 (SARS-CoV-2), the causative agent of the coronavirus disease 2019 (COVID-19) global pandemic. For example, several groups have used SARS-CoV-2 retrovirus-based pseudoviruses expressing different glycoforms of the Spike protein to elucidate the effect of blocking glycosylation on host receptor binding, viral entry, and antibody recognition [56, 57].

4.4.2 Identification and Characterization of Host Restriction

Neutralizing antibodies detection and analysis are important in characterizing the antibody response following viral infection or vaccination in any given host. Considering that neutralizing antibodies generally target the Envelope protein that is expressed on the surface of viral particles, retrovirus pseudotyping provides a convenient and safe platform to detect neutralizing antibodies against heterologous viruses. For example, HIV-1-derived pseudoviruses expressing the SARS-CoV-2 Spike protein have been produced and successfully used to measure neutralizing antibody titers against SARS-CoV-2 in the serum of individuals who have been infected as well as of those who have been vaccinated against COVID-19 [58]. Similarly, HIV-1 pseudoviruses are an important tool to measure HIV-1 neutralizing antibody titers. An important criterion of a successful HIV-1 vaccine is to induce the production of neutralizing antibodies that can protect against infection by diverse HIV-1 strains. These broadly neutralizing antibodies (bNAbs) target epitopes within the Env protein that are conserved among the circulating HIV-1 strains. Rather than using the many live infectious HIV-1 strains in circulation to identify bNAbs, panels of HIV-1 pseudoviruses expressing Env proteins derived from diverse HIV-1 strains can be used to investigate whether HIV-1 candidate vaccines successfully induce the production of bNAbs that neutralize diverse strains of HIV-1 (Table 4.2) [59].

In addition to their use in measuring neutralizing antibodies, retrovirus-based pseudotyped viruses are also commonly used to identify host restriction factors that function in the frontline antiviral defense by targeting viral entry. One example of the important role of retrovirus-based pseudoviruses in the characterization of host restriction factors is in elucidating how interferon-induced transmembrane protein family (IFITM) restricts viral replication. Following the discovery of the antiviral role of IFITM3 using an RNA interference screen to identify host facts that modulate influenza A virus (IAV) infection, it was the use of retroviruses pseudotyped with IAV hemagglutinin (HA) that led to the discovery that IFITM3 specifically targets viral entry [60]. More specifically, infection of pseudotyped virus expressing IAV HA was enhanced and inhibited upon depletion and exogenous expression of IFITM3, respectively, while retroviruses pseudotyped with Env proteins from MLV and other viruses were unaffected. These findings demonstrated that IFITM proteins restrict IAV viral entry. Since this discovery, retroviruses pseudotyped with

Table 4.2 Summary of research and clinical applications of pseudotyped retroviruses

	Application	Example
Research applications	Studying the function of viral glycoproteins in cell entry	Used to identify CCR5 as a major co-receptor required for HIV-1 cell entry (Deng et al., 1996)
	Studying post-translational modifications of viral glycoproteins	Used to study the role of protein glycosylation of SARS-CoV-2 Spike protein on host receptor binding, viral entry and antibody recognition (Yang et al., 2020; Huang et al, 2022)
	Identification and characterization of host restriction	Used to elucidate the role of IFITM3 in restricting IAV host cell entry (Feeley et al., 2011)
	Discovery of HIV-1 antivirals	Used for high-throughput screening to identify novel post-entry HIV-1 inhibitors (Garcia et al., 2009)
	Characterization of HIV-1 drug resistance	Used to investigate antiretroviral drug resistance mutations in HIV-1 RT and its role in template-switching frequency using lentiviral vectors (Nikolenko et al., 2004)
	Delivery of CRISPR-Cas9 for gene editing	Used to cleave Alk and Elm4 genes resulting in chromosomal rearrangement leading to A/k-Elm4 fusion and tumor formation in mice (Blasco et al., 2014)
Clinical applications	Delivery of therapeutic genes	Delivery of IL-2R im HSPCs ex vivo in patients with X-SCID (Cavazzana-Calvo et al, 2000; Hacein-Bey-Abina et al., 2002; Gaspar et al., 2004)
	CAR T cell therapy	Use of CAR T cell therapy to treat B cell malignancies (Meng et al., 2021)

Env proteins have been used to identify several other enveloped viruses that are also restricted by IFITM proteins [57, 61–64].

4.4.3 Discovery of Antivirals and Characterization of Drug Resistance

To date, no vaccines are available to protect against or treat HIV infection making antiretroviral therapy the only option to manage infection. While antiretroviral therapies that are currently available have significantly reduced morbidity and mortality from HIV-1 infection, the emergence of HIV strains with single or multiple resistance to antivirals currently in use has led to the need to identify novel anti-HIV drugs. Considering that lentiviral vectors encode HIV-1 proteins PR, RT, and IN and since they are essential for the successful transduction of target cells and expression of the transgene, HIV-1 pseudovirus can be useful for the discovery of HIV-1 inhibitors that target these essential viral enzymes. More specifically, lentiviral particles pseudotyped with VSV-G protein, which confers broad cell tropism, can be used to screen compounds that target postentry steps of viral replication and evaluate the efficacy of anti-PR, RT, or IN compounds in a biosafety level 2 facility [65, 66]. Several studies have used lentiviral particles pseudotyped with VSV-G that encode a reporter gene, such as a luminescent reporter gene, which can be used to evaluate pseudotyped lentiviral infection upon treatment with different compounds with potential antiviral activity. For example, Garcia et al. successfully performed a high-throughput screening to identify novel HIV-1 inhibitors using pseudotyped lentiviral particles expressing VSV-G on the cell surface and a firefly luciferase reporter gene [67].

As previously mentioned, the emergence of drug resistance HIV-1 is an important concern resulting in the failure of available treatments in some infected patients. Identifying mutations that confer drug resistance is critical to ensure that infected individuals receive treatments that effectively manage HIV-1 infection. Furthermore, understanding the mechanism involved in HIV-1 drug resistance is crucial to inform future drug development. The use of pseudotyped HIV-1 particles to develop mutant panels has played an important role in identifying HIV-1 mutations that confer drug resistance and elucidating the molecular mechanisms involved. More specifically, mutations of interest, in particular those that have been identified in HIV patients who have become resistant to antiretroviral treatments, can be inserted into either PR, RT, or IN within the packaging plasmid encoding Gag-Pol. For example, Nikolenko et al. investigated antiretroviral drug resistance mutations in HIV-1 RT and its role in template-switching frequency using lentiviral vectors [68].

In addition to its use in the identification of HIV-1 postentry inhibitors, pseudotyped lentiviral particles can also be used to screen for viral entry inhibitors against heterologous enveloped viruses in addition to HIV-1. This approach has been proven to be particularly useful in screening for antiviral drug candidates that target cell entry of emerging enveloped viruses. For example, viral entry assays based on pseudotyped particles encoding a reporter gene have been successfully used to perform high-throughput screening to identify compounds targeting entry of Ebola virus, influenza, and HIV [65, 69, 70]. More recently, lentiviral particles pseudotyped with SARS-CoV-2 Spike protein have been used to screen for viral entry inhibitors to identify drug candidates for the treatment of COVID-19 [71].

4.4.4 CRISPR-Cas9 Delivery and Gene Editing

Clustered regularly interspaced short palindromic repeats (CRISPR) and CRISPR-associated (Cas) proteins system is an important tool in gene editing that can be used as an alternative to RNA interference. While there are several Cas proteins, CRISPR-Cas9 is one of the most used systems. Briefly, the CRISPR-Cas9 system uses an RNA sequence, known as the guide RNA (gRNA), to guide the Cas9 nuclease to a specific genomic locus where it will create a double-stranded break in the DNA followed by nonhomologous end joining leading to gene deletion or the introduction of point mutations within a gene of interest with the help of a homologous template [72–74]. Lentiviral vectors have been successfully used to mediate the delivery of CRISPR-Cas9 systems into both in vitro and in vivo experimental models [75]. To deliver this system to cells, the gene encoding Cas9 and gRNA targeting the gene of interest are inserted into the transgene plasmid.

Lentivirus-mediated delivery of CRISPR-Cas9 has been used for several applications. Above all, this method is commonly used to generate gene knockout cells and animal models to investigate the function of a gene of interest in a certain experimental context. However, this method is equally useful for in vivo genetic studies to induce various mutations in a model organism to study their role in a biological process of interest. Lentivirus-mediated delivery of CRISPR-Cas9 has been used to perform high-throughput genetic screening to systemically identify genes involved in various biological processes in mammalian cells. For example, Shen and colleagues performed a loss-of-function screen by using a genome-wide gRNA library delivered by a lentiviral vector to a nonmetastatic mouse cancer cell line [76]. These mutant cells were subsequently transplanted into mice leading to the successful identification of novel genes that may be involved in accelerated metastasis and tumor growth in a mouse model of lung cancer. Lentivirus-mediated delivery of CRISPR-Cas9 can also be used to generate experimental cancer models by inducing chromosomal rearrangements. A study published by Blasco and colleagues used lentivirus-mediated delivery of CRISPR-Cas9 to cleave two genes, *Alk* and *Eml4*, resulting in a chromosomal rearrangement that lead to *Alk-Eml4* fusion and tumor formation in mice [77].

Lentivirus-mediated delivery of CRISPR-Cas9 has also been used to target viral infections, specifically those with viral genomes that integrate into the host genome. More specifically, gRNAs can be designed to target specific genes within the viral genome. For example, lentivirus-mediated CRISPR-Cas9 delivery has been successfully used to eliminate Epstein-Barr virus [78], HIV-1 [79], and hepatitis B virus [80] in animal models as well as ex vivo from cells isolated from infected patients, thereby holding the potential of curing viral infections, particularly those that integrate into the host genome. Considering that current antiviral therapies target active viral replication, lentivirus-mediated delivery of CRISPR-Cas9 provides a potential future strategy to target and deplete latently infected cells in infected individuals.

4.4.5 Gene Therapy

In addition to their important role in research, retroviral vectors have become a mainstream viral vector in gene therapy [81]. In general, gene therapy is used to treat or prevent disease by correcting the underlying genetic problem either by replacing a faulty gene with a healthy copy or by inactivating a malfunctioning and disease-causing gene. Since other viral vectors such as adenoviral vectors and

Table 4.3 Examples of ongoing clinical trials for gene therapy using lentiviral vectors as of July 2022

Condition	Gene delivered by lentiviral vector	Study phase	NCT number
X-linked SCID	Human IL2RG	I/II	NCT03601286
ADA-SCID	Human ADA	I	NCT03765632
X-linked Chronic Granulomatous Disease	GP91PHOX	I/II	NCT01855685
Sick Cell Disease	Human B-A-T87Q globin	I/II	NCT02140554
B-thalassemia	Normal human B-globin	I	NCT01639690
Wiskott-Aldrich	Wiskott-Aldrich Syndrome protein (WASp)	I/II	NCT01515462
Lysosomal Storage Disease	Arylsulfatase A(ARSA)	I/II	NCT01560182

adeno-associated viral vectors exist as episomal DNA in cells, they are lost throughout cell division. Accordingly, repeated doses are required to maintain the expression of any given transgene and hence its therapeutic efficacy. In contrast, the transgene encoded in retroviral vectors is integrated into the DNA of the transduced target cells and permanently becomes part of the cellular genome. This feature of retroviral vectors allows for the therapeutic gene to be continuously expressed throughout cell division and the lifespan of the transduced cells.

Retroviral vectors can be used in a clinical context to genetically modify a given cell type by permanently integrating a transgene into the cellular DNA that encodes a protein of interest in the context of a disease caused by a faulty gene. In the context of a disease that is caused by the expression of a given gene, retroviral vectors can also be used to deliver a noncoding RNA, such as a short hairpin RNA, that degrades cellular mRNA encoding a gene of interest to inhibit its expression. Considering that retroviral vectors derived from lentiviruses have the unique ability to infect both dividing and nondividing cells, lentiviral vectors are commonly the retroviral vector of choice for gene therapy applications. To date, the clinical applications of retroviral and lentiviral vectors are mostly restricted to ex vivo genetic modification [82]. Examples of gene therapy clinical trials using lentiviral vectors are listed in Table 4.3.

4.4.5.1 Monogenic Blood Disorders

Pseudotyped retroviruses are widely used to treat monogenic blood disorders, including both primary immunodeficiencies and blood cell diseases. Accordingly, many of the current clinical applications of retroviral and lentiviral vectors are based on genetic modification of hematopoietic stem progenitor cells (HSPC) ex vivo. An example of early clinical application of retroviral vectors in gene therapy is for the treatment of patients with X-linked severe combined immunodeficiency (X-SCID), which is caused by inactivating mutations in the interleukin-2 receptor common gamma chain gene, a key player in immune cell development and function. Due to these mutations, patients with X-SCID have very low T cells and natural killer cell counts, and impaired B cells, resulting in a compromised immune system. Clinical trials using gammaretroviral vectors encoding this receptor to transduce HSPCs ex vivo were found to improve immune function in patients suffering from X-SCID [83–85]. A similar approach in which retroviral vectors encoding a healthy copy of a mutated gene causing disease has been successfully used to treat patients with other primary immunodeficiencies in clinical trials including adenosine deaminase deficiency SCID (ADA-SCID) [86], chronic granulomatous disease [87], and Wiskott-Aldrich syndrome [88, 89]. While the use of retrovirus-based gene therapy for most of the primary immunodeficiencies discussed remains limited to clinical trials, the European Medicines Agency approved Strimvelis™, the first ex vivo HSPC gene therapy treatment for ADA-SCID, another form of the disease in which a deficient ADA gene is inherited, in May 2016 [90].

Like primary immunodeficiencies, hemoglobin disorders such as β-thalassemia and sick cell disease, both of which are monogenic blood diseases caused by mutations in the HBB gene encoding for β-globin, have also been treated using retroviral vectors. Several clinical trials investigating the clinical efficacy of SIN lentiviral vectors to deliver the wild-type HBB gene in transduced HSPCs ex vivo have found that most patients with β-thalassemia achieved transfusion independence, even in severe cases, with a safe integration profile [91–93]. Considering that the treatment for β-thalassemia was lifelong red blood cell transfusions, which adversely impact patients' quality of life, gene therapy provides a promising alternative. In 2019, the European Medicines Agency approved the first ex vivo gene therapy for β-thalassemia. Similarly, several clinical trials on the use of lentiviral vectors to express the wild-type β-globin gene in patients with sickle cell disease have demonstrated promising results [94].

4.4.5.2 Cancer Immunotherapy

Another common clinical application of retroviral vectors is their use in chimeric antigen receptor (CAR) T-cell therapy, which is a targeted cancer immunotherapy generally used to treat B cell leukemias and lymphomas [95]. CARs are genetically engineered T-cell receptors that are specific to a given patient's tumor cells. This technology redirects the patient's T cells to specifically target and destroy tumor cells, making them an important tool in personalized medicine. T cells are isolated from the patient's blood, which are subsequently transduced ex vivo with a retroviral vector containing a transgene that encodes T-cell receptors that are specific for the patient's tumor cells. The CAR T cells are then expanded in cell culture, which are then infused into the patient's bloodstream where they will selectively target cancer cells. Clinical trials have reported dramatic clinical responses and high rates of complete remission achieved in CAR T-cell therapy of B cell malignancies compared to traditional treatments, making it a promising therapy for patients with B cell lymphomas [96]. As of 2022, there are six FDA-approved CAR T-cell therapies, all of which are based on retroviral vectors.

4.4.6 Insertional Mutagenesis: Lessons Learned

While retroviral vectors have several advantages for clinical applications, it is important to note that an important concern of using retroviral vectors for gene therapy is insertional mutagenesis. Since retroviral vectors integrate into cellular DNA, there is a possibility that the retroviral vector is integrated in a manner that disrupts promoter or enhancer elements of neighboring cellular genes, leading to cell transformation and tumorigenesis, resulting in the activation of proto-oncogenes. For example, early clinical trials investigating the efficacy of a MLV vector for genetic modification of HSPCs in X-SCID [97] and Wiskott-Aldrich syndrome [98] found that patients developed acute myeloid and lymphoid leukemias several years after treatment due to activation of proto-oncogenes neighboring the sites of proviral integration. Eventually, studies showed that gammaretroviruses, such as MLV, preferentially insert near transcriptional start sites while lentiviruses, such as HIV-1, preferentially insert within transcription units [99, 100]. Accordingly, these concerning events along with the evidence highlighting the lower risk of insertional mutagenesis of lentiviral vectors led to the development of self-inactivating lentiviral vectors for gene therapy. SIN lentiviral vectors have been found to have both consistent safety and efficacy, and clinical trials have reported high efficacy for the treatment of blood disorders including immunodeficiencies and hemoglobinopathies, as well as metabolic disorders such as lysosomal storage diseases [101].

4.5 Conclusions

Retroviral vectors have several unique features that render them important tools for both research and clinical applications, including their ability to integrate a gene of interest into the host genome, their ability to transduce both dividing and nondividing cells, and their ability to deliver relatively large genes. Since their initial development, many modifications have been made to retroviral vectors to improve their transduction efficiency and safety. To date, pseudotyped retroviruses have played a critical role in understanding viral entry and the function of viral glycoproteins, elucidating the importance of viral surface protein glycosylation, the identification of host restriction factors, the discovery of antiviral therapies and characterization of HIV-1 drug resistance, and the delivery of CRISPR-Cas9 systems in model organisms. Regarding their clinical application, retroviral vectors have played an invaluable role in gene therapy, notably in monogenic blood disorders and cancer immunotherapy, although their use remains largely restricted to clinical trials.

References

1. Rous, P.: A sarcoma of the fowl transmissible by an agent separable from the tumor cells. J. Exp. Med. **13**, 397–411 (1911)
2. Gross L. 1951. "Spontaneous" Leukemia Developing in C3H Mice Following
3. Jarrett, W.F., Crawford, E.M., Martin, W.B., Davie, F.: A VIRUS-LIKE PARTICLE ASSOCIATED WITH LEUKEMIA (LYMPHOSARCOMA). Nature. **202**, 567–569 (1964)
4. Miller, J.M., Miller, L.D., Olson, C., Gillette, K.G.: Virus-like particles in phytohemagglutinin-stimulated lymphocyte cultures with reference to bovine lymphosarcoma. J. Natl. Cancer Inst. **43**, 1297–1305 (1969)
5. Craigo, J.K., Montelaro, R.C.: Equine infectious anemia virus. In: Mahy, B.W.J., Van Regenmortel, M.H.V. (eds.) Encyclopedia of virology, 3rd edn). https://doi.org/10.1016/B978-012374410-4.00395-2, pp. 167–174. Academic Press, Oxford (2008)
6. Sigurdsson, B.: Rida, a chronic encephalitis of sheep: with general remarks on infections which develop slowly and some of their special characteristics. Br. Vet. J. **110**, 341–354 (1954)
7. Daniel, M.D., Letvin, N.L., King, N.W., Kannagi, M., Sehgal, P.K., Hunt, R.D., Kanki, P.J., Essex, M., Desrosiers, R.C.: Isolation of T-cell tropic HTLV-III-like retrovirus from macaques. Science. **228**, 1201–1204 (1985)
8. Temin, H.M., Mizutami, S.: RNA-dependent DNA polymerase in virions of Rous sarcoma virus. Nature. **226**, 1211–1213 (1970)
9. Baltimore, D.: Viral RNA-dependent DNA polymerase: RNA-dependent DNA polymerase in virions of RNA tumour viruses. Nature. **226**, 1209–1211 (1970)
10. Barré-Sinoussi, F., Chermann, J.C., Rey, F., Nugeyre, M.T., Chamaret, S., Gruest, J., Dauguet, C., Axler-Blin, C., Vézinet-Brun, F., Rouzioux, C., Rozenbaum, W., Montagnier, L.: Isolation of a T-lymphotropic retrovirus from a patient at risk for acquired immune deficiency syndrome (AIDS). Science. **220**, 868–871 (1983)
11. Vogt, V.M.: Retroviral virions and genomes. In: Coffin, J.M., Hughes, S.H., Varmus, H.E. (eds.) Retroviruses. Cold Spring Harbor Laboratory Press, NewYork (1997)
12. Swanstrom, R., Wills, J.W. 1997. Synthesis, assembly, and processing of viral proteins. In Coffin JM, Hughes SH, Varmus HE (ed), Retroviruses. Cold Spring Harbor Laboratory Press New York.
13. Bell, N.M., Lever, A.M.L.: HIV gag polyprotein: processing and early viral particle assembly. Trends Microbiol. **21**, 136–144 (2013)
14. Dalton, A.K., Ako-Adjei, D., Murray, P.S., Murray, D., Vogt, V.M.: Electrostatic interactions drive membrane Association of the Human Immunodeficiency Virus Type 1 gag MA domain. J. Virol. **81**, 6434–6445 (2007)
15. Saad, J.S., Miller, J., Tai, J., Kim, A., Ghanam, R.H., Summers, M.F.: Structural basis for targeting HIV-1 gag proteins to the plasma membrane for virus assembly. Proc. Natl. Acad. Sci. U. S. A. **103**, 11364–11369 (2006)
16. Solbak, S.M., Reksten, T.R., Hahn, F., Wray, V., Henklein, P., Henklein, P., Halskau, Ø., Schubert, U., Fossen, T.: HIV-1 p6 – a structured to flexible multifunctional membrane-interacting protein. Biochim. Biophys. Acta. **1828**, 816–823 (2013)
17. Freed, E.O.: Viral late domains. J. Virol. **76**, 4679–4687 (2002)
18. Schur, F.K.M., Obr, M., Hagen, W.J.H., Wan, W., Jakobi, A.J., Kirkpatrick, J.M., Sachse, C., Kräusslich, H.-G., Briggs, J.A.G.: An atomic model of HIV-1 capsid-SP1 reveals structures regulating assembly and maturation. Science. **353**, 506–508 (2016)
19. Liang, C., Hu, J., Russell, R.S., Roldan, A., Kleiman, L., Wainberg, M.A.: Characterization of a putative alpha-helix across the capsid-SP1 boundary that is critical for the multimerization of human immunodeficiency virus type 1 gag. J. Virol. **76**, 11729–11737 (2002)
20. Checkley, M.A., Luttge, B.G., Freed, E.O.: HIV-1 envelope glycoprotein biosynthesis, trafficking, and incorporation. J. Mol. Biol. **410**, 582–608 (2011)

21. Johri, M.K., Mishra, R., Chhatbar, C., Unni, S.K., Singh, S.K.: Tits and bits of HIV tat protein. Expert. Opin. Biol. Ther. **11**, 269–283 (2011)
22. Truman, C.T., Järvelin, A., Davis, I., Castello, A.: HIV Rev-isited. Open Biol. **10**, 200320 (2020)
23. Malim, M.H., Emerman, M.: HIV-1 accessory proteins – ensuring viral survival in a hostile environment. Cell Host Microbe. **3**, 388–398 (2008)
24. Harris, R.S., Bishop, K.N., Sheehy, A.M., Craig, H.M., Petersen-Mahrt, S.K., Watt, I.N., Neuberger, M.S., Malim, M.H.: DNA deamination mediates innate immunity to retroviral infection. Cell. **113**, 803–809 (2003)
25. Harris, R.S., Petersen-Mahrt, S.K., Neuberger, M.S.: RNA editing enzyme APOBEC1 and some of its homologs can act as DNA mutators. Mol. Cell. **10**, 1247–1253 (2002)
26. Mangeat, B., Turelli, P., Caron, G., Friedli, M., Perrin, L., Trono, D.: Broad antiretroviral defence by human APOBEC3G through lethal editing of nascent reverse transcripts. Nature. **424**, 99–103 (2003)
27. Marin, M., Rose, K.M., Kozak, S.L., Kabat, D.: HIV-1 Vif protein binds the editing enzyme APOBEC3G and induces its degradation. Nat. Med. **9**, 1398–1403 (2003)
28. Sheehy, A.M., Gaddis, N.C., Choi, J.D., Malim, M.H.: Isolation of a human gene that inhibits HIV-1 infection and is suppressed by the viral Vif protein. Nature. **418**, 646–650 (2002)
29. Dubé, M., Bego, M.G., Paquay, C., Cohen, É.A.: Modulation of HIV-1-host interaction: role of the Vpu accessory protein. Retrovirology. **7**, 114 (2010)
30. Van Damme, N., Goff, D., Katsura, C., Jorgenson, R.L., Mitchell, R., Johnson, M.C., Stephens, E.B., Guatelli, J.: The interferon-induced protein BST-2 restricts HIV-1 release and is downregulated from the cell surface by the viral Vpu protein. Cell Host Microbe. **3**, 245–252 (2008)
31. Usami, Y., Wu, Y., Göttlinger, H.G.: SERINC3 and SERINC5 restrict HIV-1 infectivity and are counteracted by Nef. Nature. **526**, 218–223 (2015)
32. Rosa, A., Chande, A., Ziglio, S., De Sanctis, V., Bertorelli, R., Goh, S.L., McCauley, S.M., Nowosielska, A., Antonarakis, S.E., Luban, J., Santoni, F.A., Pizzato, M.: HIV-1 Nef promotes infection by excluding SERINC5 from virion incorporation. Nature. **526**, 212–217 (2015)
33. Fujita, M., Nomaguchi, M., Adachi, A., Otsuka, M.: SAMHD1-dependent and -independent functions of HIV-2/SIV Vpx protein. Front. Microbiol. **3** (2012)
34. Laguette, N., Sobhian, B., Casartelli, N., Ringeard, M., Chable-Bessia, C., Ségéral, E., Yatim, A., Emiliani, S., Schwartz, O., Benkirane, M.: SAMHD1 is the dendritic- and myeloid-cell-specific HIV-1 restriction factor counteracted by Vpx. Nature. **474**, 654–657 (2011)
35. Bobadilla, S., Sunseri, N., Landau, N.R.: Efficient transduction of myeloid cells by an HIV-1-derived lentiviral vector that packages the Vpx accessory protein. Gene Ther. **20**, 514–520 (2013)
36. Telesnitsky, A., Goff, S.P.: Reverse transcriptase and the generation of retroviral DNA. In: Coffin, J.M., Hughes, S.H., Varmus, H.E. (eds.) Retroviruses. Cold Spring Harbor Laboratory Press, New York (1997)
37. Coroadinha, A.S., Gama-Norton, L., Amaral, A.I., Hauser, H., Alves, P.M., Cruz, P.E.: Production of retroviral vectors: review. Curr. Gene Ther. **10**, 456–473 (2010)
38. Maetzig, T., Galla, M., Baum, C., Schambach, A.: Gammaretroviral vectors: biology, technology and application. Viruses. **3**, 677–713 (2011)
39. Merten, O.W., Hebben, M., Bovolenta, C.: Production of lentiviral vectors. Mol Ther Methods Clin Dev. **3**, 16017 (2016)
40. Blesch, A.: Lentiviral and MLV based retroviral vectors for ex vivo and in vivo gene transfer. Methods. **33**, 164–172 (2004)
41. Bona, R., Michelini, Z., Mazzei, C., Gallinaro, A., Canitano, A., Borghi, M., Vescio, M.F., Di Virgilio, A., Pirillo, M.F., Klotman, M.E., Negri, D., Cara, A.: Safety and efficiency modifications of SIV-based integrase-defective lentiviral vectors for immunization. Mol Ther Methods Clin Dev. **23**, 263–275 (2021)

42. Trobridge, G.D.: Foamy virus vectors for gene transfer. Expert. Opin. Biol. Ther. **9**, 1427–1436 (2009)
43. Erlwein, O., McClure, M.O.: Progress and prospects: foamy virus vectors enter a new age. Gene Ther. **17**, 1423–1429 (2010)
44. Wool-Lewis, R.J., Bates, P.: Characterization of Ebola virus entry by using pseudotyped viruses: identification of receptor-deficient cell lines. J. Virol. **72**, 3155–3160 (1998)
45. Yonezawa, A., Cavrois, M., Greene, W.C.: Studies of ebola virus glycoprotein-mediated entry and fusion by using pseudotyped human immunodeficiency virus type 1 virions: involvement of cytoskeletal proteins and enhancement by tumor necrosis factor alpha. J. Virol. **79**, 918–926 (2005)
46. Wrin, T., Loh, T.P., Vennari, J.C., Schuitemaker, H., Nunberg, J.H.: Adaptation to persistent growth in the H9 cell line renders a primary isolate of human immunodeficiency virus type 1 sensitive to neutralization by vaccine sera. J. Virol. **69**, 39–48 (1995)
47. Tokunaga, K., Greenberg, M.L., Morse, M.A., Cumming, R.I., Lyerly, H.K., Cullen, B.R.: Molecular basis for cell tropism of CXCR4-dependent human immunodeficiency virus type 1 isolates. J. Virol. **75**, 6776–6785 (2001)
48. Deng, H., Liu, R., Ellmeier, W., Choe, S., Unutmaz, D., Burkhart, M., Di Marzio, P., Marmon, S., Sutton, R.E., Hill, C.M., Davis, C.B., Peiper, S.C., Schall, T.J., Littman, D.R., Landau, N. R.: Identification of a major co-receptor for primary isolates of HIV-1. Nature. **381**, 661–666 (1996)
49. Lindenbach, B.D., Rice, C.M.: Unravelling hepatitis C virus replication from genome to function. Nature. **436**, 933–938 (2005)
50. Bartosch, B., Vitelli, A., Granier, C., Goujon, C., Dubuisson, J., Pascale, S., Scarselli, E., Cortese, R., Nicosia, A., Cosset, F.L.: Cell entry of hepatitis C virus requires a set of co-receptors that include the CD81 tetraspanin and the SR-B1 scavenger receptor. J. Biol. Chem. **278**, 41624–41630 (2003)
51. Douam, F., Dao Thi, V.L., Maurin, G., Fresquet, J., Mompelat, D., Zeisel, M.B., Baumert, T. F., Cosset, F.L., Lavillette, D.: Critical interaction between E1 and E2 glycoproteins determines binding and fusion properties of hepatitis C virus during cell entry. Hepatology. **59**, 776–788 (2014)
52. Owsianka, A.M., Timms, J.M., Tarr, A.W., Brown, R.J., Hickling, T.P., Szwejk, A., Bienkowska-Szewczyk, K., Thomson, B.J., Patel, A.H., Ball, J.K.: Identification of conserved residues in the E2 envelope glycoprotein of the hepatitis C virus that are critical for CD81 binding. J. Virol. **80**, 8695–8704 (2006)
53. Callens, N., Ciczora, Y., Bartosch, B., Vu-Dac, N., Cosset, F.L., Pawlotsky, J.M., Penin, F., Dubuisson, J.: Basic residues in hypervariable region 1 of hepatitis C virus envelope glycoprotein e2 contribute to virus entry. J. Virol. **79**, 15331–15341 (2005)
54. Haywood, A.M.: Membrane uncoating of intact enveloped viruses. J. Virol. **84**, 10946–10955 (2010)
55. Wang, W., Nie, J., Prochnow, C., Truong, C., Jia, Z., Wang, S., Chen, X.S., Wang, Y.: A systematic study of the N-glycosylation sites of HIV-1 envelope protein on infectivity and antibody-mediated neutralization. Retrovirology. **10**, 14 (2013)
56. Yang, Q., Hughes, T.A., Kelkar, A., Yu, X., Cheng, K., Park, S., Huang, W.C., Lovell, J.F., Neelamegham, S.: Inhibition of SARS-CoV-2 viral entry upon blocking N- and O-glycan elaboration. elife. **9** (2020)
57. Huang, H.Y., Liao, H.Y., Chen, X., Wang, S.W., Cheng, C.W., Shahed-Al-Mahmud, M., Liu, Y.M., Mohapatra, A., Chen, T.H., Lo, J.M., Wu, Y.M., Ma, H.H., Chang, Y.H., Tsai, H.Y., Chou, Y.C., Hsueh, Y.P., Tsai, C.Y., Huang, P.Y., Chang, S.Y., Chao, T.L., Kao, H.C., Tsai, Y.M., Chen, Y.H., Wu, C.Y., Jan, J.T., Cheng, T.R., Lin, K.I., Ma, C., Wong, C.H.: Vaccination with SARS-CoV-2 spike protein lacking glycan shields elicits enhanced protective responses in animal models. Sci. Transl. Med. **14**, eabm0899 (2022)

58. Tandon, R., Mitra, D., Sharma, P., McCandless, M.G., Stray, S.J., Bates, J.T., Marshall, G.D.: Effective screening of SARS-CoV-2 neutralizing antibodies in patient serum using lentivirus particles pseudotyped with SARS-CoV-2 spike glycoprotein. Sci. Rep. **10**, 19076 (2020)
59. Seaman, M.S., Janes, H., Hawkins, N., Grandpre, L.E., Devoy, C., Giri, A., Coffey, R.T., Harris, L., Wood, B., Daniels, M.G., Bhattacharya, T., Lapedes, A., Polonis, V.R., McCutchan, F.E., Gilbert, P.B., Self, S.G., Korber, B.T., Montefiori, D.C., Mascola, J.R.: Tiered categorization of a diverse panel of HIV-1 env pseudoviruses for assessment of neutralizing antibodies. J. Virol. **84**, 1439–1452 (2010)
60. Feeley, E.M., Sims, J.S., John, S.P., Chin, C.R., Pertel, T., Chen, L.M., Gaiha, G.D., Ryan, B. J., Donis, R.O., Elledge, S.J., Brass, A.L.: IFITM3 inhibits influenza a virus infection by preventing cytosolic entry. PLoS Pathog. **7**, e1002337 (2011)
61. Yu, J., Li, M., Wilkins, J., Ding, S., Swartz Talia, H., Esposito Anthony, M., Zheng, Y.-M., Freed Eric, O., Liang, C., Chen Benjamin, K., Liu, S.-L.: IFITM proteins restrict HIV-1 infection by antagonizing the envelope glycoprotein. Cell Rep. **13**, 145–156 (2015)
62. Muñoz-Moreno, R., Cuesta-Geijo, M.Á., Martínez-Romero, C., Barrado-Gil, L., Galindo, I., García-Sastre, A., Alonso, C.: Antiviral role of IFITM proteins in African swine fever virus infection. PLoS One. **11**, e0154366 (2016)
63. Prelli Bozzo, C., Nchioua, R., Volcic, M., Koepke, L., Krüger, J., Schütz, D., Heller, S., Stürzel, C.M., Kmiec, D., Conzelmann, C., Müller, J., Zech, F., Braun, E., Groß, R., Wettstein, L., Weil, T., Weiß, J., Diofano, F., Rodríguez Alfonso, A.A., Wiese, S., Sauter, D., Münch, J., Goffinet, C., Catanese, A., Schön, M., Boeckers, T.M., Stenger, S., Sato, K., Just, S., Kleger, A., Sparrer, K.M.J., Kirchhoff, F.: IFITM proteins promote SARS-CoV-2 infection and are targets for virus inhibition in vitro. Nat. Commun. **12**, 4584 (2021)
64. Wrensch, F., Winkler, M., Pöhlmann, S.: IFITM proteins inhibit entry driven by the MERS-coronavirus spike protein: evidence for cholesterol-independent mechanisms. Viruses. **6**, 3683–3698 (2014)
65. Garcia, J.-M., Gao, A., He, P.-L., Choi, J., Tang, W., Bruzzone, R., Schwartz, O., Naya, H., Nan, F.-J., Li, J., Altmeyer, R., Zuo, J.-P.: High-throughput screening using pseudotyped lentiviral particles: a strategy for the identification of HIV-1 inhibitors in a cell-based assay. Antivir. Res. **81**, 239–247 (2009)
66. Prokofjeva, M.M., Spirin, P.V., Yanvarev, D.V., Ivanov, A.V., Novikov, M.S., Stepanov, O. A., Gottikh, M.B., Kochetkov, S.N., Fehse, B., Stocking, C., Prassolov, V.S.: Screening of potential HIV-1 inhibitors/replication blockers using secure lentiviral in vitro system. Acta Nat. **3**, 55–65 (2011)
67. Garcia, J.M., Gao, A., He, P.L., Choi, J., Tang, W., Bruzzone, R., Schwartz, O., Naya, H., Nan, F.J., Li, J., Altmeyer, R., Zuo, J.P.: High-throughput screening using pseudotyped lentiviral particles: a strategy for the identification of HIV-1 inhibitors in a cell-based assay. Antivir. Res. **81**, 239–247 (2009)
68. Nikolenko, G.N., Svarovskaia, E.S., Delviks, K.A., Pathak, V.K.: Antiretroviral drug resistance mutations in human immunodeficiency virus type 1 reverse transcriptase increase template-switching frequency. J. Virol. **78**, 8761–8770 (2004)
69. Basu, A., Antanasijevic, A., Wang, M., Li, B., Mills, D.M., Ames, J.A., Nash, P.J., Williams, J.D., Peet, N.P., Moir, D.T., Prichard, M.N., Keith, K.A., Barnard, D.L., Caffrey, M., Rong, L., Bowlin, T.L.: New small molecule entry inhibitors targeting hemagglutinin-mediated influenza a virus fusion. J. Virol. **88**, 1447–1460 (2014)
70. Beck, S., Henß, L., Weidner, T., Herrmann, J., Müller, R., Chao, Y.K., Grimm, C., Weber, C., Sliva, K., Schnierle, B.S.: Identification of entry inhibitors of Ebola virus pseudotyped vectors from a myxobacterial compound library. Antivir. Res. **132**, 85–91 (2016)
71. Chen, C.Z., Xu, M., Pradhan, M., Gorshkov, K., Petersen, J.D., Straus, M.R., Zhu, W., Shinn, P., Guo, H., Shen, M., Klumpp-Thomas, C., Michael, S.G., Zimmerberg, J., Zheng, W., Whittaker, G.R.: Identifying SARS-CoV-2 entry inhibitors through drug repurposing screens of SARS-S and MERS-S Pseudotyped particles. ACS Pharmacology & Translational Science. **3**, 1165–1175 (2020)

72. Jinek, M., Chylinski, K., Fonfara, I., Hauer, M., Doudna, J.A., Charpentier, E.: A programmable dual-RNA-guided DNA endonuclease in adaptive bacterial immunity. Science. **337**, 816–821 (2012)
73. Wright, A.V., Nuñez, J.K., Doudna, J.A.: Biology and applications of CRISPR systems: harnessing Nature's toolbox for genome engineering. Cell. **164**, 29–44 (2016)
74. Hsu, P.D., Lander, E.S., Zhang, F.: Development and applications of CRISPR-Cas9 for genome engineering. Cell. **157**, 1262–1278 (2014)
75. Sanjana, N.E., Shalem, O., Zhang, F.: Improved vectors and genome-wide libraries for CRISPR screening. Nat. Methods. **11**, 783–784 (2014)
76. Chen, S., Sanjana, N.E., Zheng, K., Shalem, O., Lee, K., Shi, X., Scott, D.A., Song, J., Pan, J. Q., Weissleder, R.: Genome-wide CRISPR screen in a mouse model of tumor growth and metastasis. Cell. **160**, 1246–1260 (2015)
77. Blasco, R.B., Karaca, E., Ambrogio, C., Cheong, T.-C., Karayol, E., Minero, V.G., Voena, C., Chiarle, R.: Simple and rapid in vivo generation of chromosomal rearrangements using CRISPR/Cas9 technology. Cell Rep. **9**, 1219–1227 (2014)
78. Wang, J., Quake, S.R.: RNA-guided endonuclease provides a therapeutic strategy to cure latent herpesviridae infection. Proc. Natl. Acad. Sci. **111**, 13157–13162 (2014)
79. Kaminski, R., Chen, Y., Fischer, T., Tedaldi, E., Napoli, A., Zhang, Y., Karn, J., Hu, W., Khalili, K.: Elimination of HIV-1 genomes from human T-lymphoid cells by CRISPR/Cas9 gene editing. Sci. Rep. **6**, 22555 (2016)
80. Ramanan, V., Shlomai, A., Cox, D.B., Schwartz, R.E., Michailidis, E., Bhatta, A., Scott, D.A., Zhang, F., Rice, C.M., Bhatia, S.N.: CRISPR/Cas9 cleavage of viral DNA efficiently suppresses hepatitis B virus. Sci. Rep. **5**, 1–9 (2015)
81. Hawley, R.G., Lieu, F.H., Fong, A.Z., Hawley, T.S.: Versatile retroviral vectors for potential use in gene therapy. Gene Ther. **1**, 136–138 (1994)
82. Wang, X., Ma, C., Rodríguez Labrada, R., Qin, Z., Xu, T., He, Z., Wei, Y.: Recent advances in lentiviral vectors for gene therapy. Sci. China Life Sci. **64**, 1842–1857 (2021)
83. Cavazzana-Calvo, M., Hacein-Bey, S., GvdS, B., Gross, F., Yvon, E., Nusbaum, P., Selz, F., Hue, C., Certain, S., Casanova, J.-L.: Gene therapy of human severe combined immunodeficiency (SCID)-X1 disease. Science. **288**, 669–672 (2000)
84. Hacein-Bey-Abina, S., Le Deist, F., Carlier, F., Bouneaud, C., Hue, C., De Villartay, J.-P., Thrasher, A.J., Wulffraat, N., Sorensen, R., Dupuis-Girod, S.: Sustained correction of X-linked severe combined immunodeficiency by ex vivo gene therapy. N. Engl. J. Med. **346**, 1185–1193 (2002)
85. Gaspar, H.B., Parsley, K.L., Howe, S., King, D., Gilmour, K.C., Sinclair, J., Brouns, G., Schmidt, M., Von Kalle, C., Barington, T.: Gene therapy of X-linked severe combined immunodeficiency by use of a pseudotyped gammaretroviral vector. Lancet. **364**, 2181–2187 (2004)
86. Gaspar, H.B., Cooray, S., Gilmour, K.C., Parsley, K.L., Zhang, F., Adams, S., Bjorkegren, E., Bayford, J., Brown, L., Davies, E.G.: Hematopoietic stem cell gene therapy for adenosine deaminase–deficient severe combined immunodeficiency leads to long-term immunological recovery and metabolic correction. Sci. Transl. Med. (2011) 3:97ra80-97ra80
87. Kohn, D.B., Booth, C., Kang, E.M., Pai, S.-Y., Shaw, K.L., Santilli, G., Armant, M., Buckland, K.F., Choi, U., De Ravin, S.S.: Lentiviral gene therapy for X-linked chronic granulomatous disease. Nat. Med. **26**, 200–206 (2020)
88. Ferrua, F., Cicalese, M.P., Galimberti, S., Giannelli, S., Dionisio, F., Barzaghi, F., Migliavacca, M., Bernardo, M.E., Calbi, V., Assanelli, A.A.: Lentiviral haemopoietic stem/progenitor cell gene therapy for treatment of Wiskott-Aldrich syndrome: interim results of a non-randomised, open-label, phase 1/2 clinical study. The Lancet Haematology. **6**, e239–e253 (2019)
89. Boztug, K., Schmidt, M., Schwarzer, A., Banerjee, P.P., Díez, I.A., Dewey, R.A., Böhm, M., Nowrouzi, A., Ball, C.R., Glimm, H.: Stem-cell gene therapy for the Wiskott–Aldrich syndrome. N. Engl. J. Med. **363**, 1918–1927 (2010)

90. Aiuti, A., Roncarolo, M.G., Naldini, L.: Gene therapy for ADA-SCID, the first marketing approval of an ex vivo gene therapy in Europe: paving the road for the next generation of advanced therapy medicinal products. EMBO Mol. Med. **9**, 737–740 (2017)
91. Cavazzana-Calvo, M., Payen, E., Negre, O., Wang, G., Hehir, K., Fusil, F., Down, J., Denaro, M., Brady, T., Westerman, K.: Transfusion independence and HMGA2 activation after gene therapy of human β-thalassaemia. Nature. **467**, 318–322 (2010)
92. Marktel, S., Cicalese, M.P., Giglio, F., Scaramuzza, S., Calbi, V., Casiraghi, M., Ciotti, F., Lidonnici, M.R., Rossi, C., Masera, N., D'Angelo, E., Mirra, N., Origa, R., Tartaglione, I., Mandelli, G., Milani, R., Gattillo, S., Coppola, M., Viarengo, G., Santoleri, L., Calabria, A., Perrotta, S., Montini, E., Graziadei, G., Naldini, L., Cappellini, M.D., Ciceri, F., Aiuti, A., Ferrari, G.: Gene therapy for Beta thalassemia: preliminary results from the PHASE I/II Tiget-Bthal trial of autologous hematopoietic stem cells genetically modified with GLOBE lentiviral vector. Blood. **130**, 355–355 (2017)
93. Payen, E., Colomb, C., Negre, O., Beuzard, Y., Hehir, K., Leboulch, P.: Lentivirus vectors in β-thalassemia. Methods Enzymol. **507**, 109–124 (2012)
94. Demirci, S., Uchida, N., Tisdale, J.F.: Gene therapy for sickle cell disease: an update. Cytotherapy. **20**, 899–910 (2018)
95. Sterner, R.C., Sterner, R.M.: CAR-T cell therapy: current limitations and potential strategies. Blood Cancer J. **11**, 69 (2021)
96. Meng, J., Wu, X., Sun, Z., Xun, R., Liu, M., Hu, R., Huang, J.: Efficacy and safety of CAR-T cell products Axicabtagene Ciloleucel, Tisagenlecleucel, and Lisocabtagene Maraleucel for the treatment of hematologic malignancies: a systematic review and meta-analysis. Front. Oncol. **11** (2021)
97. Howe, S.J., Mansour, M.R., Schwarzwaelder, K., Bartholomae, C., Hubank, M., Kempski, H., Brugman, M.H., Pike-Overzet, K., Chatters, S.J., De Ridder, D.: Insertional mutagenesis combined with acquired somatic mutations causes leukemogenesis following gene therapy of SCID-X1 patients. J. Clin. Invest. **118** (2008)
98. Morris, E.C., Fox, T., Chakraverty, R., Tendeiro, R., Snell, K., Rivat, C., Grace, S., Gilmour, K., Workman, S., Buckland, K., Butler, K., Chee, R., Salama, A.D., Ibrahim, H., Hara, H., Duret, C., Mavilio, F., Male, F., Bushman, F.D., Galy, A., Burns, S.O., Gaspar, H.B., Thrasher, A.J.: Gene therapy for Wiskott-Aldrich syndrome in a severely affected adult. Blood. **130**, 1327–1335 (2017)
99. Ciuffi, A.: The benefits of integration. Clin. Microbiol. Infect. **22**, 324–332 (2016)
100. Lewinski, M.K., Yamashita, M., Emerman, M., Ciuffi, A., Marshall, H., Crawford, G., Collins, F., Shinn, P., Leipzig, J., Hannenhalli, S.: Retroviral DNA integration: viral and cellular determinants of target-site selection. PLoS Pathog. **2**, e60 (2006)
101. Kohn, D.B.: Gene therapy for blood diseases. Curr. Opin. Biotechnol. **60**, 39–45 (2019)

Chapter 5
Pseudotyped Virus for Papillomavirus

Xueling Wu, Jianhui Nie, and Youchun Wang

Abstract Papillomavirus is difficult to culture in vitro, which limits its related research. The development of pseudotyped virus technology provides a valuable research tool for virus infectivity research, vaccine evaluation, infection inhibitor evaluation, and so on. Depending on the application fields, different measures have been developed to generate various kinds of pseudotyped papillomavirus. L1-based and L2-based HPV vaccines should be evaluated using different pseudotyped virus system. Pseudotyped papillomavirus animal models need high-titer pseudotyped virus and unique handling procedure to generate robust results. This paper reviewed the development, optimization, standardization, and application of various pseudotyped papillomavirus methods.

Keywords Human papillomavirus · Pseudotyped virus · Neutralization · Cervical cancer

---

X. Wu
Cell Collection and Research Center, National Institutes for Food and Drug Control (NIFDC), and WHO Collaborating Center for Standardization and Evaluation of Biologicals, Beijing, China

J. Nie (✉)
Division of HIV/AIDS and Sex-transmitted Virus Vaccines, National Institutes for Food and Drug Control (NIFDC) and WHO Collaborating Center for Standardization and Evaluation of Biologicals, Beijing, China
e-mail: niejianhui@nifdc.org.cn

Y. Wang (✉)
Chinese Academy of Medical Sciences & Peking Union Medical College, Beijing, China

Institute of Medical Biology, Chinese Academy of Medicine Sciences & Peking Union Medical College, Kunming, China
e-mail: wangyc@nifdc.org.cn; wangyc@imbcams.com.cn

Y. Wang (ed.), *Pseudotyped Viruses*, Advances in Experimental Medicine and Biology 1407, https://doi.org/10.1007/978-981-99-0113-5_5

Abbreviations

BM	Basement membrane
BPV	Bovine papillomavirus
CFP	Crimson fluorescent protein
CPE	Cytopathic effect
CRPV	Cottontail rabbit papillomavirus
DNA	Deoxyribonucleic acid
FC	Furin-cleaved
GFP	Green fluorescent protein
Gluc	Gaussia luciferase
HPV	Human papillomavirus
HSPG	Heparan sulfate proteoglycan
ID_{50}	50% inhibitory dose
IVIS	In vivo imaging system
MSD	Merck Sharp & Dohme
N-9	Nonoxynol-9
NMIBC	Nonmuscular invasive bladder cancer
PBNA	Pseudotyped virus based neutralization assay
PsV	Pseudotyped virus
RFP	Red fluorescent protein
RNA	Ribonucleic acid
SEAP	Secretory alkaline phosphatase
SFV	Semliki Forest Virus
VLP	Viruslike particle
WHO	World Health Organization

5.1 Introduction

Papillomavirus belongs to double-stranded DNA virus, and its genome is circular double-stranded DNA with the full length of about 8000 bp. The viral genome is wrapped in the viral capsid formed by L1 and L2 proteins. The viral capsid is icosahedral, unenveloped, and about 60 nm in diameter. At present, papillomavirus has been isolated from mammals, birds, and reptiles [1]. When the DNA sequence difference of L1 is more than 10%, it is defined as a new genotype [2]. At present, there are more than 450 genotypes of human papillomavirus, which can infect human [1]. Human papillomavirus can be divided into five genera, including Alpha, Beta, Gamma, Mu, and Nupa. The HPV infecting the human mucus is mainly from the Alpha genus. According to its carcinogenicity, it can be divided into the high-risk type and low-risk type. At present, there are 13 kinds of high-risk HPV, including HPV16, HPV18, HPV31, HPV33, HPV35, HPV39, HPV45, HPV51, HPV52, HPV56, HPV58, HPV59, and HPV68. There are 11 possible high-risk types, including HPV26, HPV30, HPV34, HPV53, HPV66, HPV67, HPV69, HPV70, HPV73, HPV82, and HPV85. They can or potentially can cause cancers in the cervix, anus, penis, vagina, oropharynx, and so on [3]. Especially for cervical cancer, HPV infection is a necessary condition. This is the successful reason of the cervical cancer vaccine, that is, to prevent the occurrence of cervical cancer by preventing HPV infection. HPV6 and HPV11 are the most studied types in low-risk HPV, and the above two types account for 90% of the cases of condyloma acuminatum [4]. These two types are also included in some HPV vaccines such as the MSD products Gardasil and Gardasil9.

The replication of papillomavirus is highly tissue-specific. Viral replication is closely related to epithelial differentiation. The process is divided into two stages: the early stage of viral genome replication and the late stage of capsid protein production and assembly [5]. For papillomavirus to complete its complete life cycle, it must infect the basal columnar epithelial cells with active mitosis. Stable viral genome replication can occur in these cells. However, the replication of offspring viruses characterized by high copy number amplification will occur only after the basement membrane cells further differentiate and enter the epithelium. The assembly of L1 and L2 proteins into viruslike particles takes place in more differentiated keratinocytes. The release of papillomavirus in offspring is a passive process, through the death and exfoliation of keratinocytes in the outermost layer of stratified epithelium, and the virus particles are released outside the cells. The replication and assembly of papillomavirus depend on the terminal differentiation of stratified squamous epithelium, which makes the culture of papillomavirus very challenging. Infectious virus particles could not be obtained from conventional cell culture. Therefore, the initial idea was to simulate the life cycle of papillomavirus to achieve the culture of the virus [6], using organ rafts to provide differentiated epithelial cells [7] in vitro or xenografts of mouse skin to provide differentiated epithelial cells [8]. Although the above culture process can simulate the life cycle of the virus, these techniques are tedious, demanding long time, and unstable, and the virus yield is very low, which largely limits the availability of HPV in basic and clinical research.

When the main capsid protein L1 of HPV is expressed alone, it can be assembled into viruslike particles without nucleic acid, and its immunogenicity is similar to infectious live virus particles, which is the current preparation principle of HPV L1 viruslike particles [9]. If L1 and minor capsid protein L2 are expressed at the same time, viruslike particles packed with a nucleic acid can be formed. Both L1 [10] and L2 [11] protein sequences contain nuclear localization signals. After L1 is expressed in the cytoplasm, it aggregates quickly to form a pentamer. Through the binding of nuclear localization signals to the corresponding hand receptors, L1 pentamer and L2 proteins enter the nucleus. Most viral genomes contain packaging signals, and the structural proteins of the virus recognize these signals during assembly to selectively package nucleic acids into viral particles. However, no such sequence was identified in the HPV genome. There are positively charged sequences at the N and/or C ends of L1 and L2, which can bind to the phosphate skeleton of DNA in the form of ionic bonds, which may be the reason why capsid proteins can package nucleic acids

[12]. However, this packaging is not sequence-specific and only has the restriction of genome size. As long as it does not exceed limit of 8 kb for the circular double-stranded genome, HPV can effectively package it into virus particles [13]. Therefore, in the virus replication cycle, L1 pentamer, L2, and replicated HPV genome can be assembled in the nucleus to form initial virions. The initial virus particles further mature in the nucleus, forming disulfide bonds between L1 molecules, and the mature virus particles are denser and more stable, which can effectively resist protease digestion [14]. Based on the characteristics and mechanism of HPV particle formation, researchers express L1 and L2 proteins and corresponding specific genes to construct synthetic virions with different types and characteristics, such as viruslike particles (empty particles without genes), pseudotyped viruses (non-HPV genomes, usually reporter genes), and quasi-viruses (virions containing HPV genomes). Among them, pseudotyped virus has been most widely used in HPV infection research and immunogenicity evaluation, which is also the focus of this review.

5.2 Construction of Pseudotyped Papillomavirus

Based on the characteristics that HPV L proteins can package virus nucleic acids and form virus particles, the primary task is to package proper reporter gene to yield high-titer of pseudotyped viruses. There are two important types of the pseudotyped virus packaging system of HPV: the first system based on recombinant virus vector and the second based on plasmid transfection.

5.2.1 *Pseudotyped Virus Packaging System Based on Virus Vector*

Zhou et al. found that when the recombinant poxvirus with L1 and L2 genes infected CV-1 cells, HPV viruslike particles could be produced. These viruslike particles were HPV virus shells without nucleic acid [15]. Roden et al. used this technology to produce BPV particles containing virus genome, that is, BPHE-1 cells containing BPV genome were infected with recombinant poxvirus containing BPV L1 and L2 genes, and BPV viruses were produced (Fig. 5.1A). The structure of BPV viruses is completely the same as that of live viruses, but the production model is different, in which BPV-like viruses do not go through different stages of cell differentiation. However, the proportion of viruses with genomes produced in this way is relatively small, and most of them are still virus shells without virus genome [16]. The cytopathic effect can be produced when C127 cells are infected with BPV virus particles, and the content of the virus in the sample can be judged according to the cytopathic effect (CPE). If the virus is incubated with the antibody-containing

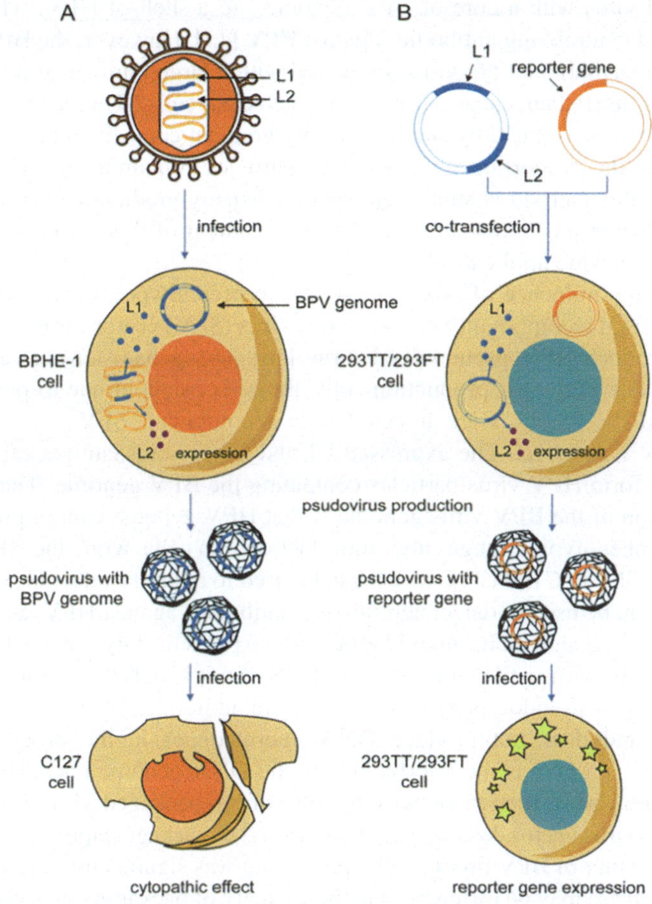

Fig. 5.1 Production of the pseudotyped papillomavirus. (**A**) Pseudotyped virus production based on virus vector. Virus vectors, containing the HPV late genes L1 and L2, are used to infect BPHE-1 cells, in which BPV genomes are consistently generated. After infection, expressed L1 and L2 proteins could spontaneously package the BPV genomes to form the pseudotyped virus with BPV genome. This type of pseudotyped virus could infect and destroy C127 cells. The amount of infected pseudotyped virus could be determined by the cytopathic effect. (**B**) Pseudotyped virus production based on plasmid transfection. The genes of HPV structural proteins L1 and L2 are cloned into the same plasmid. And reporter gene is inserted into the other plasmid, the size of which should not exceed 8 kb. The two plasmids were employed to co-transfect 293TT/293FT cells, in which the expressed L1 and L2 would self-assemble into viruslike particle packaging the reporter plasmid in it. The pseudotyped virus with reporter gene could infect the mammal cells, such as 293TT/293FT, and express the reporter gene. Through quantification of the reporter signal, the infected pseudotyped virus could be determined

sample before the virus infects the cell, the antibody content in the sample can be determined according to reduction of CPE. Using this principle, Ryndock replaced the L1 and L2 genes of BPV with the corresponding genes of HPV, resulting in a pseudotyped virus with a core of BPV genome and a shell of HPV, which can be used to detect neutralizing antibodies against HPV [17]. However, the BPV particles produced by recombinant poxvirus are mixed with a large number of poxvirus, and the poxvirus itself can cause CPE in the detected cells. Therefore, BPV virus particles should be isolated by sucrose density gradient centrifugation before detection. Because the proportion of intact virus particles containing the virus genome produced by this method is small, and sucrose density gradient centrifugation will lead to the decrease of virus stability, the virus content of this method is very low, which limits its wide application.

To avoid the influence of pox virus mixed in pseudotyped virus harvest liquid, Roden et al. used recombinant Semliki Forest virus (SFV) vector instead of poxvirus vector to produce HPV pseudotyped virus [16]. Using hamster-derived BPHE-1 cells as pseudotyped virus production cells, the cells can continue to produce BPV genomes. BPHE-1 cells were infected with recombinant SFV vectors that can express HPV L1 and L2. The expressed L1 and L2 proteins can package the BPV genome and form HPV virus particles containing the BPV genome. The reason for the application of the BPV virus genome is that HPV genome cannot produce easy quantitative phenotypic changes after introduction into cells, while the BPV genome can produce CPE in C127 cells, which can be used to quantify infected pseudotyped viruses and can be used to detect neutralizing antibodies against HPV. SFV vector is replication-defective, which can only efficiently express the target gene, but does not produce progeny virus and cytopathic effect [18]. So it avoids the disturbance on the detection of HPV pseudotyped virus infection. In addition, SFV is a positive-strand RNA virus and does not produce DNA intermediates in the process of virus replication, which avoids the interference to the DNA genome packaged by HPV, which may be part of the reason why the titer of the packaged virus is higher than that from poxvirus vector. When using this method to package papillomavirus, it was found that the titer of BPV produced by packaging was significantly higher than that of HPV, and there may be differences in the capacity of packaging genomes between the two viruses. The titer of HPV pseudotyped virus produced by this method still cannot meet the needs of large-scale serological detection, and it is difficult to achieve high-throughput detection by CPE counting to quantify neutralizing antibodies in candidate samples.

5.2.2 Pseudotyped Virus Packaging System Based on Plasmid Transfection

The packaging method of HPV pseudotyped virus based on plasmid transfection makes the production process greatly simplified and the titer of pseudotyped virus greatly increased (Fig. 5.1B). The genes of HPV structural proteins L1 and L2 are cloned into the same plasmid. And reporter gene is inserted into the other plasmid, the size of which should not exceed 8 kb. The two plasmids were employed to co-transfect 293TT/293FT cells, in which the expressed L1 and L2 would self-assemble into viruslike particle packaging the reporter plasmid in it. The pseudotyped virus with reporter gene could infect the mammal cells, such as 293TT/293FT, and express the reporter gene. Through quantification of the reporter signal, the infected pseudotyped virus could be determined.

The establishment of this efficient packaging system is accompanied by the optimization of a series of packaging conditions. The first is the optimization of codons for efficient expression of the structural protein. The codon preference of papillomavirus is very different from that of mammalian cells. Zhou et al. significantly increased the expression level of its structural proteins L1 and L2 in mammalian cell lines by codon optimization [19]. On this basis, Buck et al. further optimized the codon of L1, increased its expression level by more than 100 times, and make the expression level of HPV structural protein to meet the requirement of producing high-titer pseudotyped virus in mammalian cells [13]. In addition to the high expression of structural proteins, the level of target plasmids in packaging cells is also an important factor in determining the titer of pseudotyped viruses. To improve the replication ability of the target plasmid, the SV40 replication initiation sequence was introduced into the target plasmid, and the packaging cells were modified. Cell lines that can stably express SV40 T antigens, such as 293TT [13] or 293FT [20, 21] cells, were established. The copy number of target plasmid in packaging cells was maintained at a high level through the interaction between T antigen and the initial sequence of SV40 replication. It was found that although the packaging of the target plasmid by HPV pseudotyped virus did not have sequence specificity, it had sequence size specificity. The larger the plasmid is, the lower the packaging efficiency is. If the plasmid exceeds the size of the HPV genome (~ 8 kb), the packaging efficiency will become very low. Therefore, the modified packaging system set the size of the target plasmid at a smaller level, not more than 6 kb. For the structural gene plasmid co-transfected with the target plasmid, it should be made as large as possible. L1 and L2 are usually expressed on the same plasmid, and to reduce the packaging efficiency of the plasmid, irrelevant sequences are introduced into the plasmid to further increase the size of the plasmid. Structural gene expression plasmids are usually more than 10kb [13, 22].

In addition to the optimization of pseudotyped virus packaging cells and packaging plasmids, the optimization of pseudotyped virus harvest conditions is also very important. After the packaging of HPV pseudotyped virus is completed, it will not be automatically released into the cell supernatant, and the packaging cell needs to be lysed to release the pseudotyped virus. Through the comparison of different lysis agents and conditions, compared with ultrasound, freeze-thaw, and other decontamination agents (such as Triton Xmur100, Tween 80, NP 40, etc.), 0.5% nonionic detergent Brij 58 has the best effect. The cell lysate treated with detergent is usually adhered with pseudotyped viruses; if the cell fragments are removed directly with centrifugation, a large number of pseudotyped viruses will be lost. By adding a high concentration of the salt solution (such as 0.8 m NaCl), the pseudotyped virus can be dissociated from the fragments and the harvest liquid can be clarified. At this time, by the way of centrifuge fragments, a high concentration of pseudotyped virus solution can be obtained. For the purification of HPV pseudotyped viruses, such as the usual method of CsCl density gradient centrifugation, the titer of HPV pseudotyped viruses will be reduced by 99%. By comparison, it is found that when using nonionic low osmotic iodide (iodixanol) (trade name OptiPrep) as ultracentrifugation medium, it can effectively ensure that the pseudotyped virus titer does not decrease significantly, and usually, the purified pseudotyped virus titer can reach 50%–70% of the pre-purification [23]. The loss of the pseudotyped virus may be due to the nonspecific adsorption of the pseudotyped virus particles to the plastic purification tube [24]. Although the packaging method of HPV pseudotyped virus is still continuing to be optimized, the packaging system of HPV pseudotyped virus based on plasmid co-transfection is the most widely used at present, and it is also the cornerstone of the wide application of HPV pseudotyped virus.

5.3 Application of Pseudotyped Papillomavirus

5.3.1 The In Vitro L1 Pseudotyped Virus Based Neutralization Assay (PBNA)

PBNA is considered to be the "gold standard" for the detection of protective antibodies, and the World Health Organization also recommends this method as a reference method for evaluating vaccine-induced protective antibodies [25]. PBNA can detect all the neutralizing antibodies (such as IgM, IgA, IgG) in vitro [26].

5.3.1.1 PBNA Based on Green Fluorescent Protein (GFP) and Secretory Alkaline Phosphatase (SEAP)

These methods were originally established in John Schiller's laboratory of the National Institutes of Health [22]. The pseudotyped viruses used in this method were prepared by co-transfecting the plasmids expressing HPV structural proteins L1 and L2 with the plasmids expressing reporter genes into eukaryotic cells. The L1 and L2 proteins expressed after co-transfection with the above plasmids could be self-assembled into VLP, and the reporter gene (GFP or SEAP) plasmids were packaged into VLP to form pseudotyped viruses. Pseudotyped viruses can infect target cells (293TT or 293FT cells). After infection, reporter gene plasmids are transferred into the nucleus and express the corresponding reporter proteins. The number of cells expressing reporter genes or the signal value of reporter proteins is proportional to the number of pseudotyped viruses infected. If neutralizing antibodies or serum samples containing neutralizing antibodies are added in the process of infection, neutralizing antibodies can block pseudotyped virus to infect cells. The level of neutralizing antibodies can be calculated by detecting the signal value of the reported protein. The level of neutralizing antibody is usually expressed by the dilution of the sample when inhibiting half of the virus infection, that is, the 50% inhibitory dose (50% inhibitory dose, ID_{50}). At the beginning of the establishment of this method, in the detection of clinical samples, it was found that when the serum dilution was low (10 or 20 times dilution), the serum could nonspecifically neutralize bovine papillomavirus (BPV) pseudotyped virus. The nonspecific reaction was not changed by complement inactivation, freeze-thaw, or centrifugation of the samples. When the sample dilution reached 40 times or more, the nonspecific reaction of most samples disappeared. Therefore, the cutoff value of this method is set at 40. At the same time, to eliminate the influence of serum nonspecific reaction on the detection results, while detecting HPV specific neutralizing antibody, BPV pseudotyped virus was used as a control virus to detect nonspecific reaction in serum, which was also expressed by ID_{50}. When the ID_{50} detection value of the sample-specific virus was not less than 40 and not less than 4 times of the BPV ID_{50} detection value, it was considered that the antibody of the corresponding type was positive [22, 27].

There are two ways to read PBNA results based on GFP: microscope observation estimation and flow cytometry detection; the former is subjective and the original results are not easy to preserve, while the latter is tedious and difficult to achieve high-throughput detection. These two methods are rarely used in the present clinical sample detection. The chemiluminescence detection of PBNA based on SEAP can accurately quantify the detection results and relatively improve the throughput in a 96-well plate. This method is included in the human papillomavirus laboratory manual of the WHO [28]. However, the background of SEAP in mammalian cells is high, and the titer of pseudotyped virus is low. In the detection of SEAP, the substrates should be incubated repeatedly at 65 °C, in ice, and at room temperature, which is tedious and makes the results poorly reproducible.

5.3.1.2 PBNA Based on Secretory Membrane-Anchored Luciferase (Gaussia Luciferase, Gluc)

GLuc is a small molecular weight luciferase from marine organisms. It has a low background in mammalian cells (100 times lower than SEAP) and can be secreted to the cell supernatant after expression. Its detection procedure is simple, without repeated incubation at different temperatures. The detection time (5 minutes/96-well plate) is much shorter than that of SEAP (60 minutes/96-well plate). A Gluc-based PBNA (Gluc-PBNA) 96-well plate detection method [29] was established, which significantly improved the efficiency of the PBNA. Because of the simplicity of operation, Peter Sehr et al. combined Gluc-PBNA with PerkinElmer, PE, automation workstation to establish an automatic assay on 384-well plate, which further improved the throughput and repeatability and could be used in the detection of vaccine immune serum and natural infection serum [27].

5.3.1.3 Multiple-Color PBNA Based on Fluorescent Protein

In this method, the different types of pseudotyped viruses containing different fluorescent protein reporter genes could be generated, and neutralizing antibodies against different types of HPV could be detected in one test simultaneously. The production of the individual pseudotyped viruses in the multiple-color PBNA is identical to the GFP-based pseudotyped virus. The key part of the development of the multiple-color method is the selection of fluorescent proteins, which do not interfere with each other under a certain combination of excitation light and filter. After cross-comparison, green fluorescent protein (GFP), red fluorescent protein (RFP), and crimson fluorescent protein (CFP) without mutual interference were selected as reporter genes, and their excited/emitted light was 480/520 nm, 570/600 nm, and 630/670 nm, respectively.

In the detection of multitype antibodies, viruses with distant evolutionary relationships should be selected as a group to avoid the interaction between different types. It has been confirmed in clinical trials that high titers of HPV16 neutralizing antibodies can inhibit HPV31 and HPV33 infection, and high titers of HPV18 can inhibit HPV45 infection [30], so the above cross-protected pseudotyped viruses should not be detected together. There are many available combinations of pseudotyped viruses detected by nine-valent vaccines, such as HPV16–18-58, HPV6–33-45, HPV11–31-52, HPV16–18-58, HPV6–33-52, and HPV11–31-45. This method uses immune spot counting (FluoroSpot) instead of flow cytometry, without digesting cells or adding any reaction substrates. By directly counting the number of cells infected by pseudotyped viruses and calculating the titers of neutralizing antibodies corresponding to three types in the samples, the efficiency of the method is greatly improved, and the detection efficiency (5 min/96-well plate × 3) is greatly higher than that of flow detection (180 min/96-well plate × 3) [31].

Because the neutralization test is a detection method based on cell culture, compared with the binding test, the PBNA has a higher degree of variation, and the operation is relatively tedious, so it is difficult to carry out high-throughput detection. With the continuous updating of PBNA, the detection throughput is also gradually improved, especially the 384-well plate automatic detection method [27] based on Gluc and triple-color pseudotyped virus combined with immune spot counting method [31], which lays a foundation for the application of this method in clinical trials. There was a good correlation between the detection results of the Gluc-based method and the fluorescent protein-based method ($R^2 = 0.93$), and the sensitivity of the former was slightly higher than that of the latter [29]. When determining the results based on the fluorescent protein method, the fluorescent spot count is carried out directly without adding the detection substrate, and the detection cost is lower; the three-color detection method can detect three types of neutralizing antibodies simultaneously. The sample size is only 1/3 of that of other methods, which is more suitable for the detection of a large number of clinical samples of multivalent vaccines [31].

In addition to the evaluation of the immunogenicity of the vaccine, HPV pseudotyped virus is also applied to the effect of HPV mutation on its infectivity and antigenicity because of its easy availability and maneuverability. By comprehensively analyzing the HPV L1 gene variation, the site-directed mutation was combined with HPV pseudotyped virus to study the antigenicity of the variants. Different mutation sites were introduced into pseudotyped viruses, and the effects of these mutations on pseudotyped virus infectivity and antibody sensitivity were analyzed, and the effects of these mutations on virus epidemic characteristics and vaccine protection were predicted [32]. In addition, the method of pseudotyped virus is often used to analyze the antigenic characteristics of different lineages or sub-lineages under the same type. Different lineages or sub-lineages pseudotyped viruses were constructed and neutralized by monoclonal antibodies against epitopes. The mutation sites causing alteration of antigenicity were found and then were verified on pseudotyped viruses by site-directed mutation [33–35]. These studies are helpful to analyze the characteristics of antigenicity changes in the process of virus evolution and have important reference value for future vaccine design.

5.3.2 *In Vitro L2 Pseudotyped Virus Neutralization Test*

All the antigens of approved HPV vaccines are viruslike particles of L1, which can induce high titers of type-specific neutralizing antibodies, but the cross-neutralization reaction between different types is weak [36].If we want to expand the protective type, we can only introduce the corresponding type of L1 VLP antigen, which is the reason that the current HPV vaccine adopts a multivalent strategy. With the increase of types, the cost of vaccine preparation will be greatly increased. At present, there are 13 kinds of high-risk and 11 potential high-risk HPV. The current vaccine cannot achieve complete protection against all high-risk HPV, nor can it protect many benign warts of the skin and mucosa caused by low-risk HPV. If we only rely on the increasing types of L1 VLP vaccine, it is difficult to achieve the above protection goal. The minor capsid protein L2 of HPV contains neutralizing epitopes shared by different types, which can induce neutralizing antibodies with low titer but can cross-neutralize different types of HPV. The L2-based vaccine is an important direction of broad-spectrum HPV vaccine research [37]. However, the process of HPV pseudotyped viruses infecting monolayer mammalian cell lines is different from that of living viruses in vivo. Although pseudotyped viruses contain L2 protein, the neutralizing epitopes of L2 in the infection process of pseudotyped viruses are not fully exposed, so it is impossible to detect L2-specific neutralizing antibodies in samples. According to the characteristics of L2 neutralizing antibody epitope exposure, the original detection method was improved, and the L2 neutralizing antibody detection method based on pseudotyped virus was established.

5.3.2.1 L2-Based PBNA Mimicking In Vivo Infection

In the natural process of in vivo HPV infection, when the integrity of the cervical epithelium was destroyed, the cervical basement membrane (BM) cells were exposed. HPV binds to heparan sulfate proteoglycan (HSPG) located on the basement membrane. This binding induces a conformational change in the capsid of the virus, resulting in exposure to the Furin enzyme site at the amino-terminal of L2. Furin then digested the amino-terminal of L2, which further exposed the amino-terminal of L2, including the main cross-neutralizing epitopes. The structural change of L2 will also lead to the change of L1 conformation on the surface of virus particles. When keratinocytes migrate to exposed BM to repair broken epithelium, L1 binds to receptors on keratinocytes to complete viral infection. In the process of infection, it takes a long time, usually several hours [38]. During this period, L2 neutralizing antibodies can bind to cross-neutralizing epitopes, preventing viral particles from binding to receptors on keratinocytes, thereby preventing infection [39].

In the L1 pseudotyped virus neutralization test, the pseudotyped virus directly binds to the HSPG on the cell surface, and the cleavage of L2 occurs on the cell surface, and then L1 binds to the receptor on the cell surface quickly. The L2 neutralizing epitope is not fully exposed and the exposure time is short, so the L2 neutralizing antibody is seriously underestimated in this method. In the L2 pseudotyped virus neutralization test, the microenvironment of the extracellular matrix should be established first. MCF10A cells were added to the cell culture plate, which could secrete the extracellular matrix. After 24 hours of culture, MCF10A cells were lysed and the extracellular matrix was retained in the culture plate. Pseudotyped viruses were added to the culture plate, and the pseudotyped viruses bind to HSPG in the extracellular matrix and expose the N-terminal of L2. After that, exogenous Furin was added to cut the N-terminal of L2 to expose the neutralizing epitope of L2. After cleaning the unbound pseudotyped virus, add the sample containing L2 neutralizing antibody to the culture plate. After incubation at 37 °C for 6 hours, HSPG defective pgsa-745 cells were added and cultured for 48 hours. According to the expression of the reporter gene after pseudotyped virus infection, the level of L2-specific neutralizing antibody in the sample was calculated (Fig. 5.2) [40]. Because of the tedious operation of this method, it has not been widely used on a large scale after its establishment.

5.3.2.2 Detection of L2 Pseudotyped Virus Neutralizing Antibody Based on Furin Cleavage Intermediates

The classical L1 neutralizing antibody detection method cannot effectively detect L2 neutralizing antibodies because L2 neutralizing epitopes cannot be fully exposed during pseudotyped virus infection. If the neutralizing epitope of pseudotyped virus L2 is fully exposed by Furin cutting in advance, and then neutralizing test is carried

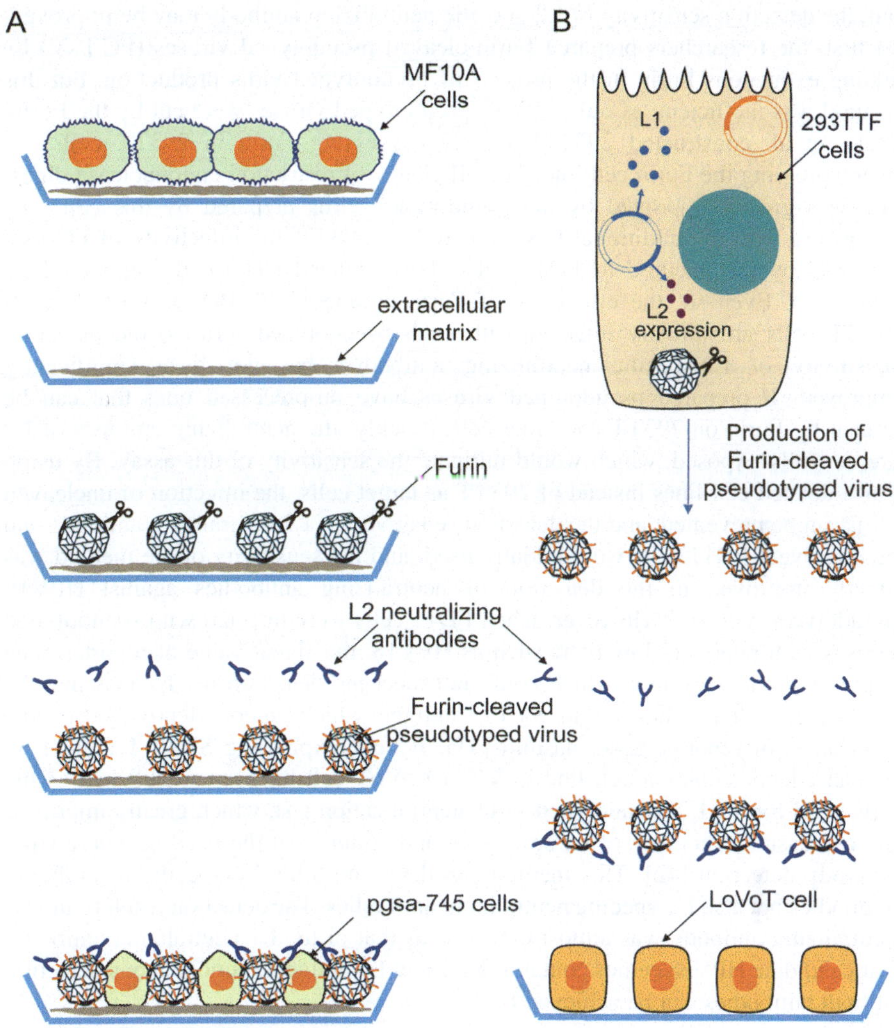

Fig. 5.2 Schematic of L2-based PBNA. (**A**) L2-based PBNA mimicking in vivo infection. MCF10A cells were added to the cell culture plate to secrete the extracellular matrix. Then, MCF10A cells were lysed and the extracellular matrix was retained in the culture plate. Pseudotyped viruses were added to the culture plate and bind to HSPG in the extracellular matrix to expose the N-terminal of L2. Exogenous Furin was added to cleave the N-terminal of L2 to expose the neutralizing epitope of L2. So, L2 neutralizing antibody could bind to the L2-specific epitope to inhibit the pseudotyped virus to infect the target pgsa-745 cell. (**B**) **L2**-based PBNA employing **Furin cleavage intermediates**. 293TTF (stably expressing Furin in 293TT cells) is used to generate the Furin-cleaved pseudotyped virus. The L2-specific neutralizing antibodies could bind to the exposed epitopes and prevent the pseudotyped virus from infecting the target LoVoT cells, which could express the SV40 T antigen to greatly improve the expression efficiency of reporter gene introduced by the pseudotyped viruses

out, the detection sensitivity of L2-specific neutralizing antibody may be improved. At first, the researchers prepared Furin-cleaved pseudotyped viruses (FC-PsV) by adding exogenous Furin in the process of pseudotyped virus production, but this method was inefficient, as only 35% of pseudotyped viruses were cut by the Furin. Wang et al. constructed 293TTF (stably expressing Furin in 293TT cells) by overexpressing the Furin cell line. The efficiency of restriction endonuclease digestion was greatly improved by the pseudotyped virus prepared by this cell line. Compared with the traditional PsV prepared by 293TT, the infectivity of FC-PsV prepared by this method to FD11 cells (Furin-deleted CHO cells) increased by 100 times. Even so, the efficiency of Furin cutting in FC-PsV is not 100%. If 293TT cells are directly infected with such pseudotyped viruses, the detection sensitivity of L2-specific neutralizing antibodies has not been significantly improved. If prepared pseudotyped viruses have un-precessed ones that can be treated by Furin on 293TT and enter cells quickly, the neutralizing epitopes of L2 are not fully exposed, which would mitigate the sensitivity of this assay. By using Furin-deleted cell lines instead of 293TT as target cells, the infection of uncleaved virus can be prevented, and the detection sensitivity of L2 neutralizing antibody can be improved. FD11 cells were initially used, and the sensitivity of the method was greatly improved in the detection of neutralizing antibodies against HPV16 pseudotyped virus L2. However, when FD11 cells were infected with pseudotyped viruses with relatively low titers such as HPV18, the signal value of reporter gene expression was very low, which could not meet the detection needs. This may be because the cell does not contain SV40 T antigen, which cannot effectively drive the replication of reporter gene plasmid [41]. By overexpressing SV40 T antigen in human adenocarcinoma cell line LoVo, a LoVoT cell line was established as the target cell for the L2 pseudotyped virus neutralization test, which greatly improved the expression efficiency of reporter gene and could meet the need of neutralizing antibody detection [42]. This method can detect not only L2-specific neutralizing antibodies but also L1-specific neutralizing antibodies. The detection result of the L1 neutralizing antibody was almost the same as that of the L1 neutralizing antibody. For candidate HPV vaccines containing L1 and L2 antigens, simultaneous detection of both antibodies can be achieved [41].

5.3.3 Animal Model of HPV Pseudotyped Virus Infection

5.3.3.1 Mouse Model of HPV Pseudotyped Virus Infection

HPV infection is species-specific, so it is impossible to establish persistent infection in animals with live viruses. Although the pseudotyped virus cannot form persistent infection, the infection can be judged by detecting the expression of reporter genes carried by the pseudotyped virus, to achieve the relative quantification of infection. The establishment of an animal infection model needs to simulate the infection process of authentic viruses in vivo. HPV infection first requires the damage of cervical epithelial cells to expose the infected target cells—basal columnar epithelial cells. For the mouse model of pseudotyped virus infection, 4–5 days before infection, the mice were injected with Depo-Provera (Pfizer), a long-acting contraceptive with medroxyprogesterone acetate, which thinned the vagina and cervical epithelium and made mice more vulnerable to damage [43]. The epithelium of mice was further damaged by physical or chemical methods before being infected with the pseudotyped virus. The former usually scrape the epithelium with a Cytobrush Plus cell collector. The latter is to add nonoxynol-9 (N-9) to the vagina of the mice, a nonionic surfactant that destroyed the vaginal and cervical epithelium of the mice. N-9 is usually added to carboxymethyl cellulose at a concentration of 4% to form a viscous liquid, which could stay longer in the vagina of mice. Chemical methods are more efficient and consistent than physical methods [44]. After 4–6 hours of epithelial treatment, the mice can be challenged intravaginally by pseudotyped virus. The pseudotyped viruses used here are similar to those in L1-PBNA, except that different reporter genes are selected according to the final detection method. If the in vivo imaging system (IVIS) is used for detection, firefly luciferase is usually used as the reporter gene. The emission wavelength of the reporter gene is about 560 nm, which can well pass through the tissue and fur. By quantifying the luminous intensity by IVIS, the amount of pseudotyped virus infection can be relatively quantified. If there is no IVIS, Gluc can also be used as a pseudotyped virus of the reporter gene. Because Gluc contains a secretory signal peptide, it can be secreted out of the cell after expression. When the pseudotyped virus is infected, the vaginal lavage fluid can be collected to detect the expression level of Gluc; thus, the amount of pseudotyped virus infection can be relatively quantified.

The mouse model can be used to evaluate the effects of HPV infection inhibitors, such as carrageenan, heparin, heparinase, furin inhibitor, and passively injected antibodies [38, 43, 45]. In addition, this model can induce long-lived tissue-resident memory CD8 T cells located in the cervicovaginal epithelium after infection with pseudotyped virus, so the model can also be applied to the study of memory T cells resident in reproductive tissues [46].

5.3.3.2 Chimeric Pseudotyped Virus Animal Model

Cottontail rabbit papillomavirus (CRPV) infection in rabbits is a classical papillomavirus research model, which is an important tool to study the pathogenic characteristics, infectious tropism, latent infection, vaccine efficacy, and immunotherapy of papillomavirus [47]. The model virus used in this model is that CRPV cannot fully reflect the characteristics of HPV; especially it cannot directly evaluate the protective effect of the HPV vaccine. After the development of pseudotyped virus technology, taking advantage of the non-sequence specificity of genome selection during the packaging of HPV pseudotyped virus, the structural gene expression plasmid of HPV and CRPV genome were co-transfected into eukaryotic cells to form HPV pseudotyped virus particles packed with CRPV genome. The pseudotyped virus can infect the skin tissue of New Zealand white rabbits, and the CRPV genome released after infection can form papilloma on the skin of New Zealand white rabbits which becomes an animal model of benign disease [48]. The animal model can directly evaluate the preventive effect of HPV antibodies, vaccines, and infection inhibitors on diseases [49].

5.3.4 Other Applications of Pseudotyped Papillomavirus

Pseudotyped papillomavirus is also used in gene therapy and tumor vaccine development because of the non-sequence specificity of the papillomavirus pseudotyped virus packaging genome. For example, suicide genes packaged by HPV pseudotyped viruses are used to treat nonmuscular invasive bladder cancer (NMIBC). In the mouse bladder cancer model, HPV pseudotyped virus carrying suicide gene is more than ten times more infectious to bladder cancer cells than ordinary bladder cells. It can specifically introduce suicide genes into cancerous cells, lead to cell death, and produce tumor cell-specific CD8 T cell response, which significantly inhibits tumor growth and prolongs the life span of tumor-bearing mice [50]. HPV pseudotyped virus can also be used as a vector carrying gene editing plasmid for gene therapy. The plasmid carrying cervical cancer oncogene E6-gRNA and Cas9 was packaged into HPV16 pseudotyped virus, which was used to infect cervical cancer cell SiHa, which could effectively inhibit tumor growth and metastasis in vitro and in vivo [51]. In a therapeutic study of a mouse model of colorectal cancer, the intestinal cancer protein gene packaged by BPV showed a strong tumor-killing effect by oral immunization, which was achieved by inducing innate immunity, while the packaged oncoprotein gene could induce an adaptive immune response and effectively prevent tumor recurrence. The life span of tumor-bearing mice could be prolonged by three times through immunization. It has become a very promising potential strategy for the treatment of colorectal cancer [52].

5.4 Conclusion

Papillomavirus pseudotyped virus technology continues to develop with the deepening of the understanding of the life cycle, pathogenic mechanism, and immune characteristics of papillomavirus, which continues to enrich the tools for papillomavirus research. It has been more and more widely used in viral infection characteristics, antigen characteristics, vaccine evaluation, and treatment research and even directly used as a therapeutic tool in the treatment of diseases. With the development of technology, the method of papillomavirus pseudotyped virus is also upgrading, which will be more widely used in the future, not limited to the research field of papillomavirus.

Acknowledgments This work was supported by the General Program of National Natural Science Foundation of China [grant number 82172244&82073621] and a major project of the Study on Pathogenesis and Epidemic Prevention Technology System [2021YFC2302500].

References

1. McBride, A.A.: Human papillomaviruses: diversity, infection and host interactions. Nat. Rev. Microbiol. **20**, 95–108 (2022)
2. de Villiers, E.M., Fauquet, C., Broker, T.R., Bernard, H.U., zur Hausen, H.: Classification of papillomaviruses. Virology. **324**, 17–27 (2004)
3. de Martel, C., Plummer, M., Vignat, J., Franceschi, S.: Worldwide burden of cancer attributable to HPV by site, country and HPV type. Int. J. Cancer. **141**, 664–670 (2017)
4. Garland, S.M., et al.: Natural history of genital warts: analysis of the placebo arm of 2 randomized phase III trials of a quadrivalent human papillomavirus (types 6, 11, 16, and 18) vaccine. J. Infect. Dis. **199**, 805–814 (2009)
5. Doorbar, J., Egawa, N., Griffin, H., Kranjec, C., Murakami, I.: Human papillomavirus molecular biology and disease association. Rev. Med. Virol. **25**(Suppl 1), 2–23 (2015)
6. Ryndock, E.J., Biryukov, J., Meyers, C.: Replication of human papillomavirus in culture. Methods Mol. Biol. **1249**, 39–52 (2015)
7. Meyers, C., Frattini, M.G., Hudson, J.B., Laimins, L.A.: Biosynthesis of human papillomavirus from a continuous cell line upon epithelial differentiation. Science. **257**, 971–973 (1992)
8. McBride, A.A., Dlugosz, A., Baker, C.C.: Production of infectious bovine papillomavirus from cloned viral DNA by using an organotypic raft/xenograft technique. Proc. Natl. Acad. Sci. U. S. A. **97**, 5534–5539 (2000)
9. Kirnbauer, R., Booy, F., Cheng, N., Lowy, D.R., Schiller, J.T.: Papillomavirus L1 major capsid protein self-assembles into virus-like particles that are highly immunogenic. Proc. Natl. Acad. Sci. U. S. A. **89**, 12180–12184 (1992)
10. Nelson, L.M., Rose, R.C., Moroianu, J.: Nuclear import strategies of high risk HPV16 L1 major capsid protein. J. Biol. Chem. **277**, 23958–23964 (2002)
11. Darshan, M.S., Lucchi, J., Harding, E., Moroianu, J.: The l2 minor capsid protein of human papillomavirus type 16 interacts with a network of nuclear import receptors. J. Virol. **78**, 12179–12188 (2004)
12. El Mehdaoui, S., et al.: Gene transfer using recombinant rabbit hemorrhagic disease virus capsids with genetically modified DNA encapsidation capacity by addition of packaging sequences from the L1 or L2 protein of human papillomavirus type 16. J. Virol. **74**, 10332–10340 (2000)
13. Buck, C.B., Pastrana, D.V., Lowy, D.R., Schiller, J.T.: Efficient intracellular assembly of papillomaviral vectors. J. Virol. **78**, 751–757 (2004)
14. Buck, C.B., Thompson, C.D., Pang, Y.Y., Lowy, D.R., Schiller, J.T.: Maturation of papillomavirus capsids. J. Virol. **79**, 2839–2846 (2005)
15. Zhou, J., Sun, X.Y., Stenzel, D.J., Frazer, I.H.: Expression of vaccinia recombinant HPV 16 L1 and L2 ORF proteins in epithelial cells is sufficient for assembly of HPV virion-like particles. Virology. **185**, 251–257 (1991)
16. Roden, R.B., et al.: In vitro generation and type-specific neutralization of a human papillomavirus type 16 virion pseudotype. J. Virol. **70**, 5875–5883 (1996)
17. Sapp, M., Selinka, H.C.: Generation and applications of HPV pseudovirions using vaccinia virus. Methods Mol. Med. **119**, 463–482 (2005)
18. Liljestrom, P., Garoff, H.: A new generation of animal cell expression vectors based on the Semliki Forest virus replicon. Biotechnology (N. Y). **9**, 1356–1361 (1991)

19. Zhou, J., Liu, W.J., Peng, S.W., Sun, X.Y., Frazer, I.: Papillomavirus capsid protein expression level depends on the match between codon usage and tRNA availability. J. Virol. **73**, 4972–4982 (1999)
20. Wu, X., et al.: Detection of HPV types and neutralizing antibodies in Gansu province, China. *J Med Virol.* **81**, 693–702 (2009)
21. Wu, X.L., Zhang, C.T., Zhu, X.K., Wang, Y.C.: Detection of HPV types and neutralizing antibodies in women with genital warts in Tianjin City, China. *Virol Sin.* **25**, 8–17 (2010)
22. Pastrana, D.V., et al.: Reactivity of human sera in a sensitive, high-throughput pseudovirus-based papillomavirus neutralization assay for HPV16 and HPV18. Virology. **321**, 205–216 (2004)
23. Zolotukhin, S., et al.: Recombinant adeno-associated virus purification using novel methods improves infectious titer and yield. Gene Ther. **6**, 973–985 (1999)
24. Yeager, M.D., et al.: Neutralization of human papillomavirus (HPV) pseudovirions: a novel and efficient approach to detect and characterize HPV neutralizing antibodies. Virology. **278**, 570–577 (2000)
25. WHO Expert Committee on Biological Standardization: Recommendations to assure the quality, safety and efficacy of recombinant human papillomavirus virus-like particle vaccines. World Health Organ. Tech. Rep. Ser. **999**, 151–233 (2016)
26. Frazer, I.H.: Measuring serum antibody to human papillomavirus following infection or vaccination. Gynecol. Oncol. **118**, S8–S11 (2010)
27. Sehr, P., et al.: High-throughput pseudovirion-based neutralization assay for analysis of natural and vaccine-induced antibodies against human papillomaviruses. PLoS One. **8**, e75677 (2013)
28. World Health Organization: Human papillomavirus laboratory manual. First edition. **84–94** (2009)
29. Nie, J., Huang, W., Wu, X., Wang, Y.: Optimization and validation of a high throughput method for detecting neutralizing antibodies against human papillomavirus (HPV) based on pseudovirons. J. Med. Virol. **86**, 1542–1555 (2014)
30. Wheeler, C.M., et al.: Cross-protective efficacy of HPV-16/18 AS04-adjuvanted vaccine against cervical infection and precancer caused by non-vaccine oncogenic HPV types: 4-year end-of-study analysis of the randomised, double-blind PATRICIA trial. Lancet Oncol. **13**, 100–110 (2012)
31. Nie, J., Liu, Y., Huang, W., Wang, Y.: Development of a triple-color Pseudovirion-based assay to detect neutralizing antibodies against human papillomavirus. Viruses. **8**, 107 (2016)
32. Ning, T., et al.: Naturally occurring single amino acid substitution in the L1 major capsid protein of human papillomavirus type 16: alteration of susceptibility to antibody-mediated neutralization. J. Infect. Dis. **216**, 867–876 (2017)
33. Godi, A., et al.: Impact of naturally occurring variation in the human papillomavirus 52 capsid proteins on recognition by type-specific neutralising antibodies. J. Gen. Virol. **100**, 237–245 (2019)
34. Godi, A., et al.: Comprehensive assessment of the antigenic impact of human papillomavirus lineage variation on recognition by neutralizing monoclonal antibodies raised against lineage a major capsid proteins of vaccine-related genotypes. J. Virol. **94** (2020)
35. Godi, A., et al.: Impact of naturally occurring variation in the human papillomavirus 58 capsid proteins on recognition by type-specific neutralizing antibodies. J. Infect. Dis. **218**, 1611–1621 (2018)
36. Bissett, S.L., et al.: Pre-clinical immunogenicity of human papillomavirus alpha-7 and alpha-9 major capsid proteins. Vaccine. **32**, 6548–6555 (2014)
37. Schellenbacher, C., Roden, R.B.S., Kirnbauer, R.: Developments in L2-based human papillomavirus (HPV) vaccines. Virus Res. **231**, 166–175 (2017)
38. Kines, R.C., Thompson, C.D., Lowy, D.R., Schiller, J.T., Day, P.M.: The initial steps leading to papillomavirus infection occur on the basement membrane prior to cell surface binding. Proc. Natl. Acad. Sci. U. S. A. **106**, 20458–20463 (2009)

39. Day, P.M., et al.: In vivo mechanisms of vaccine-induced protection against HPV infection. Cell Host Microbe. **8**, 260–270 (2010)
40. Day, P.M., et al.: A human papillomavirus (HPV) in vitro neutralization assay that recapitulates the in vitro process of infection provides a sensitive measure of HPV L2 infection-inhibiting antibodies. Clin. Vaccine Immunol. **19**, 1075–1082 (2012)
41. Wang, J.W., et al.: Measurement of neutralizing serum antibodies of patients vaccinated with human papillomavirus L1 or L2-based immunogens using furin-cleaved HPV Pseudovirions. PLoS One. **9**, e101576 (2014)
42. Wang, J.W., et al.: Production of Furin-Cleaved Papillomavirus Pseudovirions and Their Use for In Vitro Neutralization Assays of L1- or L2-Specific Antibodies. *Curr Protoc Microbiol.* **38**(14B 15), 11–26 (2015)
43. Roberts, J.N., et al.: Genital transmission of HPV in a mouse model is potentiated by nonoxynol-9 and inhibited by carrageenan. Nat. Med. **13**, 857–861 (2007)
44. Cuburu, N., Cerio, R.J., Thompson, C.D., Day, P.M.: Mouse model of cervicovaginal papillomavirus infection. Methods Mol. Biol. **1249**, 365–379 (2015)
45. Ujma, S., et al.: Surfactant protein a impairs genital HPV16 Pseudovirus infection by innate immune cell activation in a murine model. Pathogens. **8** (2019)
46. Cuburu, N., et al.: Intravaginal immunization with HPV vectors induces tissue-resident CD8+ T cell responses. J. Clin. Invest. **122**, 4606–4620 (2012)
47. Christensen, N.D., Budgeon, L.R., Cladel, N.M., Hu, J.: Recent advances in preclinical model systems for papillomaviruses. Virus Res. **231**, 108–118 (2017)
48. Culp, T.D., et al.: Papillomavirus particles assembled in 293TT cells are infectious in vivo. J. Virol. **80**, 11381–11384 (2006)
49. Mejia, A.F., et al.: Preclinical model to test human papillomavirus virus (HPV) capsid vaccines in vivo using infectious HPV/cottontail rabbit papillomavirus chimeric papillomavirus particles. J. Virol. **80**, 12393–12397 (2006)
50. Hojeij, R., et al.: Immunogenic human papillomavirus Pseudovirus-mediated suicide-gene therapy for bladder cancer. Int. J. Mol. Sci. **17** (2016)
51. Cheng, Y.X., Chen, G.T., Yang, X., Wang, Y.Q., Hong, L.: Effects of HPV Pseudotype virus in cutting E6 gene selectively in SiHa cells. Curr Med Sci. **38**, 212–221 (2018)
52. Zhong, Z., Zhai, Y., Bu, P., Shah, S., Qiao, L.: Papilloma-pseudovirus eradicates intestinal tumours and triples the lifespan of Apc(min/+) mice. Nat. Commun. **8**, 15004 (2017)

Chapter 6
Pseudotyped Viruses for *Marburgvirus* and *Ebolavirus*

Li Zhang, Shou Liu, and Youchun Wang

Abstract Marburg virus (MARV) and Ebola virus (EBOV) of the *Filoviridae* family are the most lethal viruses in terms of mortality rate. However, the development of antiviral treatment is hampered by the requirement for biosafety level-4 (BSL-4) containment. The establishment of BSL-2 pseudotyped viruses can provide important tools for the study of filoviruses. This chapter summarizes general information on the filoviruses and then focuses on the construction of replication-deficient pseudotyped MARV and EBOV (e.g., lentivirus system and vesicular stomatitis virus system). It also details the potential applications of the pseudotyped viruses, including neutralization antibody detection, the study of infection mechanisms, the evaluation of antibody-dependent enhancement, virus entry inhibitor screening, and glycoprotein mutation analysis.

Keywords EBOV · MARV · Lentivirus · VSV neutralization · Drug screening · Mutation analysis

Abbreviations

aa	Amino acids
ADE	Antibody-dependent enhancement
ASMase	Acid sphingomyelinase
BDBV	Bundibugyo virus

L. Zhang · S. Liu
Division of HIV/AIDS and Sex-transmitted Virus Vaccines, National Institutes for Food and Drug Control (NIFDC) and WHO Collaborating Center for Standardization and Evaluation of Biologicals, Beijing, China

Y. Wang (✉)
Chinese Academy of Medical Sciences & Peking Union Medical College, Beijing, China

Institute of Medical Biology, Chinese Academy of Medicine Sciences & Peking Union Medical College, Kunming, China
e-mail: wangyc@nifdc.org.cn; wangyc@imbcams.com.cn

Y. Wang (ed.), *Pseudotyped Viruses*, Advances in Experimental Medicine and Biology 1407, https://doi.org/10.1007/978-981-99-0113-5_6

BLI	Bioluminescent imaging
BOMV	Bombali virus
BSL	Biosafety level
CatB	Cathepsin B
CatL	Cathepsin L
DC-SIGN	Dendritic Cell-Specific Intercellular Adhesion Molecule 3-Grabbing Non-integrin
DC-SIGNR	DC-SIGN-related molecule
DRC	Democratic Republic of Congo
EBOV	*Zaire ebolavirus*
E-S-FLU	EBOV-GP-pseudotyped influenza virus
EVD	Ebola virus disease
FRα	Folate Receptor Alpha
GAGs	Glycosaminoglycans
GP	Glycoprotein
GPCR	G protein-coupled receptor
HER2	Human Epidermal Growth Factor Receptor 2
HOS	Human osteosarcoma
HTS	High-throughput screening
ID50	50% inhibition dilution
IFL	Internal fusion loop
Luc	Luciferase
MARCH8	Membrane-Associated RING-CH-Type 8
MARV	Marburg virus
MBL	Mannose-Binding Lectin
MLD	Mucin-like domain
MVA	Modified vaccinia Ankara
MVD	Marburg virus disease
NP	Nucleoprotein
NPC1	Niemann Pick C1 Protein
RAVV	Ravn virus
RBD	Receptor-binding domain
RBS	Receptor-binding site
RdRp/L	RNA-dependent RNA polymerase
RESTV	Reston virus
RTK	Receptor Tyrosine Kinase
sGP	Secreted Forms of the GP Protein
SM	Sphingomyelin
SUDV	Sudan virus
TAFV	Taï Forest virus
TAM	Tyro3/Axl/Mer
TIM-1/4	T-Cell Immunoglobulin Mucin Domain-1/4
trVLPs	Transcription-competent viruslike particles
VP35/40/30/24	Viral Protein 35/40/30/24 kDa
VSV	Vesicular stomatitis virus
β-gal	β-Galactosidase

6.1 Introduction

The genera *Marburgvirus* and *Ebolavirus* belong to the *Filoviridae* family, which comprises four other genera: *Cuevavirus*, *Dianlovirus*, *Striavirus*, and *Thamnovirus* [1, 2]. The genus *Marburgvirus* includes only one species, *Marburg marburgvirus*, which has two members: Marburg virus (MARV) and Ravn virus (RAVV) [1]. The genus *Ebolavirus* includes six species, each comprising one virus: Ebola virus (EBOV) of the species *Zaire ebolavirus*, Sudan virus (SUDV) of the species *Sudan ebolavirus*, Bundibugyo virus (BDBV) of the species *Bundibugyo ebolavirus*, Taï Forest virus (TAFV) of the species *Taï Forest ebolavirus*, Reston virus (RESTV) of the species *Reston ebolavirus*, and Bombali virus (BOMV) of the species *Bombali ebolavirus* [3]. Among the viruses of the *Filoviridae* family, only six viruses from the genera *Marburgvirus* and *Ebolavirus* are pathogenic to humans (MARV, RAVV, BDBV, SUDV, TAFV, and EBOV) [4], of which MARV and EBOV have a much higher prevalence rate than the other viruses. Therefore, we focus on MARV and EBOV in this chapter.

6.2 The Biological Characteristics of *Marburgvirus* and *Ebolavirus*

MARV was the first discovered filovirus in 1967 in Germany and Yugoslavia [5]. Up until 2021, 470 reported cases of MARV infection in humans had been reported [6]. However, EBOV is the main causative agent of human outbreaks and the most virulent type of filovirus [7]. Ebola virus disease (EVD) and Marburg virus disease (MVD) both cause hemorrhagic fever in humans but also show nonspecific common symptoms, such as fever, malaise, headache, diarrhea, and vomiting [8]. The average case fatality rates for EVD and MVD are 65.0 and 53.8%, respectively, but these range from 24% to 90% [9, 10]. Both EVD and MVD are listed on the World Health Organization Priority Diseases List [11]. Fruit bats act as a reservoir of MARV and EBOV [12], primates are the main source of zoonotic spillover, and the leading cause of emergence is reported to be eating and hunting habits [7].

6.2.1 Morphology and Genome Structure

Marburgvirus and *Ebolavirus* are negative-strand, non-segmented RNA viruses [13]. Their morphology is commonly observed as long, filamentous viral particles, which are 80 nm in width and up to 1400 nm in length [14]. The virion particles of EBOV are thought to be longer than those of MARV [15]. Their genomes are approximately 19 kilobases and encode seven structural proteins, including the viral nucleoprotein (NP), viral protein of 35 kDa (VP35), VP40, a type I transmembrane glycoprotein (GP), VP30, VP24, and RNA-dependent RNA polymerase (RdRp, L) [16]. The viral RNA genome forms a strong complex with NP and VP30, constituting the central nucleocapsid core along with VP35 and RdRp. VP40 and VP24 reside in the matrix surrounding the nucleocapsid core and line the inner leaflet of the envelope [17]. The virus particles are surrounded by a host-derived lipid envelope, studded by GP (Fig. 6.1) [18]. The mature filoviral GP is composed of GP1 and GP2, which are cleaved by furin-like proteases from the P0 precursor [19]. The position of the conserved cleavage site is different between MARV and EBOV, but the gene structure and functional organization are the same [20, 21]. The heterodimers of GP1/GP2 are disulfide-linked and form chalice-like shaped trimers [22]. GP2 serves as the base, and GP1 serves as the cup [18]. Viruses of the *Ebolavirus* genus, but not the *Marburgvirus* genus, also produce secreted forms of GP (sGP) [23, 24]. The function of sGP may be related to viral immune evasion [25].

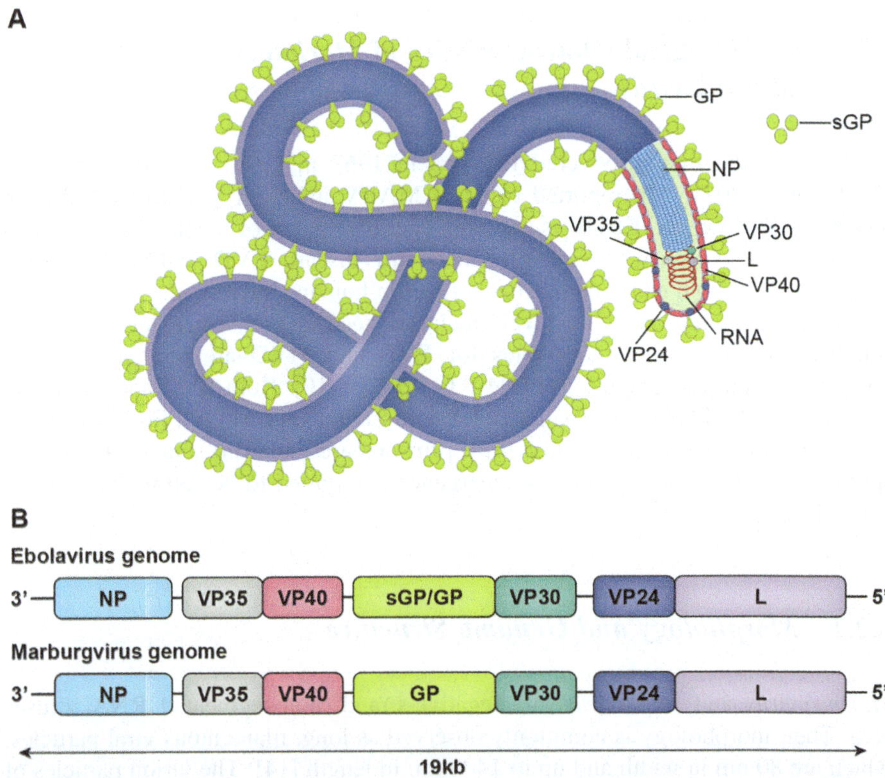

Fig. 6.1 Morphology and genome structure of EBOV and MARV. (**a**) Schematic illustration of a filovirus particle. (**b**) Schematic representation of the EBOV and MARV genomes

6.2.2 Virus Entry

Virus entry into cells is mediated by GP [18]. GP1 is required for receptor binding, while GP2 is required for membrane fusion [18]. Mature GP1 has four distinct subdomains: base, receptor-binding domain (RBD), glycan cap, and mucin-like domain (MLD) [14].

The MLD is heavily glycosylated with N-linked glycans and O-linked glycans but is yet to be structurally defined [16, 26]. *Ebolavirus* hold their MLD as a single unit on GP1, whereas the MLD of MARV is separated into two parts: one in GP1 and the other in GP2 [16]. These glycans shield the GP ectodomain from neutralizing antibodies [27].

The entry of filoviruses is also mediated by several host proteins, including the carbohydrate-binding receptors and phosphatidylserine receptors [14]. The carbohydrate-binding receptors include C-type lectins (such as hMGL, ASGPR, DC-SIGN, L-SIGN, and L-SECtin) and glycosaminoglycans (GAGs, such as heparin, heparin sulfate, chondroitin, and chondroitin sulfate), which are expected to interact with glycans on the MLD and the glycan cap of GP1. The phosphatidylserine (PS) receptors include the Tyro3/Axl/Mer (TAM) family receptors and T-cell immunoglobulin mucin domain-1 and 4 (TIM-1 and TIM-4), which target the PS on the outer leaflet of the filovirus membrane [14]. In addition, β1-integrins and α5β1-integrin may also positively regulate the entry of filoviruses [28, 29].

The uptake of filoviruses has been proposed to involve different endocytic pathways, including the clathrin-dependent and lipid raft-/caveolin-dependent pathways [30]. Recent data suggest that macropinocytosis plays a clear and predominant role [31]. After internalization into an endosomal compartment, GP1 and GP2 are sequentially processed by the cysteine proteases, cathepsin B (CatB) and/or cathepsin L (CatL), under acidic pH and reducing conditions and interact with the Niemann Pick C1 protein (NPC1) [14]. NPC1 is a crucial receptor for filoviruses, which potentially determines cell tropism and host species susceptibility [32].

6.3 Drugs and Vaccines for *Marburgvirus* and *Ebolavirus*

To date, more than 70 clinical trials for therapeutics targeting EBOV have been registered and carried out [33]. Four vaccines have been approved by the FDA in the USA (ERVEBO®, rVSV-ZEBOV-GP, v920), China (Ad5-EBOV), Russia (GamEvac-Combi), and the European Union (Zabdeno and Mvabea, Ad26. ZEBOV, MVABN-Filo) [33]. Several vaccines against filoviruses are also under development, including EBOV, SUDV, and MARV (https://clinicaltrials.gov/). The most advanced vaccine approaches include vesicular stomatitis virus (VSV), human (Ad5, Ad26) and chimpanzee (ChAd3) adenovirus, modified vaccinia Ankara (MVA), and DNA platforms [34]. Although several monoclonal antibody-based therapeutics (such as ZMapp, mAb114, and REGN-EB3) and small-molecular drugs (such as remdesivir) combined with supportive care, including intravenous fluid replacement to maintain blood volume, blood pressure, and electrolyte balance, as well as nutritional supplementation and analgesics, were used in clinical trials during the 2018–2019 Democratic Republic of Congo (DRC) MVD outbreak and showed statistically significant survival benefits in patients with early infection or mild disease, the case fatality rates for patients with a high viral load or severe disease are still unacceptably high [35]. Recently, research has focused on small-molecular inhibitors that target RNA synthesis, virus entry, or virus egress. Several promising pan-filovirus therapeutics have been demonstrated to be effective in cell culture and animal experiments, including RNA synthesis inhibitors (such as BXC4430, favipiravir, 3-deazaneplanocin A, β-d-N4-hydroxycytidine, 6-azauridine, benzoquinoline, GSK983, okadaic acid, and tolcapone), entry inhibitors (such as compound 7, LJ001, arbidol, ethylisopropylamiloride, E64, E64a, E64d, K11777, CA074, imipramine, U18666A, clomiphene, toremifene-mefloquine-posaconazole, verapamil, nimodipine and diltiazem, tetrandrine, apilimod, and AG1478), and egress inhibitors (such as Tsg101 and Nedd4) [36].

6.4 Construction of Pseudotyped *Marburgvirus* and *Ebolavirus*

Owing to their extreme virulence, live *Marburgvirus* and *Ebolavirus* should only be studied in Biosafety Level 4 (BSL4) laboratories, which limits investigators in terms of direct access to the live virus and makes the study of antiviral therapies challenging [37]. However, the development of BSL-2-safe systems targeting specific steps of the virus replication cycle has greatly promoted the study of filoviruses [36]. These BSL-2-safe systems comprise replication-defective pseudotyped viruses that express filovirus GP based on a VSV or lentiviral system, which are focused on the virus entry step [38, 39]; viruslike particles (VLPs), which are self-assembled VP40 alone or VP40 and GP [40]; a "minigenome," which can recapitulate the filovirus RNA synthesis machinery by co-transfection of four viral proteins (NP, VP35, VP30, and L) with a model viral genomic RNA [41]; and the replication- and transcription-competent viruslike particles (trVLPs), which also include the viral genes (VP40, GP, and VP24), allowing for most steps of the viral lifecycle to be analyzed, including entry, budding, and RNA synthesis, under BSL-2 conditions [42]. This review mainly focuses on the replication-defective pseudotyped filoviruses based on GP, and it summarizes the technical details of pseudotyped virus construction, the application of pseudotyped viruses in neutralization antibody evaluation and viral entry small molecule drug screening, and the investigation of filovirus GP mutations using pseudotyped viruses (Tables 6.1, 6.2, and 6.3).

Table 6.1 Summary of lentiviral-based pseudotyped viruses (HIV)

Back bone	Package cell	Backbone vectors	Reporter	Plasmid ratio	Transfection reagents	Application
HIV	293 T	pNL4-3. Luc. $R^{-}E^{-}$	Luc	HIV: GP = 1:1	PEI	Infection mechanism [43]
				/		Drug screening [44], neutralization antibody test (vaccine) [45, 46]
				HIV: GP = 2:5	Calcium phosphate	Neutralization antibody test (serum, mAbs) [47–50], infection mechanism [51]
				/		
				HIV: GP = 1:1	Lipofectamine 2000/3000	Neutralization antibody test (vaccine) [52, 53], infection mechanism [54, 55], drug screening [56]
				/	Lipofectamine	Infection mechanism [57, 58]
				/	/	Neutralization antibody test (serum) [59], infection mechanism (tropism, receptor, fusion) [60–64], drug screening [65]
		pNL4-3 Δenv	BlaM-Vpr	/	/	Infection mechanism [66]
		pCD/NL-BH*DDD	GFP	HIV:GFP: GP = 93: 125:1	TransIT-LT1	Mutation study [67]
		pSG3-Fluc	Luc	HIV: GP = 2:1	Lipofectamine 3000	Drug screening [68–70]
				HIV: GP = 2000: 285	PEI	Neutralization antibody test [71]

(continued)

Table 6.1 (continued)

Back bone	Package cell	Backbone vectors	Reporter	Plasmid ratio	Transfection reagents	Application
		pCMVΔR8.2	Luc/ GFP	HIV:Luc/ GFP: GP = 10: 24:5	Lipofectamine 2000	Entry mechanism study [72]
			β-Gal	HIV:β-gal: GP = 1:2:3	Calcium phosphate	Infection mechanism (tropism) [73]
				HIV:β-gal: GP = 2:4:1	CalPhos kit	Pseudovirus optimization [74]
			Luc	HIV:Luc: GP = 7: 7:400	Calcium phosphate	Neutralization antibody test (vaccine) [75]
		pCMV-Δ8.91	Luc	HIV:Luc: GP = 2:3:2	Fugene	Neutralization antibody test (serum) [76]
		p8.9QV	GFP	/	/	Infection mechanism (RTK) [77]

6.4.1 VSV-Based Pseudotyped Marburgvirus *and* Ebolavirus

VSV is a member of the *Rhabdoviridae* family within the taxonomic order of *Mononegavirales* [39]. It has a similar genome structure to the filoviruses [39]. In 1997, a replication-deficient VSV-based pseudotyped RESTV, with GFP as a reporter, was first reported by Takada et al. [90] Briefly, 293 T cells were transfected with pCAGGS-ResGP (plasmid expressing RESTV GP) using Lipofectamine [90]. Then, after 36 h, the cells were infected with VSVΔG*-G at a multiplicity of infection (MOI) of 1 and were incubated for 1 h at 37 °C [90]. Cells were then washed with PBS three times, and fresh medium was added. The VSV-RESTV-GP pseudotyped virus-containing supernatants were collected 24 h later, centrifuged or filtered (to remove cell debris), and were then used directly, temporarily stored at 4 ° C for up to 1 week, or aliquoted and stored at −80 °C until use [90]. The detailed method for producing VSV-ΔG*-G is described in the manuscript [90]. BHK cells infected with vTF7-3 (to obtain T7 RNA polymerase activity) were co-transfected with pVSV-ΔG* (a full-length cDNA clone of the VSV genome, with the G protein coding region replaced with the GFP gene) and VSV nucleocapsid protein-, phosphoprotein-, polymerase protein-, and glycoprotein-expressing plasmids at weight ratios of 10, 3, 5, 1, and 3 μg, respectively [90]. VSVΔG*-G-containing supernatants were collected 48 h later [90].

Table 6.2 Summary of lentiviral-based pseudotyped viruses (FIV, SIV, and MLV)

Back bone	Package cell	Backbone vectors	Reporter	Plasmid ratio	Transfection reagents	Application
FIV	293 T	FIV (gag-pol)	β-Gal/ Luc	FIV: Luc/β-gal: GP = 2:3:1	Calcium phosphate	Mutation study [78], infection mechanism (RTK) [79]
			β-Gal	FIV:b-gal: GP = 2:4:1	CalPhos kit	Pseudotyped virus optimization [74]
SIV	293 T	pSIV-12 (gag-pol-rev)	GFP	SIV:EGFP: GP = 3:3:1	/	Lentiviral pseudovirus in primary cells [80]
MLV	293 T	MLV (gag/pol)	GFP	MLV:GFP: GP = 1:1:2	Calcium phosphate	Mutation study [81]
				MLV:GFP: GP = 1:1:1	Calcium phosphate	Infection mechanism (GP1) [82]
				/	Calcium phosphate	Infection mechanism (other receptors) [83]
			LacZ	MLV: LacZ: GP = 4:4:5	/	Infection mechanism (RTK) [84]
			β-Gal	/	Calcium phosphate	Infection mechanism (cell tropism) [85]
				MLV:β-gal: GP = 1:4:2	Calcium phosphate	Entry mechanism (GP1) [86]
			Luc	MLV:Luc: GP = 10:10:1	Polyethylenimine	Mutation study [87]
			/	MLV: GP = 2:1	Calcium phosphate	Infection mechanism [88]
	MLV stable expressing cell		GFP	/	/	Infection mechanism (RTK) [89]

The VSV-based pseudotyped filoviruses have been widely used by several groups worldwide [39]. The construction procedures are all similar, although they vary in detail. Most filoviruses have been successfully constructed into pseudotyped VSV-GP, including EBOV, BDBV, SUDV, TAFV, LLOV, MARV, and RAVV [91]. Different eukaryotic expression vectors are used for expressing GP (e.g., pcDNA3.1, pCAGGS) [38, 91]. The transfection reagents also vary (including calcium phosphate [92], Lipofectamine [90], polyethylenimine (PEI) [39], and TransIT LT-1 [93]). The GFP reporter gene can be replaced by other fluorescence proteins or luciferases [39]. The MOI and the time of infection or transduction of

Table 6.3 Summary of VSV-based pseudotyped viruses

Back bone	Package cell	Backbone virus	Transfection reagents	Reporter	MOI (VSVΔG)	Application
VSV	293 T	VSV*ΔG	Lipofectamine	GFP	1	Pseudotyped virus construction [90], infection mechanism (RTK, NPC1, sGP) [32, 89, 119–121, 123, 126, 127, 131, 145], drug screening [44], ADE [116–118]
			Lipofectamine 2000		/	Infection mechanism (e.g., NPC1) [122, 125, 129, 130]
			PEI	β-Gal / Luc	1-3	Pseudotyped virus construction [39], infection mechanism [132]
				GFP		
			Fugene HD	GFP	3	Neutralization antibody test (vaccine, mAbs) [102, 112]
				Luc		
			Trans-IT LT1	GFP	1	Infection mechanism (e.g., GP1) [136, 137]
				Luc	0.1	Mutation study [93, 148]
			Calcium phosphate/ Lipofectamine 2000	GFP/ Luc	3	Mutation study [92], entry mechanism (e.g., enzyme) [91, 124]
			/	GFP	/	Neutralization antibody test (vaccine, mAbs) [104, 115]

VSV-ΔG also vary between different studies [39]. Furthermore, VSVΔG-stocks encoding a reporter gene (GFP, dsRed, or luciferase) are commercially available from Kerafast (http://www.kerafast.com/c-310-delta-g-vsv-pseudotyping-system.aspx) [39].

When large stocks of pseudotyped viruses are needed, fresh DMEM can be added to cells after the first collection, and collection can be repeated at 48 h following transduction [39]. The collected supernatants can then be pooled and concentrated by overnight centrifugation in a high-speed centrifuge (5400×g for at least 16 h at 4 °C) and/or concentrated and purified via ultracentrifugation (80,000×g for 2 h at 4 °C) [39]. For the latter method, PBS containing 20% sucrose is first added to the ultracentrifuge tube [39].

6.4.2 *Lentiviral-Based Pseudotyped* Marburgvirus *and* Ebolavirus

Pseudotyped, lentiviral-based filoviruses are another widely used type of BSL-2-safe pseudotyped viruses and include HIV (human immunodeficiency virus)-based [38, 67, 68, 72], SIV (simian immunodeficiency virus)-based [80], MLV (murine leukemia virus)-based [81, 87], and FIV (feline immunodeficiency virus)-based pseudotyped viruses [78]. Standard lentiviral-based pseudotyped filovirus production involves transient transfection of viral component expression plasmids into package cells, followed by the collection and centrifugation of the supernatant to purify and concentrate the viron [38]. The 293 T or 293FT cells are commonly used as packaging cells, which are transfected with viral glycoprotein plasmid (e.g., EBOV-GP) and lentiviral backbone plasmids [38, 67, 68, 72, 78, 80, 81, 87, 94]. There are several types of lentiviral backbones, which are composed of three plasmids (e.g., gag/pol, rev, and HIV-CMV-eGFP), two plasmids (e.g., gag/pol and HR'-CMV-Luc), or one plasmid (e.g., pSG3.ΔEnv.CMV.Fluc) [38, 67, 68, 72, 78, 80, 81, 87, 94]. The ratios of lentiviral package plasmids, glycoprotein plasmid, and transfection reagents vary between different studies [38, 67, 68, 72, 78, 80, 81, 87, 94]. Pseudotyped virus-containing supernatants can be collected every 12–24 h, up until 72 h post-transfection [38, 67, 68, 72, 78, 80, 81, 87, 94]. The combined supernatants are stored, and concentrated or purified, similar to the VSV-based pseudotyped viruses [38].

6.4.3 *Influenza-Based Pseudotyped* Marburgvirus *and* Ebolavirus

The EBOV-GP-pseudotyped influenza virus (E-S-FLU) was constructed by removing the endogenous HA sequence (S-FLU) and incorporating EBOV-GP into the envelope [95, 96]. Briefly, the lentiviral vector expressing the full-length EBOV-GP gene was transduced into MDCK-SIAT1 cells [96]. Cells with high expression of EBOV-GP were sorted and stored as the E-SIAT cell line [96]. S-FLU virus coated in the HA of influenza virus was added and propagated in E-SIAT cells to produce the EBOV-GP-pseudotyped S-FLU virus [96]. After 48 h, culture supernatants were harvested that contained adequate pseudovirions to be used for the screening of inhibitory drugs or antibodies against EBOV [96, 97].

6.4.4 *Comparison of Different Types of Pseudotyped Filoviruses and Authentic Filoviruses*

Urbanowicz et al. compared the pseudotyped HIV system with the MLV system. Their results revealed that pseudotyped EBOV and RESTV based on the

HIV-backbone had stronger signals compared with the MLV-backbone [98]. In another study, Steeds et al. compared the VSV system with the HIV system and found that VSV-based pseudotyped EBOV was superior to HIV-based pseudotyped viruses for the assessment of neutralizing antibodies [99].

A high degree of correlation was shown between Ebola virus BSL-4 plaque reduction neutralization assays and pseudotyped VSV BSL-2 fluorescence reduction assays ($R^2 = 0.96$) when serum samples from a cohort of healthy survivors of the Sierra Leone EBOV outbreak of 2013–2016 were tested and when EBOV protein-immunized guinea pig serum samples were tested [71, 100]. The 50% inhibition dilution (ID50) was similar between the authentic virus and the pseudotyped virus [100]. However, another study indicated that when the same monoclonal antibodies (mAbs) were tested by both pseudotyped and authentic filoviruses, the antibody appeared to neutralize authentic filoviruses less efficiently than VSV-based pseudotyped filoviruses (e.g., EBOV, BDBV, and MARV) [101].

6.5 Application of the Pseudotyped Viruses

6.5.1 Neutralization Antibody Detection

Pseudotyped viruses are widely used for neutralizing antibody evaluation against highly pathogenic filoviruses, which is convenient to evaluate the efficacy of filovirus vaccines and antibody therapies and the protection provided against natural infection [47–50, 52–55, 59–61, 73, 75, 76, 82, 85, 86, 102–123].

6.5.1.1 Vaccine Efficacy Evaluation

VSV-based and HIV-based pseudotyped viruses are used to evaluate the neutralization antibodies elicited by various EBOV vaccine platforms, including recombinant GP vaccine, the gene vector vaccine, the vaccinia virus Tiantan strain harboring EBOV, and filovirus VLPs [52, 53, 102, 103]. Using HIV-based pseudotyped viruses, Sullivan et al. suggested that single low-dose adenovirus vectors encoding modified EBOV-GPs could provide immune protection in nonhuman primates against Ebola virus [75]. Using VSV-based pseudotyped viruses, modified rabies virus-vectored Ebola vaccines have been proven to induce efficacious neutralization antibodies in mice and dogs [104].

6.5.1.2 Therapeutic Antibody Analysis

Convalescent-phase human serum, polyclonal immune serum, and mAbs have been used in the treatment of Ebola virus infections in animal experiments and in clinics [97, 105–108]. The neutralization titers are representative of the protective effect.

Although mAbs with neutralizing activity (mAb144, 2G4, 4G7, FVM04, CA45, MR72, MR78, MR82, MR191) are usually more protective than non-neutralizing antibodies [109–111], several experimentally or clinically proven protective mAbs do not show strong neutralization of filoviruses (e.g., 13C6, 1H3, 6D8,13F6, MR228). In addition, other studies suggest that mAbs with in vitro neutralization activity cannot protect against filovirus infection in vivo (e.g., KZ52) [48, 109, 112–114]. HIV- or VSV-based pseudotyped viruses are commonly used to evaluate the neutralizing sensitivity of mAbs [48, 112, 115]. Because pseudotyped viruses are easily manipulated by mutagenesis, they have also been used to study the detailed target epitopes for the mAbs [48, 112, 115] and to identify cross-neutralizing mAbs against pan-filoviruses [112].

6.5.1.3 Investigation of the Kinetics of Convalescent Sera

The specificity and kinetics of the neutralizing antibody response of convalescent sera after filoviruses infection were analyzed using pseudotyped viruses, especially the HIV-based pseudotyped viruses. One study analyzed three convalescent sera during the 2014 EVD outbreak and showed that EBOV-GP-specific neutralizing activity increases up to 9 months postinfection [59]. The antibody responses in a cohort of healthy survivors of the Sierra Leone EBOV outbreak of 2013–2016 were investigated using HIV-based pseudotyped viruses and suggested that EBOV antibody reactivity declines over the 0.5–2 years following recovery [71]. Another study examined the serum neutralization activity of 14 survivors from the 1976 EBOV outbreak using HIV-based pseudotyped viruses, and their results indicated that 4 of 14 serum samples could still neutralize pseudotyped virus 40 years after the initial infection, although the neutralization titer was relatively low (1:50) [49]. The neutralization breadth of 15 survivors from the 2014 EBOV outbreak was also tested using HIV-based pseudotyped viruses, and the results suggested that some survivors of naturally acquired Ebola virus infection mount not only a pan-*Ebolavirus* response (5/15) but also less frequently (3/15) a pan-filovirus neutralizing response [50]. Bombali virus can be cross-neutralized by EBOV antibodies and convalescent plasma when tested with HIV-based pseudotyped viruses [76]. In addition, using HIV-based pseudotyped viruses to test the neutralization antibodies of 2430 serum samples, Steffen et al. reported that the prevalence of Ebola infection in Equatorial Africa during 1997–2012 was 2–3.5% [47].

6.5.2 Antibody-Dependent Enhancement (ADE) Evaluation

VSV-pseudotyped filoviruses were used in ADE evaluation. Takada et al. found that antisera produced by EBOV-GP DNA immunization enhanced the infectivity of VSV-pseudotyped EBOV [116]. This enhancement was weaker for RESTV-immunized sera against VSV-pseudotyped RESTV [116]. The group further generated an anti-EBOV-GP mAb that enhanced viral infectivity [116] and mapped the epitopes on the GP, which is required for ADE of EBOV infection, using chimeric VSV-pseudotyped filoviruses [117]. The same group also investigated different MARV variants using VSV-pseudotyped MARV and indicated that ADE activity depended on epitopes in the MLD and amino acid G547, which is present in the Angola EBOV-GP but absent in the Musoke EBOV-GP [118].

6.5.3 *Study of the Mechanisms of Filovirus Infection*

A large number of studies on the filovirus entry mechanism have been performed using pseudotyped filoviruses. Using different pseudotyped platforms, researchers have examined the cell tropism of filoviruses and discovered the receptor of filovirus, mapped the key amino acids on GP, analyzed the host proteolytic enzymes, and compared the differences between different strains of filoviruses. Below we summarize the main findings regarding the mechanisms of action of filoviruses, as determined using pseudotyped viruses.

6.5.3.1 Mapping the Key Domains and Amino Acids on the GP

Most of the key domains and amino acids in the GP were discovered using pseudotyped filoviruses. The GP1 of EBOV [amino acids (aa) 54–201] and MARV (aa 38–188) were proposed to be the receptor-binding domains based on studies using MLV-pseudotyped EBOV and MARV [82]. Using site-directed mutagenesis and MLV-pseudotyped EBOV, 15 conserved residues in the N-terminus of GP1 were examined, and the results suggested that the phenylalanine residues at positions 88 and 159 were critical for cell entry [86]. A putative fusion domain (aa 524–539) on GP2 of EBOV-GP was proven by mutational analysis based on VSV-pseudotyped filoviruses [119]. The mutations I532R, F535R, G536A, and P537R almost completely abolished the infectivity of pseudotyped virus in 293 cells, whereas mutations G528R, L529A, L529R, I532A, and F535A reduced the infectivity of the pseudotyped viruses by approximately one-half [119]. Using a HIV-based pseudotyping system, 41 amino acid mutations on GP1 of EBOV were tested. R64 and K95 were found to be involved in receptor binding, while I170 was important for viral entry but not binding [60]. Using HIV-pseudotyped filoviruses, Manicassamy et al. constructed more than 100 mutations to comprehensively analyze the role of GP1 in viral entry and identified 6 amino acids (L57, L63, R64, F88, K95, and I170) likely involved in receptor recognition and/or post-binding events [54]. Using HIV-pseudotyped filoviruses and mutational analysis, the role of the charged residues in the GP2 helical regions was also examined and most of the mutations were found to greatly impair efficient viral infection [55]. Furthermore, using VSV-pseudotyped sGP of EBOV, one study reported that sGP can substitute for GP1, forming a sGP-GP2 complex and conferring infectivity [120].

6.5.3.2 Cell Tropism Examination

Pseudotyped platforms can also be applied to the examination of cell tropism. MLV-pseudotyped EBOV and VSV-pseudotyped EBOV were both used to examine the cell tropism of Ebola virus in several cell lines of different hosts [85, 121]. The cell lines most susceptible to MLV-pseudotyped EBOV were 293 T and Vero cells, and those most susceptible to VSV-pseudotyped EBOV were 293 T cells, which may be because of the high protein expression characteristic of these cell lines. Although hepatocytes and endothelial cells are both targets of filoviruses, cell lines from these corresponding tissues are nonpermissive to VSV-pseudotyped EBOV [121]. In addition to 293 T cells, Chan et al. reported that human osteosarcoma (HOS), HeLa, HepG2, and primary HUVEC cells show a high luciferase signal after HIV-pseudotyped EBOV inoculation, whereas no signal was detected in human astrocyte U87 cells, murine NIH-3 T3 cells, or human Jurkat cells [61]. The authors of the study suggested that the lack of signal may be caused by the compromised luciferase gene promoter in U87 and NIH-3 T3 cells rather than a failure of entry [61]. In addition, HIV-pseudotyped EBOV was proven to efficiently transduce airway epithelial cells [73]. Finally, all of the MLV-, VSV-, and HIV-pseudotyped virus systems lacked susceptibility to lymphocytes [61, 85, 121].

6.5.3.3 Proteolytic Enzyme Analysis

Using VSV-pseudotyped EBOV, Chandran et al. found that the activity of endosomal cysteine proteases (CatB and CatL) is necessary for EBOV infection [122], which was confirmed by Schornberg et al. [123] and Hernández et al. [124]. However, using HIV-pseudotyped Ebola viruses and MARV, Gnirss et al. suggested that CatB and CatL activate Ebola but not Marburg virus GPs for efficient entry into the 293 T-cell line and primary macrophages [51]. Furthermore, using VSV-pseudotyped filoviruses, Misasi et al. suggested that CatB is more important for EBOV infection than for infection of SUDV, RESTV, and MARV, which correlated with the sequence polymorphisms at residues 47 in GP1 and 584 in GP2 [125].

6.5.3.4 Discovery and Analysis of Receptor NPC1

Using genome-wide haploid genetic screening and VSV-pseudotyped EBOV, the cholesterol transporter NPC1 was proven to be indispensable for EBOV infection [32]. Subsequently, NPC1 was proven to be the critical filovirus receptor, which was the first known viral receptor that recognizes its ligand within an intracellular compartment and not at the plasma membrane [126]. VSV pseudotypes bearing GPs derived from EBOV, SUDV, and MARV were used in these studies [126]. A large panel of VSV-pseudotyped viruses bearing mutant GPs was used to map the

EBOV-GP endosomal receptor-binding site (RBS) at molecular resolution (T83, I113, K114, and K115) [127]. In addition, using VSV-pseudotyped EBOV, F88A was demonstrated to impair GP binding to human NPC1 but had little impact on GP binding to mouse NPC1 [128]. Another study investigated the binding of EBOV-GP and NPC1 and found that NPC1 is required for fusogenic triggering of EBOV-GP [129]. They also showed that VSV-EBOV-GPΔMuc particles bearing three mutations in the NPC1-binding site (T83M in the trough, K114E/K115E in the crest) abolish the GP1-NPC1 interaction and viral infectivity [129]. Mittler et al. reported an in situ assay to monitor binding of the EBOV-GP to its receptor NPC1 in intact infected cells based on proximity ligation using VSV-pseudotyped EBOV and verified the previously found key amino acids for the GP and NPC1 interaction, which is a useful tool to delineate receptor-viral GP interactions [130].

6.5.3.5 Receptor Tyrosine Kinase (RTK)-Related Studies

By screening a kinase inhibitor library using MLV-pseudotyped EBOV, Stewart et al. found that EBOV triggers RTK-dependent signaling to traffic to intracellular vesicles that contain the receptor NPC1 [84]. Tyro3 receptor tyrosine kinase family members, Axl, Dtk, and Mer, were shown to enhance the infection of MLV-pseudotyped and VSV-pseudotyped filoviruses [89]. The mechanism of Axl-mediated Ebola virus infection was further analyzed using HIV-pseudotyped EBOV and a series of Axl mutants [77]. In addition, FIV-pseudotyped EBOV was used in the study of Axl and micropinocytosis [79]. Receptor tyrosine kinase human epidermal growth factor receptor 2 (HER2) was also found to mediate enhancement of filovirus entry, which was confirmed by VSV-pseudotyped EBOV [131].

6.5.3.6 Glycosylation and Acylation Analysis

Using HIV-pseudotyped EBOV, N-glycosylation sites of GP were studied, and the results suggested that N563 and N618 are essential for GP function [43]. Two cysteine residues were conserved among all filoviruses and posttranslationally acylated with palmitic acids. However, using VSV-pseudotyped mutant GPs, acylation was found not to be required for EBOV-GP function [121].

6.5.3.7 Studies of Other Host Factors

TIM-family proteins have been reported to promote MLV-pseudotyped EBOV and MARV infection by binding to PS on the cell membrane, indicating that TIM-family proteins and related PS play an important role in filovirus infection [83]. The role of acid sphingomyelinase (ASMase) and its substrate, lipid sphingomyelin (SM), in EBOV infection was investigated using VSV-pseudotyped EBOV, and the results confirmed that ASMase activity and the presence of SM are necessary for efficient infection of cells by EBOV [132]. The binding of filoviruses with DC-SIGN (dendritic cell-specific intercellular adhesion molecule 3-grabbing non-integrin), DC-SIGNR (DC-SIGN-related molecule) [62, 133], and mannose-binding lectin (MBL) to HIV-/VSV-pseudotyped viruses has also been reported [57]. Mechanistic studies with HIV-pseudotyped EBOV suggested that MBL binds to N-linked glycan epitopes on viral surfaces and mediates lipid raft-dependent micropinocytosis, which is necessary for enhanced infection [58]. However, based on studies using HIV-pseudotyped viruses, the role of folate receptor alpha (FRα) and caveolae in EBOV infection is controversial. Chan et al. proposed that FRα is a cofactor for filoviruses cellular entry [63], while Simmon et al. suggested that FRα is not necessary for EBOV infection [64]. A study of GP-mediated entry and fusion using HIV-pseudotyped EBOV suggested the involvement of cytoskeletal proteins (microtubules and microfilaments) and the enhancement of infection by tumor necrosis factor-alpha [66].

Using HIV-pseudotyped EBOV, membrane-associated RING-CH-type 8 (MARCH8) was found to block EBOV-GP incorporation by downregulating its cell surface expression [134]. Using beta-lactamase-containing VLP, pseudotyped with EBOV-GP, the authors discovered that cytokine interleukin-10 pretreatment significantly enhanced filovirus entry into monocyte-derived macrophage cells [135].

6.5.3.8 Comparisons Between Filoviruses

EBOV and RESTV are two filoviruses with opposite levels of lethality in humans, and therefore they are regularly compared by researchers. Through mutagenesis of GP2 based on VSV-pseudotyped EBOV and RESTV, the transmembrane-anchored subunit of GP2 (aa 502–527) was found to determine the different levels of infectivity between EBOV and RESTV, with H516 being critical for EBOV infectivity [136]. In addition, the low infectivity of RESTV mediated by the interaction between GPs and MGL was dramatically increased when the N-terminal 18 amino acids (aa 33–50) were replaced with those of ZEBOV [137].

Using VSV-based pseudotyped viruses harboring the GPs of EBOV, SUDV, TAFV, BDBV, RESTV, MARV, and LLOV, Hoffmann et al. compared the infectivity of these pseudotyped viruses in several cell lines of different hosts [91]. Filoviruses employ the same host cell factors for entry into human, nonhuman primate, and fruit bat cell lines, although the efficiency of the usage of these factors may differ between filovirus species [91].

6.5.4 *Virus Entry Inhibitor Screening*

Virus entry into host cells is an attractive target for antiviral therapy. Because they are easy to manipulate and highly stable, pseudotyped filoviruses based on GP can mimic the viral entry process and can be used in high-throughput screening (HTS) of anti-filovirus drugs [138].

6.5.4.1 In Vitro Screening

Using HIV-pseudotyped EBOV, SUDV, MARV, and RAVV, a compound library of 2635 compounds composed of approved drugs, mechanistic probes, and elite compounds (e.g., dietary supplements and vitamins) was screened. In total, 80 candidate drugs were discovered, including selective estrogen receptor modulators, antihistamines, calcium channel blockers, and antidepressants [44]. Using HIV-based pseudotyped viruses expressing EBOV/MARV-GP, Cheng et al. screened the Prestwick Chemical Library, which contains 1200 FDA-approved drugs, and discovered that 24 of 29 effective compounds were G protein-coupled receptor (GPCR) antagonists targeting different GPCRs, including histamine receptors, 5-HT (serotonin) receptors, muscarinic acetylcholine receptor, and adrenergic receptor [45]. They further screened 1220 antihistamine-focused chemical libraries and identified 2 novel small molecules with potent anti-filoviral properties [46]. Our group screened 767 chemical compounds against HIV-pseudotyped MARV and proposed that the candidates can be clustered into categories similar to those identified for anti-EBOV [68]. Compound 8a and its derivatives were examined using HIV-pseudotyped EBOV and MARV through a library of 237 small molecules [56]. Traditional Chinese medicines were also screened using HIV-based pseudotyping platforms, and several Chinese medicinal herbs showed anti-Ebola activity (e.g., *Prunella vulgaris* L., ellagic acid, aloperine) [69, 139, 140]. Madrid et al. screened 1012 FDA-approved drugs with the help of pseudotyped viruses and identified chloroquine, which was shown to protect against EBOV challenge both in vitro and in vivo [141]. In addition, an in silico-designed peptide inhibitor, targeting the interaction between GP1 and NPC-1, was confirmed using HIV-pseudotyped EBOV [65].

6.5.4.2 In Vivo Verification

In vivo bioluminescent imaging (BLI) mouse models for EBOV and MARV were developed based on the pseudotyped virus and were used to verify the candidates from in vitro screening [68, 70, 94]. Pseudotyped viruses with reporter genes are convenient for real-time analysis of the infection process in small animal models, without the need for euthanization [68, 94]. Furthermore, the distribution of pseudotyped filoviruses among organs can be observed by ex vivo imaging of necropsied tissues [68, 94]. In addition, the BLI models could also be used to evaluate protective neutralization antibodies [68]. A detailed protocol for pseudotyped virus-based in vivo BLI for MARV was published in Lei et al. [142]

6.5.5 Analysis of Mutations Within the GP

RNA viruses have high mutation rates [143]. The resulting amino acid changes, especially changes to the GP, may significantly alter viral infectivity and antigenicity [144]. The uses of pseudotyped filovirus in mutation analyses are summarized below according to the mutation site.

-A82V: The Makona variant with an A82V mutation in the GP was detected during the 2013–2016 EVD epidemic in western Africa and rapidly became the predominant EBOV of the epidemic. A82V is located at the NPC1-binding interface of the GP [67]. Using pseudotyped lentiviruses, Diehl et al. found that the enhanced infectivity of Makona strain was associated with the A82V mutation [67]. Using a VSV-based pseudotyped virus, Fels et al. showed that the A82V mutation promotes more efficient GP processing by CatL, which then leads to faster viral fusion kinetics and higher levels of infectivity [145]. Wang et al. used MLV particle pseudotyped EBOV GPs to reveal that Makona-82 V is more resistant to an NPC1-targeting inhibitor than Makano-82A and that the A82V mutation increases activation of EBOV-GP by reducing the threshold for NPC1 activation by the conformational change in GP2 [81]. Using the VLP of EBOV, which was composed of GP, VP40, NP, L, VP35, and VP30-luc, Dietzel et al. also demonstrated that the A82V mutation increases the efficiency of virus entry into target cells [146].

T544I: The T544I mutation has been detected during several EVD outbreaks [93]. The 544 residue is located on the internal fusion loop (IFL), which plays a key role in membrane fusion [147]. Ueda et al. found that pseudotyped VSV-GP-T544I could infect Huh 7 cells more efficiently (4.3-fold) than the ancestral Makona variant, while VSV-GP-A82V could increase the viral infectivity by only 1.8-fold [147]. The infectivity of VSV-GP-T544I + A82V was increased 8.3-fold compared with the ancestral Makona variant [147]. The enhanced virus infectivity resulting from the T544I mutation was also proven by Fels et al. using VSV-GP pseudotyped virus in Vero cells [145]. Hoffman et al. also used VSV-GP to demonstrate that the reduced infectivity of 2014-EBOV(Makona) compared with 1976-EBOV(Mayinga) in NHP-derived cell lines and certain human target cells (monocyte-derived macrophages and dendritic cells) was because of the 544 T mutation, not the 82 V mutation [92]. Using MLV-GPs, Wang et al. showed that, as for A82V, T544I was more resistant to NPC1 inhibitor 3.47 as a result of the reduced stability of the prefusion conformation of EBOV-GP [81]. Using VSV-pseudotyped EBOV and RESTV, 10 cell lines from nine mammalian species were compared and the 544I mutation was found to increase viral infectivity in all host species, whereas the 82 V mutation modulates viral infectivity, depending on the viral and host species [148].

Other mutations: Using MLV-pseudotyped EBOV, a panel of pseudotyped viruses was constructed based on the earliest reported isolate with amino acid substitutions introduced that defined viral lineages and infected different human and bat cell lines [87]. Several specific amino acid substitutions in the EBOV-GP were found to have increased tropism for human cells (e.g., R29K, T206M, T230A) while reducing tropism for bat cells [87]. Using VSV-based pseudotyped viruses, the

E545D mutation of variant Makona EBOV, but not the D552N mutation, was proven to be an escape mutant in response to the selective pressure from mAb CA45 treatment [149]. To identify potentially important mutations, Brindley et al. compared the representative GPs from EBOV and MARV outbreaks and focused on five completely conserved regions [78]. In total, they evaluated 63 mutants using FIV-based pseudotyped viruses for their transduction efficiency, receptor binding, and ability to be cleaved by cathepsins L and B[53]. They identified residues G87/F88, K114/K115, K140, G143, and Y162 as being important for receptor binding, while residues P146A/C147A, F153A/H154A, F159A, and F160A were inefficiently cleaved by cathepsins [78].

6.6 Summary

There are several pseudotyped virus systems for filoviruses, including lentiviral-based and VSV-based platforms, which are easy to construct and manipulate. During a filovirus pandemic, the pseudotyped virus can overcome the limitation of requiring a BSL4-level laboratory and thereby facilitate the study of filoviruses. A neutralization assay based on pseudotyped virus is a rapid and convenient method for the evaluation of vaccines and therapeutic antibodies. The high-throughput potential and high stability of the pseudovirus system make the large-scale screening of entry-inhibiting drugs possible. The ease with which the system can be manipulated and transformed makes it convenient in terms of exploring the mechanism of filovirus infection, especially for determining the key amino acids of GPs. Finally, the ease of manipulating specific amino acids makes it convenient for real-time monitoring of the infectivity and antigenicity changes of emerging mutants or variants during a pandemic.

References

1. Kuhn, J.H.: Guide to the correct use of filoviral nomenclature. Curr. Top. Microbiol. Immunol. **411**, 447–460 (2017). https://doi.org/10.1007/82_2017_7
2. Yang, X.L., et al.: Characterization of a filovirus (Mengla virus) from Rousettus bats in China. Nat. Microbiol. **4**, 390–395 (2019). https://doi.org/10.1038/s41564-018-0328-y
3. Euren, J., et al.: Human interactions with bat populations in Bombali, Sierra Leone. Ecohealth. **17**, 292–301 (2020). https://doi.org/10.1007/s10393-020-01502-y
4. Hoenen, T., Groseth, A., Feldmann, H.: Therapeutic strategies to target the Ebola virus life cycle. Nat. Rev. Microbiol. **17**, 593–606 (2019). https://doi.org/10.1038/s41579-019-0233-2
5. Brauburger, K., Hume, A.J., Muhlberger, E., Olejnik, J.: Forty-five years of Marburg virus research. Viruses. **4**, 1878–1927 (2012). https://doi.org/10.3390/v4101878
6. Horimoto, T., Nakayama, K., Smeekens, S.P., Kawaoka, Y.: Proprotein-processing endoproteases PC6 and furin both activate hemagglutinin of virulent avian influenza viruses. J. Virol. **68**, 6074–6078 (1994). https://doi.org/10.1128/jvi.68.9.6074-6078.1994

7. Ponce, L., Kinoshita, R., Nishiura, H.: Exploring the human-animal interface of Ebola virus disease outbreaks. Math. Biosci. Eng. **16**, 3130–3143 (2019). https://doi.org/10.3934/mbe.2019155
8. Dixon, M.G., Schafer, I.J., Centers for Disease, C. & Prevention: Ebola viral disease outbreak--West Africa, 2014. MMWR Morb. Mortal. Wkly Rep. **63**, 548–551 (2014)
9. Anthony, S.M., Bradfute, S.B.: Filoviruses: one of these things is (not) like the other. Viruses. **7**, 5172–5190 (2015). https://doi.org/10.3390/v7102867
10. Nyakarahuka, L., et al.: How severe and prevalent are Ebola and Marburg viruses? A systematic review and meta-analysis of the case fatality rates and seroprevalence. BMC Infect. Dis. **16**, 708 (2016). https://doi.org/10.1186/s12879-016-2045-6
11. Volchkov, V.E., et al.: Proteolytic processing of Marburg virus glycoprotein. Virology. **268**, 1–6 (2000). https://doi.org/10.1006/viro.1999.0110
12. Leroy, E.M., et al.: Fruit bats as reservoirs of Ebola virus. Nature. **438**, 575–576 (2005). https://doi.org/10.1038/438575a
13. Ascenzi, P., et al.: Ebolavirus and Marburgvirus: insight the filoviridae family. Mol. Asp. Med. **29**, 151–185 (2008). https://doi.org/10.1016/j.mam.2007.09.005
14. Davey, R.A., et al.: Mechanisms of filovirus entry. Curr. Top. Microbiol. Immunol. **411**, 323–352 (2017). https://doi.org/10.1007/82_2017_14
15. Geisbert, T.W., Jahrling, P.B.: Differentiation of filoviruses by electron microscopy. Virus Res. **39**, 129–150 (1995). https://doi.org/10.1016/0168-1702(95)00080-1
16. Martin, B., Hoenen, T., Canard, B., Decroly, E.: Filovirus proteins for antiviral drug discovery: a structure/function analysis of surface glycoproteins and virus entry. Antivir. Res. **135**, 1–14 (2016). https://doi.org/10.1016/j.antiviral.2016.09.001
17. Gordon, T.B., Hayward, J.A., Marsh, G.A., Baker, M.L., Tachedjian, G.: Host and viral proteins modulating Ebola and Marburg virus egress. Viruses. **11** (2019). https://doi.org/10.3390/v11010025
18. Emanuel, J., Marzi, A., Feldmann, H.: Filoviruses: ecology, molecular biology, and evolution. Adv. Virus Res. **100**, 189–221 (2018). https://doi.org/10.1016/bs.aivir.2017.12.002
19. Volchkov, V.E., Feldmann, H., Volchkova, V.A., Klenk, H.D.: Processing of the Ebola virus glycoprotein by the proprotein convertase furin. Proc. Natl. Acad. Sci. U. S. A. **95**, 5762–5767 (1998). https://doi.org/10.1073/pnas.95.10.5762
20. Maruyama, J., et al.: Characterization of the envelope glycoprotein of a novel filovirus, lloviu virus. J. Virol. **88**, 99–109 (2014). https://doi.org/10.1128/JVI.02265-13
21. Manicassamy, B., et al.: Characterization of Marburg virus glycoprotein in viral entry. Virology. **358**, 79–88 (2007). https://doi.org/10.1016/j.virol.2006.06.041
22. Lee, J.E., et al.: Structure of the Ebola virus glycoprotein bound to an antibody from a human survivor. Nature. **454**, 177–182 (2008). https://doi.org/10.1038/nature07082
23. Mehedi, M., et al.: A new Ebola virus nonstructural glycoprotein expressed through RNA editing. J. Virol. **85**, 5406–5414 (2011). https://doi.org/10.1128/JVI.02190-10
24. Negredo, A., et al.: Discovery of an ebolavirus-like filovirus in europe. PLoS Pathog. **7**, e1002304 (2011). https://doi.org/10.1371/journal.ppat.1002304
25. Bhattacharyya, S.: Mechanisms of immune evasion by Ebola virus. Adv. Exp. Med. Biol. **1313**, 15–22 (2021). https://doi.org/10.1007/978-3-030-67452-6_2
26. Jeffers, S.A., Sanders, D.A., Sanchez, A.: Covalent modifications of the ebola virus glycoprotein. J. Virol. **76**, 12463–12472 (2002). https://doi.org/10.1128/jvi.76.24.12463-12472.2002
27. Lennemann, N.J. et al.: Comprehensive functional analysis of N-linked glycans on Ebola virus GP1. mBio **5**, e00862–00813 (2014). https://doi.org/10.1128/mBio.00862-13
28. Takada, A., et al.: Downregulation of beta1 integrins by Ebola virus glycoprotein: implication for virus entry. Virology. **278**, 20–26 (2000). https://doi.org/10.1006/viro.2000.0601
29. Schornberg, K.L., et al.: Alpha5beta1-integrin controls ebolavirus entry by regulating endosomal cathepsins. Proc. Natl. Acad. Sci. U. S. A. **106**, 8003–8008 (2009). https://doi.org/10.1073/pnas.0807578106

30. Rasmussen, A.L.: Host factors in Ebola infection. Annu. Rev. Genomics Hum. Genet. **17**, 333–351 (2016). https://doi.org/10.1146/annurev-genom-083115-022446
31. Aleksandrowicz, P., et al.: Ebola virus enters host cells by macropinocytosis and clathrin-mediated endocytosis. J. Infect. Dis. **204**(Suppl 3), S957–S967 (2011). https://doi.org/10.1093/infdis/jir326
32. Carette, J.E., et al.: Ebola virus entry requires the cholesterol transporter Niemann-pick C1. Nature. **477**, 340–343 (2011). https://doi.org/10.1038/nature10348
33. Sharma, A.R., Lee, Y.H., Nath, S., Lee, S.S.: Recent developments and strategies of Ebola virus vaccines. Curr. Opin. Pharmacol. **60**, 46–53 (2021). https://doi.org/10.1016/j.coph.2021.06.008
34. Matz, K.M., Marzi, A., Feldmann, H.: Ebola vaccine trials: progress in vaccine safety and immunogenicity. Expert Rev. Vaccines. **18**, 1229–1242 (2019). https://doi.org/10.1080/14760584.2019.1698952
35. Iversen, P.L., et al.: Recent successes in therapeutics for Ebola virus disease: no time for complacency. Lancet Infect. Dis. **20**, e231–e237 (2020). https://doi.org/10.1016/S1473-3099(20)30282-6
36. Edwards, M.R., Basler, C.F.: Current status of small molecule drug development for Ebola virus and other filoviruses. Curr. Opin. Virol. **35**, 42–56 (2019). https://doi.org/10.1016/j.coviro.2019.03.001
37. Kerper, M., Puckett, Y.: In StatPearls (2022)
38. Sinn, P.L., Coffin, J.E., Ayithan, N., Holt, K.H., Maury, W.: Lentiviral vectors Pseudotyped with filoviral glycoproteins. Methods Mol. Biol. **1628**, 65–78 (2017). https://doi.org/10.1007/978-1-4939-7116-9_5
39. Brouillette, R.B., Maury, W.: Production of filovirus glycoprotein-Pseudotyped vesicular stomatitis virus for study of filovirus entry mechanisms. Methods Mol. Biol. **1628**, 53–63 (2017). https://doi.org/10.1007/978-1-4939-7116-9_4
40. Sun, Y., et al.: Protection against lethal challenge by Ebola virus-like particles produced in insect cells. Virology. **383**, 12–21 (2009). https://doi.org/10.1016/j.virol.2008.09.020
41. Hoenen, T., Groseth, A., de Kok-Mercado, F., Kuhn, J.H., Wahl-Jensen, V.: Minigenomes, transcription and replication competent virus-like particles and beyond: reverse genetics systems for filoviruses and other negative stranded hemorrhagic fever viruses. Antivir. Res. **91**, 195–208 (2011). https://doi.org/10.1016/j.antiviral.2011.06.003
42. Hoenen, T., Watt, A., Mora, A., Feldmann, H.: Modeling the lifecycle of Ebola virus under biosafety level 2 conditions with virus-like particles containing tetracistronic minigenomes. J. Vis. Exp. **52381** (2014). https://doi.org/10.3791/52381
43. Wang, B., et al.: Mechanistic understanding of N-glycosylation in Ebola virus glycoprotein maturation and function. J. Biol. Chem. **292**, 5860–5870 (2017). https://doi.org/10.1074/jbc.M116.768168
44. Johansen, L.M., et al.: A screen of approved drugs and molecular probes identifies therapeutics with anti-Ebola virus activity. Sci. Transl. Med. **7**, 290ra289 (2015). https://doi.org/10.1126/scitranslmed.aaa5597
45. Cheng, H., et al.: Inhibition of Ebola and Marburg virus entry by G protein-coupled receptor antagonists. J. Virol. **89**, 9932–9938 (2015). https://doi.org/10.1128/JVI.01337-15
46. Cheng, H., et al.: Identification of a coumarin-based antihistamine-like small molecule as an anti-filoviral entry inhibitor. Antivir. Res. **145**, 24–32 (2017). https://doi.org/10.1016/j.antiviral.2017.06.015
47. Steffen, I., et al.: Serologic prevalence of Ebola virus in equatorial Africa. Emerg. Infect. Dis. **25**, 911–918 (2019). https://doi.org/10.3201/eid2505.180115
48. Davidson, E., et al.: Mechanism of binding to Ebola virus glycoprotein by the ZMapp, ZMAb, and MB-003 cocktail antibodies. J. Virol. **89**, 10982–10992 (2015). https://doi.org/10.1128/JVI.01490-15

49. Rimoin, A.W., et al.: Ebola virus neutralizing antibodies detectable in survivors of theYambuku, Zaire outbreak 40 years after infection. J. Infect. Dis. **217**, 223–231 (2018). https://doi.org/10.1093/infdis/jix584
50. Bramble, M.S., et al.: Pan-filovirus serum neutralizing antibodies in a subset of Congolese ebolavirus infection survivors. J. Infect. Dis. **218**, 1929–1936 (2018). https://doi.org/10.1093/infdis/jiy453
51. Gnirss, K., et al.: Cathepsins B and L activate Ebola but not Marburg virus glycoproteins for efficient entry into cell lines and macrophages independent of TMPRSS2 expression. Virology. **424**, 3–10 (2012). https://doi.org/10.1016/j.virol.2011.11.031
52. Wu, F., et al.: A chimeric Sudan virus-like particle vaccine candidate produced by a recombinant baculovirus system induces specific immune responses in mice and horses. Viruses. **12** (2020). https://doi.org/10.3390/v12010064
53. Xie, L., Zai, J., Yi, K., Li, Y.: Intranasal immunization with recombinant vaccinia virus Tiantan harboring Zaire Ebola virus gp elicited systemic and mucosal neutralizing antibody in mice. Vaccine. **37**, 3335–3342 (2019). https://doi.org/10.1016/j.vaccine.2019.04.070
54. Manicassamy, B., Wang, J., Jiang, H., Rong, L.: Comprehensive analysis of ebola virus GP1 in viral entry. J. Virol. **79**, 4793–4805 (2005). https://doi.org/10.1128/JVI.79.8.4793-4805.2005
55. Jiang, H., et al.: The role of the charged residues of the GP2 helical regions in Ebola entry(). Virol. Sin. **24**, 121–135 (2009). https://doi.org/10.1007/s12250-009-3015-6
56. Yermolina, M.V., Wang, J., Caffrey, M., Rong, L.L., Wardrop, D.J.: Discovery, synthesis, and biological evaluation of a novel group of selective inhibitors of filoviral entry. J. Med. Chem. **54**, 765–781 (2011). https://doi.org/10.1021/jm1008715
57. Ji, X., et al.: Mannose-binding lectin binds to Ebola and Marburg envelope glycoproteins, resulting in blocking of virus interaction with DC-SIGN and complement-mediated virus neutralization. J. Gen. Virol. **86**, 2535–2542 (2005). https://doi.org/10.1099/vir.0.81199-0
58. Brudner, M., et al.: Lectin-dependent enhancement of Ebola virus infection via soluble and transmembrane C-type lectin receptors. PLoS One. **8**, e60838 (2013). https://doi.org/10.1371/journal.pone.0060838
59. Luczkowiak, J., et al.: Specific neutralizing response in plasma from convalescent patients of Ebola virus disease against the West Africa Makona variant of Ebola virus. Virus Res. **213**, 224–229 (2016). https://doi.org/10.1016/j.virusres.2015.12.019
60. Wang, J., Manicassamy, B., Caffrey, M., Rong, L.: Characterization of the receptor-binding domain of Ebola glycoprotein in viral entry. Virol. Sin. **26**, 156–170 (2011). https://doi.org/10.1007/s12250-011-3194-9
61. Chan, S.Y., Speck, R.F., Ma, M.C., Goldsmith, M.A.: Distinct mechanisms of entry by envelope glycoproteins of Marburg and Ebola (Zaire) viruses. J. Virol. **74**, 4933–4937 (2000). https://doi.org/10.1128/jvi.74.10.4933-4937.2000
62. Simmons, G., et al.: DC-SIGN and DC-SIGNR bind ebola glycoproteins and enhance infection of macrophages and endothelial cells. Virology. **305**, 115–123 (2003). https://doi.org/10.1006/viro.2002.1730
63. Chan, S.Y., et al.: Folate receptor-alpha is a cofactor for cellular entry by Marburg and Ebola viruses. Cell. **106**, 117–126 (2001). https://doi.org/10.1016/s0092-8674(01)00418-4
64. Simmons, G., et al.: Folate receptor alpha and caveolae are not required for Ebola virus glycoprotein-mediated viral infection. J. Virol. **77**, 13433–13438 (2003). https://doi.org/10.1128/jvi.77.24.13433-13438.2003
65. Li, Q., et al.: Novel cyclo-peptides inhibit Ebola pseudotyped virus entry by targeting primed GP protein. Antivir. Res. **155**, 1–11 (2018). https://doi.org/10.1016/j.antiviral.2018.04.020
66. Yonezawa, A., Cavrois, M., Greene, W.C.: Studies of ebola virus glycoprotein-mediated entry and fusion by using pseudotyped human immunodeficiency virus type 1 virions: involvement of cytoskeletal proteins and enhancement by tumor necrosis factor alpha. J. Virol. **79**, 918–926 (2005). https://doi.org/10.1128/JVI.79.2.918-926.2005

67. Diehl, W.E., et al.: Ebola virus glycoprotein with increased infectivity dominated the 2013-2016 epidemic. Cell. **167**, 1088–1098 e1086 (2016). https://doi.org/10.1016/j.cell.2016.10.014
68. Zhang, L., et al.: A bioluminescent imaging mouse model for Marburg virus based on a pseudovirus system. Hum. Vaccin. Immunother. **13**, 1811–1817 (2017). https://doi.org/10.1080/21645515.2017.1325050
69. Zhang, X., et al.: Discovery and evolution of aloperine derivatives as novel anti-filovirus agents through targeting entry stage. Eur. J. Med. Chem. **149**, 45–55 (2018). https://doi.org/10.1016/j.ejmech.2018.02.061
70. Zhang, L., et al.: Screening and identification of Marburg virus entry inhibitors using approved drugs. Virol. Sin. **35**, 235–239 (2020). https://doi.org/10.1007/s12250-019-00184-3
71. Adaken, C., et al.: Ebola virus antibody decay-stimulation in a high proportion of survivors. Nature. **590**, 468–472 (2021). https://doi.org/10.1038/s41586-020-03146-y
72. Manicassamy, B., Rong, L.: Expression of ebolavirus glycoprotein on the target cells enhances viral entry. Virol. J. **6**, 75 (2009). https://doi.org/10.1186/1743-422X-6-75
73. Kobinger, G.P., Weiner, D.J., Yu, Q.C., Wilson, J.M.: Filovirus-pseudotyped lentiviral vector can efficiently and stably transduce airway epithelia in vivo. Nat. Biotechnol. **19**, 225–230 (2001). https://doi.org/10.1038/85664
74. Medina, M.F., et al.: Lentiviral vectors pseudotyped with minimal filovirus envelopes increased gene transfer in murine lung. Mol. Ther. **8**, 777–789 (2003). https://doi.org/10.1016/j.ymthe.2003.07.003
75. Sullivan, N.J., et al.: Immune protection of nonhuman primates against Ebola virus with single low-dose adenovirus vectors encoding modified GPs. PLoS Med. **3**, e177 (2006). https://doi.org/10.1371/journal.pmed.0030177
76. Bentley, E.M., et al.: Cross-neutralisation of novel Bombali virus by Ebola virus antibodies and convalescent plasma using an optimised Pseudotype-based neutralisation assay. Trop Med Infect Dis. **6** (2021). https://doi.org/10.3390/tropicalmed6030155
77. Shimojima, M., Ikeda, Y., Kawaoka, Y.: The mechanism of Axl-mediated Ebola virus infection. J. Infect. Dis. **196**(Suppl 2), S259–S263 (2007). https://doi.org/10.1086/520594
78. Brindley, M.A., et al.: Ebola virus glycoprotein 1: identification of residues important for binding and postbinding events. J. Virol. **81**, 7702–7709 (2007). https://doi.org/10.1128/JVI.02433-06
79. Hunt, C.L., Kolokoltsov, A.A., Davey, R.A., Maury, W.: The Tyro3 receptor kinase Axl enhances macropinocytosis of Zaire ebolavirus. J. Virol. **85**, 334–347 (2011). https://doi.org/10.1128/JVI.01278-09
80. Sandrin, V., et al.: Lentiviral vectors pseudotyped with a modified RD114 envelope glycoprotein show increased stability in sera and augmented transduction of primary lymphocytes and CD34+ cells derived from human and nonhuman primates. Blood. **100**, 823–832 (2002). https://doi.org/10.1182/blood-2001-11-0042
81. Wang, M.K., Lim, S.Y., Lee, S.M., Cunningham, J.M.: Biochemical basis for increased activity of Ebola glycoprotein in the 2013-16 epidemic. Cell Host Microbe. **21**, 367–375 (2017). https://doi.org/10.1016/j.chom.2017.02.002
82. Kuhn, J.H., et al.: Conserved receptor-binding domains of Lake Victoria marburgvirus and Zaire ebolavirus bind a common receptor. J. Biol. Chem. **281**, 15951–15958 (2006). https://doi.org/10.1074/jbc.M601796200
83. Jemielity, S., et al.: TIM-family proteins promote infection of multiple enveloped viruses through virion-associated phosphatidylserine. PLoS Pathog. **9**, e1003232 (2013). https://doi.org/10.1371/journal.ppat.1003232
84. Stewart, C.M., et al.: Ebola virus triggers receptor tyrosine kinase-dependent signaling to promote the delivery of viral particles to entry-conducive intracellular compartments. PLoS Pathog. **17**, e1009275 (2021). https://doi.org/10.1371/journal.ppat.1009275

85. Wool-Lewis, R.J., Bates, P.: Characterization of Ebola virus entry by using pseudotyped viruses: identification of receptor-deficient cell lines. J. Virol. **72**, 3155–3160 (1998). https://doi.org/10.1128/JVI.72.4.3155-3160.1998
86. Mpanju, O.M., Towner, J.S., Dover, J.E., Nichol, S.T., Wilson, C.A.: Identification of two amino acid residues on Ebola virus glycoprotein 1 critical for cell entry. Virus Res. **121**, 205–214 (2006). https://doi.org/10.1016/j.virusres.2006.06.002
87. Urbanowicz, R.A., et al.: Human adaptation of Ebola virus during the west African outbreak. Cell. **167**, 1079–1087 e1075 (2016). https://doi.org/10.1016/j.cell.2016.10.013
88. Miao, C., Li, M., Zheng, Y.M., Cohen, F.S., Liu, S.L.: Cell-cell contact promotes Ebola virus GP-mediated infection. Virology. **488**, 202–215 (2016). https://doi.org/10.1016/j.virol.2015.11.019
89. Shimojima, M., et al.: Tyro3 family-mediated cell entry of Ebola and Marburg viruses. J. Virol. **80**, 10109–10116 (2006). https://doi.org/10.1128/JVI.01157-06
90. Takada, A., et al.: A system for functional analysis of Ebola virus glycoprotein. Proc. Natl. Acad. Sci. U. S. A. **94**, 14764–14769 (1997). https://doi.org/10.1073/pnas.94.26.14764
91. Hoffmann, M., Gonzalez Hernandez, M., Berger, E., Marzi, A., Pohlmann, S.: The glycoproteins of all filovirus species use the same host factors for entry into bat and human cells but entry efficiency is species dependent. PLoS One. **11**, e0149651 (2016). https://doi.org/10.1371/journal.pone.0149651
92. Hoffmann, M., et al.: A polymorphism within the internal fusion loop of the Ebola virus glycoprotein modulates host cell entry. J. Virol. **91** (2017). https://doi.org/10.1128/JVI.00177-17
93. Ueda, M.T., et al.: Functional mutations in spike glycoprotein of Zaire ebolavirus associated with an increase in infection efficiency. Genes Cells. **22**, 148–159 (2017). https://doi.org/10.1111/gtc.12463
94. Liu, Q., et al.: Antibody-dependent-cellular-cytotoxicity-inducing antibodies significantly affect the post-exposure treatment of Ebola virus infection. Sci. Rep. **7**, 45552 (2017). https://doi.org/10.1038/srep45552
95. Powell, T.J., Silk, J.D., Sharps, J., Fodor, E., Townsend, A.R.: Pseudotyped influenza a virus as a vaccine for the induction of heterotypic immunity. J. Virol. **86**, 13397–13406 (2012). https://doi.org/10.1128/JVI.01820-12
96. Xiao, J.H., et al.: Characterization of influenza virus Pseudotyped with ebolavirus glycoprotein. J. Virol. **92** (2018). https://doi.org/10.1128/JVI.00941-17
97. Rijal, P., et al.: Therapeutic monoclonal antibodies for Ebola virus infection derived from vaccinated humans. Cell Rep. **27**, 172–186 e177 (2019). https://doi.org/10.1016/j.celrep.2019.03.020
98. Urbanowicz, R.A., et al.: Novel functional hepatitis C virus glycoprotein isolates identified using an optimized viral pseudotype entry assay. J. Gen. Virol. **97**, 2265–2279 (2016). https://doi.org/10.1099/jgv.0.000537
99. Steeds, K., et al.: Pseudotyping of VSV with Ebola virus glycoprotein is superior to HIV-1 for the assessment of neutralising antibodies. Sci. Rep. **10**, 14289 (2020). https://doi.org/10.1038/s41598-020-71225-1
100. Konduru, K., Shurtleff, A.C., Bavari, S., Kaplan, G.: High degree of correlation between Ebola virus BSL-4 neutralization assays and pseudotyped VSV BSL-2 fluorescence reduction neutralization test. J. Virol. Methods. **254**, 1–7 (2018). https://doi.org/10.1016/j.jviromet.2018.01.003
101. Ilinykh, P.A., et al.: Chimeric filoviruses for identification and characterization of monoclonal antibodies. J. Virol. **90**, 3890–3901 (2016). https://doi.org/10.1128/JVI.00101-16
102. Warfield, K.L., et al.: Role of antibodies in protection against Ebola virus in nonhuman primates immunized with three vaccine platforms. J. Infect. Dis. **218**, S553–S564 (2018). https://doi.org/10.1093/infdis/jiy316

103. Yang, R., et al.: Neutralizing antibody Titer test of Ebola recombinant protein vaccine and gene vector vaccine pVR-GP-FC. Biomed. Environ. Sci. **31**, 721–728 (2018). https://doi.org/10.3967/bes2018.097
104. Shuai, L., et al.: Genetically modified rabies virus-vectored Ebola virus disease vaccines are safe and induce efficacious immune responses in mice and dogs. Antivir. Res. **146**, 36–44 (2017). https://doi.org/10.1016/j.antiviral.2017.08.011
105. Mupapa, K., et al.: Treatment of Ebola hemorrhagic fever with blood transfusions from convalescent patients. International scientific and technical committee. J. Infect. Dis. **179**(Suppl 1), S18–S23 (1999). https://doi.org/10.1086/514298
106. Gupta, M., Mahanty, S., Bray, M., Ahmed, R., Rollin, P.E.: Passive transfer of antibodies protects immunocompetent and imunodeficient mice against lethal Ebola virus infection without complete inhibition of viral replication. J. Virol. **75**, 4649–4654 (2001). https://doi.org/10.1128/JVI.75.10.4649-4654.2001
107. Jahrling, P.B., et al.: Passive immunization of Ebola virus-infected cynomolgus monkeys with immunoglobulin from hyperimmune horses. Arch. Virol. Suppl. **11**, 135–140 (1996). https://doi.org/10.1007/978-3-7091-7482-1_12
108. Audet, J., et al.: Molecular characterization of the monoclonal antibodies composing ZMAb: a protective cocktail against Ebola virus. Sci. Rep. **4**, 6881 (2014). https://doi.org/10.1038/srep06881
109. Saphire, E.O., Schendel, S.L., Gunn, B.M., Milligan, J.C., Alter, G.: Antibody-mediated protection against Ebola virus. Nat. Immunol. **19**, 1169–1178 (2018). https://doi.org/10.1038/s41590-018-0233-9
110. Brannan, J.M., et al.: Post-exposure immunotherapy for two ebolaviruses and Marburg virus in nonhuman primates. Nat. Commun. **10**, 105 (2019). https://doi.org/10.1038/s41467-018-08040-w
111. Flyak, A.I., et al.: Mechanism of human antibody-mediated neutralization of Marburg virus. Cell. **160**, 893–903 (2015). https://doi.org/10.1016/j.cell.2015.01.031
112. Howell, K.A., et al.: Antibody treatment of Ebola and Sudan virus infection via a uniquely exposed epitope within the glycoprotein receptor-binding site. Cell Rep. **15**, 1514–1526 (2016). https://doi.org/10.1016/j.celrep.2016.04.026
113. Oswald, W.B., et al.: Neutralizing antibody fails to impact the course of Ebola virus infection in monkeys. PLoS Pathog. **3**, e9 (2007). https://doi.org/10.1371/journal.ppat.0030009
114. Ilinykh, P.A., et al.: Non-neutralizing antibodies from a Marburg infection survivor mediate protection by fc-effector functions and by enhancing efficacy of other antibodies. Cell Host Microbe. **27**, 976–991 e911 (2020). https://doi.org/10.1016/j.chom.2020.03.025
115. Koellhoffer, J.F., et al.: Two synthetic antibodies that recognize and neutralize distinct proteolytic forms of the ebola virus envelope glycoprotein. Chembiochem. **13**, 2549–2557 (2012). https://doi.org/10.1002/cbic.201200493
116. Takada, A., Watanabe, S., Okazaki, K., Kida, H., Kawaoka, Y.: Infectivity-enhancing antibodies to Ebola virus glycoprotein. J. Virol. **75**, 2324–2330 (2001). https://doi.org/10.1128/JVI.75.5.2324-2330.2001
117. Takada, A., Ebihara, H., Feldmann, H., Geisbert, T.W., Kawaoka, Y.: Epitopes required for antibody-dependent enhancement of Ebola virus infection. J. Infect. Dis. **196**(Suppl 2), S347–S356 (2007). https://doi.org/10.1086/520581
118. Nakayama, E., et al.: Antibody-dependent enhancement of Marburg virus infection. J. Infect. Dis. **204**(Suppl 3), S978–S985 (2011). https://doi.org/10.1093/infdis/jir334
119. Ito, H., Watanabe, S., Sanchez, A., Whitt, M.A., Kawaoka, Y.: Mutational analysis of the putative fusion domain of Ebola virus glycoprotein. J. Virol. **73**, 8907–8912 (1999). https://doi.org/10.1128/JVI.73.10.8907-8912.1999
120. Iwasa, A., Shimojima, M., Kawaoka, Y.: sGP serves as a structural protein in Ebola virus infection. J. Infect. Dis. **204**(Suppl 3), S897–S903 (2011). https://doi.org/10.1093/infdis/jir313

121. Ito, H., Watanabe, S., Takada, A., Kawaoka, Y.: Ebola virus glycoprotein: proteolytic processing, acylation, cell tropism, and detection of neutralizing antibodies. J. Virol. **75**, 1576–1580 (2001). https://doi.org/10.1128/JVI.75.3.1576-1580.2001
122. Chandran, K., Sullivan, N.J., Felbor, U., Whelan, S.P., Cunningham, J.M.: Endosomal proteolysis of the Ebola virus glycoprotein is necessary for infection. Science. **308**, 1643–1645 (2005). https://doi.org/10.1126/science.1110656
123. Schornberg, K., et al.: Role of endosomal cathepsins in entry mediated by the Ebola virus glycoprotein. J. Virol. **80**, 4174–4178 (2006). https://doi.org/10.1128/JVI.80.8.4174-4178.2006
124. Gonzalez-Hernandez, M., Muller, A., Hoenen, T., Hoffmann, M., Pohlmann, S.: Calu-3cells are largely resistant to entry driven by filovirus glycoproteins and the entry defect can be rescued by directed expression of DC-SIGN or cathepsin L. Virology. **532**, 22–29 (2019). https://doi.org/10.1016/j.virol.2019.03.020
125. Misasi, J., et al.: Filoviruses require endosomal cysteine proteases for entry but exhibit distinct protease preferences. J. Virol. **86**, 3284–3292 (2012). https://doi.org/10.1128/JVI.06346-11
126. Miller, E.H., et al.: Ebola virus entry requires the host-programmed recognition of an intracellular receptor. EMBO J. **31**, 1947–1960 (2012). https://doi.org/10.1038/emboj.2012.53
127. Bornholdt, Z.A. et al.: Host-primed Ebola virus GP exposes a hydrophobic NPC1 receptor-binding pocket, revealing a target for broadly neutralizing antibodies. mBio. **7**, e02154–02115 (2016). https://doi.org/10.1128/mBio.02154-15
128. Martinez, O., et al.: A mutation in the Ebola virus envelope glycoprotein restricts viral entry in a host species- and cell-type-specific manner. J. Virol. **87**, 3324–3334 (2013). https://doi.org/10.1128/JVI.01598-12
129. Spence, J.S., Krause, T.B., Mittler, E., Jangra, R.K., Chandran, K.: Direct visualization of Ebola virus fusion triggering in the endocytic pathway. MBio. **7**, e01857–e01815 (2016). https://doi.org/10.1128/mBio.01857-15
130. Mittler, E., Alkutkar, T., Jangra, R.K., Chandran, K.: Direct intracellular visualization of Ebola virus-receptor interaction by in situ proximity ligation. MBio. **12** (2021). https://doi.org/10.1128/mBio.03100-20
131. Kuroda, M., Halfmann, P., Kawaoka, Y.: HER2-mediated enhancement of Ebola virus entry. PLoS Pathog. **16**, e1008900 (2020). https://doi.org/10.1371/journal.ppat.1008900
132. Miller, M.E., Adhikary, S., Kolokoltsov, A.A., Davey, R.A.: Ebolavirus requires acid sphingomyelinase activity and plasma membrane sphingomyelin for infection. J. Virol. **86**, 7473–7483 (2012). https://doi.org/10.1128/JVI.00136-12
133. Alvarez, C.P., et al.: C-type lectins DC-SIGN and L-SIGN mediate cellular entry by Ebola virus in cis and in trans. J. Virol. **76**, 6841–6844 (2002). https://doi.org/10.1128/jvi.76.13.6841-6844.2002
134. Yu, C., et al.: MARCH8 inhibits Ebola virus glycoprotein, human immunodeficiency virus type 1 envelope glycoprotein, and avian influenza virus H5N1 hemagglutinin maturation. MBio. **11** (2020). https://doi.org/10.1128/mBio.01882-20
135. Stantchev, T.S., et al.: Cytokine effects on the entry of filovirus envelope Pseudotyped virus-like particles into primary human macrophages. Viruses. **11** (2019). https://doi.org/10.3390/v11100889
136. Usami, K., et al.: Involvement of viral envelope GP2 in Ebola virus entry into cells expressing the macrophage galactose-type C-type lectin. Biochem. Biophys. Res. Commun. **407**, 74–78 (2011). https://doi.org/10.1016/j.bbrc.2011.02.110
137. Fujihira, H., et al.: A critical domain of ebolavirus envelope glycoprotein determines Glycoform and infectivity. Sci. Rep. **8**, 5495 (2018). https://doi.org/10.1038/s41598-018-23357-8
138. Basu, A., Mills, D.M., Bowlin, T.L.: High-throughput screening of viral entry inhibitors using pseudotyped virus. Curr. Protoc. Pharmacol.. **Chapter 13**, Unit 13B 13. (2010). https://doi.org/10.1002/0471141755.ph13b03s51

139. Yang, Y., et al.: A cell-based high-throughput protocol to screen entry inhibitors of highly pathogenic viruses with traditional Chinese medicines. J. Med. Virol. **89**, 908–916 (2017). https://doi.org/10.1002/jmv.24705
140. Cui, Q., et al.: Identification of ellagic acid from plant Rhodiola rosea L. as an anti-Ebola virus entry inhibitor. Viruses. **10** (2018). https://doi.org/10.3390/v10040152
141. Madrid, P.B., et al.: A systematic screen of FDA-approved drugs for inhibitors of biological threat agents. PLoS One. **8**, e60579 (2013). https://doi.org/10.1371/journal.pone.0060579
142. Lei, S., Huang, W., Wang, Y., Liu, Q.: In vivo bioluminescent imaging of Marburg virus in a rodent model. Methods Mol. Biol. **2081**, 177–190 (2020). https://doi.org/10.1007/978-1-4939-9940-8_12
143. Duffy, S.: Why are RNA virus mutation rates so damn high? PLoS Biol. **16**, e3000003 (2018). https://doi.org/10.1371/journal.pbio.3000003
144. Li, Q., et al.: The impact of mutations in SARS-CoV-2 spike on viral infectivity and antigenicity. Cell. **182**, 1284–1294.e1289 (2020). https://doi.org/10.1016/j.cell.2020.07.012
145. Fels, J.M., et al.: A glycoprotein mutation that emerged during the 2013-2016 Ebola virus epidemic alters proteolysis and accelerates membrane fusion. MBio. **12** (2021). https://doi.org/10.1128/mBio.03616-20
146. Dietzel, E., Schudt, G., Krahling, V., Matrosovich, M., Becker, S.: Functional characterization of adaptive mutations during the west African Ebola virus outbreak. J. Virol. **91** (2017). https://doi.org/10.1128/JVI.01913-16
147. Gregory, S.M., et al.: Structure and function of the complete internal fusion loop from ebolavirus glycoprotein 2. Proc. Natl. Acad. Sci. U. S. A. **108**, 11211–11216 (2011). https://doi.org/10.1073/pnas.1104760108
148. Kurosaki, Y., et al.: Different effects of two mutations on the infectivity of Ebola virus glycoprotein in nine mammalian species. J. Gen. Virol. **99**, 181–186 (2018). https://doi.org/10.1099/jgv.0.000999
149. Banadyga, L., et al.: Atypical Ebola virus disease in a nonhuman primate following monoclonal antibody treatment is associated with glycoprotein mutations within the fusion loop. MBio. **12** (2021). https://doi.org/10.1128/mBio.01438-20

Chapter 7
Pseudotyped Viruses for Coronaviruses

Meiyu Wang, Jianhui Nie, and Youchun Wang

Abstract Seven coronaviruses have been identified that can infect humans, four of which usually cause mild symptoms, including HCoV-229E, HCoV-NL63, HCoV-OC43, and HCoV-HKU1, three of which are lethal coronaviruses, named severe acute respiratory syndrome coronavirus, Middle East respiratory syndrome coronavirus, and severe acute respiratory syndrome coronavirus 2. Pseudotyped virus is an important tool in the field of human coronavirus research because it is safe, easy to prepare, easy to detect, and highly modifiable. In addition to the application of pseudotyped viruses in the study of virus infection mechanism, vaccine, and candidate antiviral drug or antibody evaluation and screening, pseudotyped viruses can also be used as an important platform for further application in the prediction of immunogenicity and antigenicity after virus mutation, cross-species transmission prediction, screening, and preparation of vaccine strains with better broad spectrum and antigenicity. Meanwhile, as clinical trials of various types of vaccines and post-clinical studies are also being carried out one after another, the establishment of a

---

Meiyu Wang and Jianhui Nie have equally contributed equally to this work.

M. Wang
Division of HIV/AIDS and Sex-transmitted Virus Vaccines, National Institutes for Food and Drug Control (NIFDC) and WHO Collaborating Center for Standardization and Evaluation of Biologicals, Beijing, China

Units Department of Laboratory Medicine, Sichuan Academy of Medical Sciences & Sichuan Provincial People's Hospital, School of Medicine, University of Electronic and Technology of China, Chengdu, China

J. Nie
Units Department of Laboratory Medicine, Sichuan Academy of Medical Sciences & Sichuan Provincial People's Hospital, School of Medicine, University of Electronic and Technology of China, Chengdu, China

Y. Wang (✉)
Chinese Academy of Medical Sciences & Peking Union Medical College, Beijing, China

Institute of Medical Biology, Chinese Academy of Medicine Sciences & Peking Union Medical College, Kunming, China
e-mail: wangyc@nifdc.org.cn; wangyc@imbcams.com.cn

Y. Wang (ed.), *Pseudotyped Viruses*, Advances in Experimental Medicine and Biology 1407, https://doi.org/10.1007/978-981-99-0113-5_7

high-throughput and fully automated detection platform based on SARS-CoV-2 pseudotyped virus to further reduce the cost of detection and manual intervention and improve the efficiency of large-scale detection is also a demand for the development of SARS-CoV-2 pseudotyped virus.

Keywords Pseudotyped virus · SARS-CoV · MERS-CoV · SARS-CoV-2 · Vaccine · Monoclonal antibodies · Neutralization assay

Abbreviations

6-HB	6-helix bundle
ACE2	Angiotensin-converting enzyme 2
BSL	Biosafety level
COVID-19	Coronavirus disease 2019
CP	Cytoplasmic domain
CRFK	Crandell-Rees feline kidney
CTD	C-terminal domain
DLS	Dynamic light scattering
DPP4	Dipeptidyl peptidase 4
ELISA	Enzyme-linked immunoassay
fluc	Firefly luciferase
FP	Fusion peptide
GFP	Green fluorescent protein
hCoV	Human coronavirus
HCoV-229E	Human coronavirus 229E
HCoV-HKU1	Human coronavirus HKU1
HCoV-NL63	Human coronavirus NL63
HCoV-OC43	Human coronavirus OC43
HIV	Human immunodeficiency virus
HR1	Heptad repeat 1
HR2	Heptad repeat 2
LTR	Long terminal repeat
MERS-CoV	Middle East respiratory syndrome coronavirus
MLV	Murine leukemia virus
NSP	Nonstructural protein
NT_{50}	50% neutralization titer
NTA	Nanoparticle tracking analysis
NTD	N-terminal domain
ORF	Open reading frame
QDs	Quantum dots
RBD	Receptor-binding domain
SARS-CoV	Severe acute respiratory syndrome coronavirus
SARS-CoV-2	Severe acute respiratory syndrome coronavirus 2

SP	Signal peptide
TEM	Transmission electron microscopy
TM	Transmembrane domain
TMPRSS2	Transmembrane serine protease 2
UH	Upstream helix
VLPs	Viruslike particles
VOC	Variant of concern
VOI	Variant of interest
Vpr	Viral protein r
VSV	Vesicular stomatitis virus

7.1 Biological Characteristics of Coronavirus

Coronavirus was first isolated from chickens in the 1930s and has been widely reported in humans, other mammals, and birds, causing acute and persistent infections [1]. Coronavirus is classified as one of the subfamilies (*Coronavirinae*) in the family *Coronaviridae*, which is the biggest taxon among the *Nidovirales*. The *Coronavirinae* subfamily can be further divided into three *genera* based on phylogenetic clustering, Alpha Coronavirus, Beta Coronavirus, and Gamma Coronavirus (Fig. 7.1). Seven coronaviruses have been identified that can infect humans, four of which usually cause mild symptoms, such as the common cold, including human coronavirus 229E (HCoV-229E) and human coronavirus NL63 (HCoV-NL63), which are Alpha Coronavirus, and human coronavirus OC43 (HCoV-OC43) and human coronavirus HKU1 (HCoV-HKU1), which are Beta Coronavirus. However, since 2002, three lethal coronaviruses have been identified in succession, named severe acute respiratory syndrome coronavirus (SARS-CoV), Middle East respiratory syndrome coronavirus (MERS-CoV), and severe acute respiratory syndrome coronavirus 2 (SARS-CoV-2). They are all classified as Beta Coronavirus and

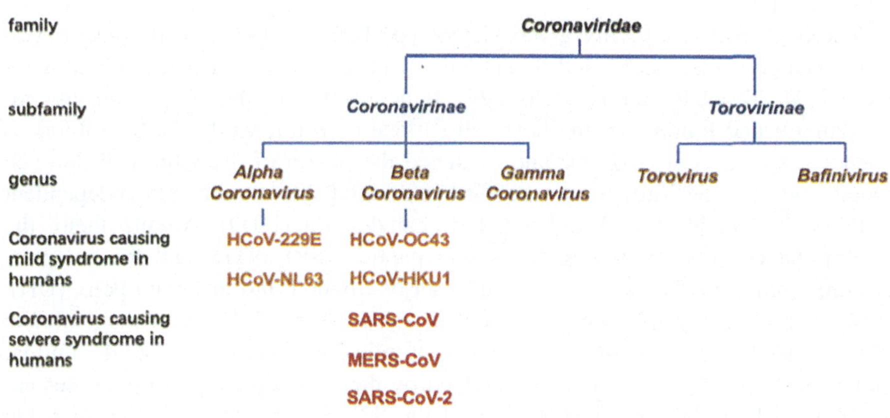

Fig. 7.1 Taxonomy of the family *Coronaviridae*

mainly infect the human lower respiratory tract and cause a severe acute respiratory syndrome in humans [2, 3]. Because of their highly infectious and pathogenic nature, viral culture and animal infection experiments should be performed in biosafety level 3 (BSL-3) laboratories, which pose an obstacle to the study of viruses. Because of the significant impact of these three coronaviruses on human society, this chapter will concentrate exclusively on the SARS-CoV, MERS-CoV, and, especially, SARS-CoV-2, which is ravaging the world.

7.1.1 Structure of Coronaviruses

Like other coronaviruses, SARS-CoV, MERS-CoV, and SARS-CoV-2 all have ~30-kb genomes of positive single-stranded non-segmented RNA containing a typical 5′ cap and a poly (a)-3′ tail [4]. The 5′ end of the SARS-CoV-2 genome contains two large open reading frames (ORFs) – ORF1a and ORF1b genes (also named as rep 1a and 1b in MERS) – which account for about two-thirds of the entire genome. They together encode 15 nonstructural proteins (NSPs), including NSP1 to NSP10 and NSP12 to NSP16 [5]. The remaining one-third of the genome encodes eight accessory proteins (3a, 3b, p6, 7a, 7b, 8b, 9b, and orf14) and four structural proteins (N, M, S, and E), which are components of mature viruses and are essential to the structural and functional integrity of the virus [6]. The genome of coronavirus is wrapped by the viral N protein to form a helical nucleocapsid, which is coated with a viral envelope in which the M, E, and Spike protein are embedded [7]. The M protein (~25 to 30 kDa) exists as a dimer in the virus and is able to both promote membrane bending and bind to the capsid through different conformational changes. E proteins (~8 to 12 kDa) are often thought to contribute to the assembly and release of viral particles and have ion channel roles [8]. S protein (~150 kDa) is exposed on the cell surface and is a key protein for virus recognition and infection of cells, making it a major target for postinfection neutralizing antibody/antiviral drug and vaccine design [9, 10].

The S protein is a highly glycosylated type I fusion protein that contains two functional subunits, the S1 and S2 subunits, with a protease cleavage site between them [11]. The S1 subunit is located at the distal end and is responsible for recognition and binding to the host cell surface receptor, while the S2 subunit is anchored to the viral envelope and mediated the fusion of the virus with the cell membrane. The S1 subunit can be further divided into a relatively independent N-terminal domain (NTD) and a C-terminal domain (CTD). Among them, the N-terminal domain starts with the signal peptide (SP), NTD, and the receptor-binding domain (RBD) [12]. S2 subunit can be divided into upstream helix (UH), fusion peptide (FP), heptad repeat 1 (HR1), heptad repeat 2 (HR2), transmembrane domain (TM), and intracellular region cytoplasmic domain (CP) [11]. Under normal conditions, the FP of S proteins is shielded by their UH domains, so proteases are required to hydrolyze the UH domain at the S2' end to activate the irreversible conformational change of S2 and initiate S protein-mediated membrane fusion [11] (Fig. 7.2).

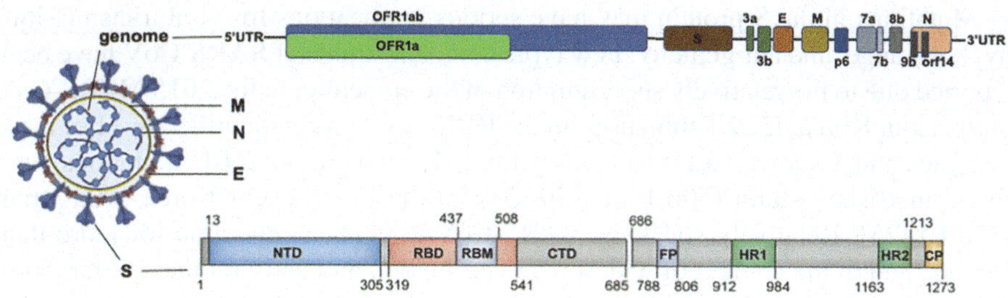

Fig. 7.2 Schematic illustration of the structure of SARS-CoV-2, genome of SARS-CoV-2, and construction of spike protein

7.1.2 *Infection of Coronaviruses*

The initial step in coronavirus infection is the specific binding of S proteins to host cell surface receptors, and the histophilic and pathogenic properties of the virus correlate with the tissue distribution of these receptors. The cellular receptors for infecting human coronaviruses are now relatively well investigated and include angiotensin-converting enzyme 2 (ACE2; HCoV-NL63, SARS-CoV, and SARS-CoV-2) and dipeptidyl peptidase 4 (DPP4; MERS-CoV) and human aminopeptidase N (APN; HCoV-229E) [13]. When infection of SARS-COV-2 occurs, the RBD of the S protein first binds to the cell surface receptor, followed by cleavage by the host protease; S1 detaches from S2, exposing a hydrophobic FP that is inserted into the host cell membrane; and HR1 and HR2 fold over each other to form a 6-helix bundle (6-HB) and bring the virus and host membrane close to fusion [14–17]. Compared with other coronaviruses, the S protein of SARS-CoV-2 has 12 extra nucleotides upstream, forming a new furin cleavage site, which may facilitate the initiation of the S protein and may increase the efficiency of SARS-CoV-2 transmission [10, 18, 19].

7.1.3 *Diversity for Each Coronavirus*

Coronaviruses are the largest RNA viruses in the genome, and their rate of mutation is lower than other RNA viruses such as influenza viruses and human immunodeficiency viruses. This is partly due to the fact that the coronavirus-encoded nonstructural protein 14 (NSP14) has N-terminal ribonuclease activity that can repair some mutations in the viral replication process [20–22] and mutates only one-tenth the frequency of influenza viruses [23]. Although the mutation rate of the coronavirus is low, the long-term widespread of SARS-CoV-2 has led to the emergence and accumulation of various variants. The constant emergence, replacement, and simultaneous circulation of new variants have made the prevention and control of coronavirus disease 2019 (COVID-19) pandemic more complex and unpredictable.

Mutations in the S protein may have serious implications for viral transmissibility, host range, and antigenicity. Few typical mutant strains of SARS-CoV have been reported due to the relatively short duration of the epidemic. In the 2015 MERS-CoV outbreak in Korea, I529 T mutations in the RBD region were identified in 11 of these genomes and 1 showed a D510 G mutation [24]. In addition, RBD mutations were found in Bishal strain (506 L and 509G) and the more recent Korea a002 strain (530 L) [25]. Relatively speaking, SARS-CoV-2 has been endemic for more than 3 years, and in the context of widespread vaccination and antiviral therapy targeting the S protein worldwide, SARS-CoV-2 has evolved multiple mutations during the epidemic, which in turn have evolved into epidemic strains that are widely prevalent in specific regions [26, 27]. Therefore, the WHO continuously conducts surveillance for mutations in SARS-CoV-2 and classifies SARS-CoV-2 epidemic mutant strains as VOC (variant of concern) and VOI (variant of interest). VOC refers to mutant strains with evidence of the following characteristics: increased transmissibility, which can lead to more severe disease. VOIs refer to prevalent strains with specific mutations that may affect binding to receptors or may reduce the neutralization of antibodies from previous infections or vaccines, reduce the efficacy of treatments or vaccines, have a potential diagnostic impact, or predict increased transmissibility or disease severity. The identification of VOC and VOI is constantly changing as viruses continue to evolve, and as of April 1, 2022, the prevalent strains defined as VOC include Delta and Omicron [28]. The currently circulating Omicron variants are also constantly evolving into new sub-lineages, such as BA.4, BA.5, and BA.2.75. The antigenicity of these new sub-lineages has been significantly different from that of the original Omicron [29], and they are more infectious, becoming the dominant strains [30], posing a serious challenge to the prevention and control of the pandemic.

7.2 Construction of Pseudotyped Viruses for Coronaviruses

The isolation of viral strains is important for research on the pathogenesis of viruses and the development of corresponding vaccines, therapeutic drugs, and diagnostic reagents. However, due to the high pathogenicity and transmission risk of SARS-CoV, MERS-CoV, and SARS-CoV-2, the acquisition and cultivation of virulent strains have always been limited by the level of laboratory protection and personnel capacity, which seriously hinder the research work. The pseudotyped virus is a synthetic chimeric virus, which is an infectious viruslike particle composed of a lipid envelope or capsid protein of the target virus wrapped around a heterologous nucleic acid. The pseudotyped virus can replace authentic viruses and has become an important tool for CoV research. Currently, all seven human coronaviruses can be successfully synthesized as pseudotyped viruses, with the most commonly studied pseudotyped viruses being the highly pathogenic SARS-CoV, MERS-CoV, and SARS-CoV-2. The technical key to human coronavirus (hCoV) pseudotyped virus preparation is to obtain infectious particles with higher titer and safer and easier detection, which needs to be achieved based on the selection of a suitable pseudotyped virus packaging system and optimization of parameters for hCoV pseudotyped virus preparation. HCoV pseudotyped viruses have been reported based both on viral vector systems specifically targeting enveloped viruses and on protein-autonomous packaging systems for viruslike particles [31].

7.2.1 CoV Pseudotyped Virus Based on Vesicular Stomatitis Virus (VSV)

Currently, pseudotyped viruses based on VSV have been successfully constructed for SARS-CoV, MERS-CoV, and SARS-CoV-2. That is, BHK21 cells were infected by recombinant poxvirus expressing T7 RNA polymerase and cotransfected with the pVSV-ΔG-luciferase plasmid. Alternatively, 293 T cells were infected with T7 polymerase poxvirus or transfected with pT7 NA polymerase, which is deficient in G protein and can express luciferase, and four helper plasmids expressing N, P, G, and L. Subsequently, G*ΔG-VSV was obtained by passaging on cells expressing VSV G protein [32, 33]. Using the above G*ΔG-VSV to infect HEK-293 T expressing hCoV S protein, high titers of hCoV pseudotyped virus can be obtained by harvesting the supernatant usually after 24h [33, 34]. Quantification of pseudotyped viruses is mainly achieved by quantifying VSV genes by real-time fluorescence PCR [35, 36]. In addition, Li et al. [37] constructed a replication-complete SARS-CoV pseudotyped virus based on the VSV packaging system. They constructed a backbone plasmid by replacing the G protein gene of VSV with a gene expressing the full length of the S protein of SARS-CoV-2 and inserted the reporter gene green fluorescent protein (GFP) or firefly luciferase (fluc) in front of its N gene and performed viral rescue by reverse genetic method, i.e., backbone plasmid and supporting plasmid expressing VSV structural proteins and T7 polymerase were cotransfected with HEK-293 T cells. Virus titration and neutralization experiments were performed using Vero cells and observed by phosphorescence or GFP fluorescence or luminescence values after the addition of substrate. The replicative pseudotyped virus was amplified by successive passages on Vero cells, and the pseudotyped virus titer obtained by amplification could reach 10^6 FFU/ml.

7.2.2 CoV Pseudotyped Virus Based on Human Immunodeficiency Virus (HIV)

Lentiviral packaging systems are genetically modified expression vectors based on lentiviruses, which can express target genes while retaining the cis-acting elements necessary for viral transcription, packaging, and integration. HIV, as one of the most well-researched lentiviruses, is often the packaging vector of choice for pseudotyped virus preparation [38]. Unlike VSV vector construction, which requires a complex viral rescue process, HIV lentiviral packaging vectors are in the form of plasmids (backbone plasmids) that can be cotransfected with SARS-CoV-2 S protein expression plasmids to form pseudotyped virus particles expressing SARS-CoV-2 S protein on the envelope. The current pseudotyped viruses based on the HIV packaging system are all replication-deficient. The main genes provided by the HIV packaging system are *gag* and *pol*, with *gag* providing the structural proteins p15, p18, and p24 and *pol* providing the integrase and reverse transcriptase. The reporter genes can be expressed in species-specific cells by integration in the HIV vector plasmid or as a separate expression plasmid cotransfected with the HIV vector plasmid and a membrane plasmid expressing the S protein, i.e., the reporter genes are designed to be expressed on a separate plasmid from the packaging signals (ψ, psi) normally associated with the HIV vector, as well as other packaging elements [31]. However, HIV packaging vectors from different laboratories also have some differences due to their different design concepts. To avoid potential reply mutation or integration risks and to improve the safety of the HIV packaging system, different fragments of HIV packaging vectors may be constructed in different plasmids for cotransfection use, too. The most frequently reported backbone plasmid for hCoV pseudotyped virus preparation is pNL4–3.Luc.R-E, which is a replication-deficient plasmid with HIV-based *env* deletion and carrying a *fluc* reporter gene [39–42]. Quantification of HIV-based hCoV pseudotyped viruses is usually performed by enzyme-linked immunoassay (ELISA) to detect the expression of p24 antigen carried by the HIV backbone [43]. Unlike the VSV system, harvesting of HIV-based hCoV pseudotyped viruses usually takes place 48 h after plasmid transfection [41, 44].

Compared with the VSV-based system, pseudotyped virus generated using the HIV backbone usually showed relatively lower titers. All phase III clinical trials of the COVID-19 vaccine approved in the United States have been evaluated for immunogenicity using the HIV-based SARS-CoV-2 pseudotyped virus neutralization assays. It is mainly done by two measures: one is performed in Duke University and one at Monogram Biosciences. The two methods are slightly different, but both employ similar mechanism to improve the infectivity of the fake virus, thus ensuring the repeatability and durability of the method. The former introduces transmembrane serine protease 2 (TMPRSS2) during the pseudotyped virus packaging process. The target cells infected by the pseudovirus are 293 T cells that overexpress the ACE2 receptor. The target cells of the latter are highly expressed TMPRSS2 and ACE2 receptors at the same time [45].

7.2.3 Construction of CoV Pseudotyped Virus Based on Other Packaging Systems

The murine leukemia virus (MLV) packaging system, another lentivirus packaging system, has likewise been reported for the preparation of CoV pseudotyped virus.

Compared to HIV, MLV itself cannot infect humans, so the MLV pseudotyped virus packaging system is relatively safer. The MLV packaging vector, also in plasmid form, lacks the MLV envelope protein but provides the core viral genes *gag* and *pol*. CoV pseudotyped viruses based on MLV packaging systems have been reported including MERS-CoV, SARS-CoV, and SARS-CoV-2, but relatively few MLV-based CoV pseudotyped viruses have been reported compared to VSV and HIV vectors [46, 47]. Millet et al. [47] employed a three-plasmid cotransfection strategy, combining the MLV vector, transfer vector carrying the fluc reporter gene flanked by retroviral regulatory long terminal repeat (LTR) region, a packing signal, and a plasmid expressing SARS-CoV or MERS-CoV S protein. Upon pseudotyped virus particle formation, the reporter gene plasmid is integrated into the nascent particle assembled from the MLV capsid protein and subsequently wrapped around the heterologous CoV S protein. When infecting sensitive cells, the RNA of the pseudotyped virus is released in the cell, reverse transcribed, and integrated into the target cell genome, which ensures the safety of pseudotyped virus use since the integrated genes do not contain MLV genes but only reporter genes [47]. However, it has also been reported that the pseudotyped virus titers constructed with the MLV packaging system were lower than those with the HIV lentivirus packaging system, even though the same optimized S protein was used as the membrane plasmid in the preparation of SARS-CoV-2 [48].

7.2.4 Construction of CoV Pseudotyped Virus Based on Protein-Autonomous Packaging Systems for Virus like Particles

In addition to the traditional pseudotyped virus packaging systems for chimeric viruses, new systems based on other principles are constantly being used to prepare CoV viruslike particles for subsequent studies. Gorshkov, Kirill, et al. constructed a pseudotyped virus based on fluorescent quantum dots (QDs) conjugated to the RBD region of SARS-CoV-2 recombinant S protein as a multifunctional imaging probe [49]. It can mimic infection by binding to the cell surface ACE2 receptor, thereby enabling biochemical energy transfer and thus endocytosis, and this same infection can also be inhibited by SARS-CoV-2 neutralizing antibodies and ACE2-Fc. Viruslike particles (VLPs) are multi-protein capsids structurally and functionally similar to infectious viral particles, which are formed from structural viral proteins with self-assembling properties when overexpressed in a suitable host cell and which are commonly used for packaging envelope-free viruses. Like other pseudoviral forms, these capsids not only mimic the morphology of the parental virus but also specifically transduce permissive cells. Unlike vector packaging systems, VLP does not carry the viral genome, which makes it safer and can be performed even in BSL-1 level laboratories. However, VLP systems have also recently been used for packaging CoV pseudotyped viruses, and, depending on the packaging strategy used, VLP for CoV can also be classified as homotypic VLP consisting of only coronavirus proteins and as VLP consisting of S proteins bound to a heterologous viral scaffold, such as poxvirus or phage. There are also pseudotyped viruses without coronavirus proteins but, with coronavirus nucleic acids encapsulated within them, used for diagnostic quality control. The quantification and characterization of CoV VLP can be done by nanoparticle tracking analysis (NTA), dynamic light scattering (DLS), and transmission electron microscopy (TEM) [50]. Xie et al. reported an alternative type of pseudotyped virus based on a reverse genetic system by constructing seven complementary DNA fragments of the SARS-CoV-2 genome assembled into a genome-wide cDNA with retrotranscribed RNA that can infect cells by electrotransformation, which was found to produce a cellular infection similar to that of a live virus infection in terms of spotophilic morphology, viral RNA profile, and replication kinetics. It provides new ideas for further studies on viral clearance as well as inhibition of viral replication [51].

7.3 Application of the HCoV Pseudotyped Viruses

The application of pseudotyped viruses breaks through the barriers to research caused by their strong infectiousness, pathogenicity, and scarcity of sources. CoV pseudotyped viruses have been widely used in pathogenesis research, mutant strain research, clinical disease process monitoring, vaccine development and clinical evaluation, antiviral drug/monoclonal antibody candidate evaluation, and many other fields.

7.3.1 *Infectivity of Highly Pathogenic hCoV and the Possibility of Cross-Species Transmission*

The coronavirus S protein plays a key role in recognizing host cell receptors and mediating the fusion of viruses with cell membranes and is the most critical viral protein for cross-species transmission and infection [52]. Clarification of the specific cellular receptors for S proteins is key to the study of their infectivity and infection mechanisms. In addition, the interaction of S proteins with homologous receptors of another species is a prerequisite for interspecies transmission [53]. Therefore, the comparison of coronavirus S proteins of different species and the study of their interaction with homologous receptors of different species and cross-immune protection are important for the study of virus transmission and evolution across species. CoV pseudotyped virus particles are embedded with S proteins on the envelope, thus enabling the study of the infection mechanism of highly pathogenic hCoV such as SARS-CoV, SARS-CoV-2, and MERS-CoV under limited experimental conditions. Several research teams have used hCoV pseudotyped viruses as tools to conduct studies on the infectivity of highly pathogenic hCoV and its highly homologous animal-derived viruses as well as cross-species scenarios. Hoffmann et al. [54] found that similar to SARS-CoV, cell entry of SARS-CoV-2 was also dependent on the receptor ACE2 but not on the DPP4 receptor by constructing pseudotyped viruses expressing the full length of the S proteins of SARS-CoV-2, SARS-CoV, MERS-CoV, and 229E-CoV, respectively, based on the VSV packaging system, for infection of cells expressing multiple potential viral receptors. Zhu et al. [55] suggested that the S protein of SARS-CoV-2 has higher cell fusion-mediating activity relative to SARS-CoV, which is important for the understanding of SARS-CoV-2 transmissibility and pathogenesis and also reveals targets for therapeutic intervention. Bats are thought to be natural reservoirs of a and β coronaviruses and are an evolutionary source of hCoV. Pseudotyped virus-based studies have shown that bat isolates of SARS-associated coronaviruses can infect cells from humans and many other animal species using ACE2 from humans, civets, and bats as their cellular receptors. Both MERS-CoV and BAT CoV-HKU4 can use BAT and human DPP4 to infect cells derived from humans, camels, bats, and other animal species [56]. MERS-CoV has a wider range of cells than SARS-CoV and 229E, which also corroborates with the evolutionary chain of epidemiology [57]. By constructing pseudotyped viruses of civet-derived coronaviruses, their ability to infect human cells was found to be much lower than that of SARS-CoV, and most sera immunized with civet-derived S proteins were found to have strong neutralizing activity against this strain, but no cross-protection against SARS-CoV [58]. RaTG13 of bat origin is by far the most relevant virus for SARS-CoV-2 phylogeny. By constructing the pseudotyped virus of RaTG13, SARS-CoV-2 was found to induce strong cross-reactive antibodies against RaTG13, and the receptor binding and host adaptation mechanisms of RaTG13 were elucidated in combination with the results of the interaction between RaTG13 and ACE2 of different host origin. hDPP4 can be used by HKU25-spike to enter hDPP4-expressing cells and suggest that bat CoV spike proteins may have evolved to bind to hDPP4 [59]. The above studies highlight the importance of continuous monitoring of animal host-borne coronaviruses to prevent CoV re-spillover and provide important insights into the evolution and cross-species transmission of hCoV [52, 53].

Studies have shown that the role of host proteases on CoV infection is significantly associated with the zoonotic potential of coronaviruses. The host cell protease activates the S protein to initiate infection by targeting the specific enzymatic cleavage site of the coronavirus S protein [54]. The S protein, a highly glycosylated protein, has 22 glycosylation sites. By preparing SARS-CoV-2 pseudotyped viruses modified with S protein glycosylation sites, our laboratory found that the infectivity of glycosylation-deleted N331Q and N343Q pseudotyped viruses appeared significantly reduced, which in turn illustrated the importance of glycosylation for S protein-mediated virus infection [60]. Moreover, pseudotyped viruses can also be used to study the dynamic tracking and visualization of the infection process of virus particles in cells. Ma, Yingxin, et al. [61] prepared a SARS-CoV-2 pseudotyped virus based on the HIV system and labeled its envelope with a lipophilic dye DiO, while fusing the genomic RNA-associated viral protein r (Vpr) of the viral core with mCherry fluorescent protein, and used this two-color labeled pseudotyped virus to infect cells to achieve real-time kinetic simulation tracking of viral infection. It was found that the SARS-CoV-2 pseudotyped virus entered respiratory-associated susceptible cells through cellular endocytosis and released the viral core into the cytoplasm [62].

7.3.2 Study of Highly Pathogenic hCoV Mutant Strains

The prevalence of coronaviruses as RNA viruses is accompanied by mutations, and the infectivity of mutated viruses and the effectiveness of established prophylactic and antiviral therapies are a major concern. However, isolating and obtaining specific viral strains for further studies is relatively difficult. Therefore, pseudotyped viruses carrying specific mutation sites that are already present or predicted to be present in nature are obtained by targeted mutation of the membrane plasmid S gene and used to study the effect of mutation on infectivity and to further evaluate the effectiveness of existing antiviral drugs, monoclonal antibody drugs, and plasma after vaccine immunization. Due to the relatively short duration and limited prevalence of SARS-CoV and MERS-CoV, studies of pseudotyped viruses applied to mutations have been more focused on SARS-CoV-2. Our laboratory constructed a pseudotyped virus library of hundreds of SARS-CoV-2 mutant strains and endemic strains by continuously tracking the reported mutations based on the above method and systematically investigated the infectivity and antigenicity of the reported S protein mutation loci. It was found that the D614G mutation led to a significant enhancement of viral infectivity compared with the wild type, and the mutations affecting neutralization susceptibility were mainly concentrated in the RBD region of the S protein [60]. We also found mutant strains such as 501Y. V2, although similar in infectivity to D614G, had relatively reduced sensitivity to both monoclonal and serum polyclonal antibodies [63], a finding that was also verified in live virus studies [64]. Studies based on SARS-CoV-2 pseudotyped viruses have further targeted the importance of K417N, E484K, or N501Y mutations, as they reduce or eliminate the neutralizing activity of most of the reportedly effective monoclonal antibodies as well as provide a degree of diminished protection against vaccine immunization sera, and recovery sera, and these mutations may have an important role in virus escape [65, 66]. Animal immunization with a pseudotyped virus mutant strain revealed that the E484K mutation had a large effect on the antigenicity of the virus [36]. Kemp [67] et al. tracked strain mutations in patients receiving plasma therapy by second-generation sequencing and subsequently constructed pseudoviral mutant strains using a pseudoviral approach to mutations they found to be predominantly S protein D796H and ΔH69/ΔV70 deletions. In vitro studies of these pseudotyped virus mutant strains revealed no change in infectivity of the pseudotyped viruses carrying the mutations compared to the wild type, but a reduced susceptibility to neutralization in recovery plasma. In addition, SARS-CoV-2 replicating pseudotyped viruses were passaged in the presence of monoclonal antibodies or recovery sera to simulate virus replication under immune stress and thus predict the possible mutant strains. Using this approach, researchers found that the K417N, E484K, or N501Y mutations were also present in replicating pseudotyped viruses in the presence of vaccine-initiated monoclonal antibodies [66], thus completing the cross-correlation with previous studies on the broad spectrum of mutations.

7.3.3 *Pseudotyped Virus for Quantifying hCoV Neutralizing Antibodies*

Natural infection, vaccine immunization efficacy evaluation, and antiviral drug screening are essential for the development of prevention and control strategies for coronaviruses, especially SARS-CoV, MERS-CoV, and SARS-CoV-2 [44, 68]. The procedure of the CoV pseudotyped virus neutralization assay to detect neutralizing antibodies has been reported by several laboratories. The antibody-pseudotyped virus-target cell mixture is incubated for 24–72 h, and the expression of the reporter gene is quantified to calculate the 50% neutralization titer (NT_{50}) [48, 69].

Cells that have been reported for highly pathogenic hCoV pseudotyped virus neutralization assays are usually naturally or transfected with cells expressing the ACE2 receptor (DPP4 for MERS), including Huh7, Vero, 293 T/ 293FT/BHK21/ HeLa transiently or stably transfected with human ACE2 (DPP4), etc. [39, 46, 70]. Overexpression of TMPRSS2 protease based on ACE2-expressing cells can further increase the susceptibility of cells to SARS-CoV and SARS-CoV-2 pseudotyped virus infection [71] [72]. Furthermore, a Japanese team reported that in the preparation of SARS-CoV-2 pseudotyped virus based on the HIV system, despite the use of a unmodified S protein expression plasmid, Crandell-Rees feline kidney (CRFK) overexpressing ACE2 was tenfold more infectious when used as target cells compared to 293 T cells overexpressing ACE2 [37].

7.3.3.1 Natural Infection

Viral infection is followed by the production of specific antibody responses by the immune system against the virus, and the type and concentration of these antibodies change dynamically with the progression of infection. Neutralizing antibodies, as functional antibodies that prevent viral infection of target cells, are important for disease awareness and outbreak control strategies by tracking their relationship with disease progression and the persistence of immunity in patients after healing. Pseudotyped virus-based neutralization assays offer the possibility of the high-volume patient, multicenter follow-up in clinical practice due to their rapid and safe operation. The pseudotyped virus-based neutralizing antibody assay revealed that the neutralizing antibody levels in SARS-CoV-2 infected patients gradually increased with the duration of infection and then slowly decreased for up to 8 months or more. In addition, the levels of neutralizing antibodies were correlated with the duration and severity of clinical symptoms, but not with the age of the patients. The more severe the disease and the longer the duration of symptoms, the higher the titer of neutralizing antibodies in general [73–75], and in addition, there is a correlation between late neutralizing antibody deficiency and fatal outcome in severely ill patients [76].

7.3.3.2 Pseudotyped Virus for CoV Vaccine Development and Clinical Evaluation

Vaccines are the most effective means of dealing with highly pathogenic infectious diseases. Because the SARS-CoV epidemic has lasted just over a year, there is still no marketed preventive vaccine for SARS-CoV, although research teams in various countries have developed multiple vaccine candidates in the laboratory based on the induction of neutralizing antibodies, including inactivated whole viruses, viral vector vaccines, and DNA vaccines, some of which have demonstrated their efficacy in animal models. Because MERS-CoV has been a case epidemic, there are still no vaccine candidates in clinical studies, and multiple vaccine candidates, including inactivated whole virus, live attenuated virus, viral vector vaccines, subunit vaccines, and DNA vaccines are in laboratory stage development [77, 78]. Based on these prior research efforts, when the SARS-CoV-2 epidemic began, researchers responded quickly to begin vaccine development. Vaccines based on different approaches, including mRNA, adenoviral vectors, protein subunits, and whole cell inactivated virus vaccines, have been developed for the ongoing widespread SARS-CoV-2 epidemic, and some of them have been urgently approved in many countries [79]. In vaccine development and clinical trials, the level of immune-induced neutralizing antibodies to a vaccine is key to its evaluation of protective efficacy. The conduct of large-scale phase III efficacy trials will become increasingly difficult due to the ethical issues of placebo control in the presence of an effective vaccine available. Therefore, neutralizing antibody levels have been used as an alternative evaluation index for immune protection [80], which has allowed pseudotyped virus-based neutralizing antibody assays to be widely used in vaccine development and clinical evaluation. Pseudotyped virus-based neutralizing antibody titers have been used in almost all preclinical and clinical trials of current vaccine candidates as well as in the follow-up monitoring of immune efficacy. In all the COVID-19 vaccine phase III trials sponsored by US government, neutralizing antibody responses were assessed using one of the pseudotyped virus-based neutralization assays developed by Duke University or Monogram Biosciences [45]. Four of these vaccines have been approved in United States (https://www.fda.gov/emergency-preparedness-and-response/counterterrorism-and-emerging-threats/coronavirus-disease-2019-covid-19). The pseudotyped virus-based neutralization assay has also been employed to investigate the correlation of protection of COVID-19 vaccine, in which the neutralization titers were determined as the strongest marker of protection [81].

7.4 Summary

Pseudotyped viruses are safe, easy to prepare, easy to detect, and highly accessible and have become an important tool in the field of coronavirus research. Due to the use of report gene testing, the pseudotyped-based assay is more likely to achieve the high throughput, which is essential for clinical testing with large-volume samples, so it is also becoming more widely used. Since pseudotyped viruses can be site-directed modified, they have great advantages in terms of infectivity and antigenicity analyses of variants. In particular, it has been widely used in the evaluation of the preventive effect of vaccines on circulating variants. Results from different laboratories or generated in different pseudotyped virus systems may produce different results, and the standardization of the method is particularly important. The application of international standards or traceable internal controls can effectively improve repeatability of the assays and the comparability of results.

Acknowledgments This work was supported by the General Program of National Natural Science Foundation of China [grant number 82172244&82073621] and major project of Study on Pathogenesis and Epidemic Prevention Technology System [2021YFC2302500].

References

1. Najimudeen, S.M., Hassan, M., Cork, S.C., Abdul-Careem, M.F.: Pathogens infectious bronchitis coronavirus infection in chickens: multiple system disease with immune suppression. Pathogens. **9** (2020)
2. Peng, R., Wu, L., Wang, Q., Qi, J., Gao, G.: Cell entry of SARS-CoV-2. Trends Biochem. Sci. **46** (2021). https://doi.org/10.1016/j.tibs.2021.06.001
3. Roujian et al.: Genomic characterisation and epidemiology of 2019 novel coronavirus: implications for virus origins and receptor binding
4. Neuman, B.W., et al.: Ultrastructure of SARS-CoV, FIPV, and MHV Revealed by Electron Cryomicroscopy (2006)
5. Wu, A., et al.: Genome composition and divergence of the novel coronavirus (2019-nCoV) originating in China. Cell Host Microbe. **27**, 325–328 (2020). https://doi.org/10.1016/j.chom.2020.02.001
6. Romano, M., Ruggiero, A., Squeglia, F., Maga, G., Berisio, R.: A structural view of SARS-CoV-2 RNA replication machinery: RNA synthesis, proofreading and final capping. Cells. **9**, 1267 (2020)
7. Asselah, T., Durantel, D., Pasmant, E., Lau, G., Schinazi, R.F.: COVID-19: discovery, diagnostics and drug development. J. Hepatol. **74**, 168–184 (2021). https://doi.org/10.1016/j.jhep.2020.09.031
8. Fehr, A.R., Perlman, S.: In Coronaviruses: Methods and Protocols (eds Helena Jane Maier, Erica Bickerton, & Paul Britton) 1–23 (Springer New York, 2015)
9. Walls, A.C., et al.: Structure, function and antigenicity of the SARS-CoV-2 spike glycoprotein. Cell. **181** (2020)
10. Vilcek, S.: SARS-CoV-2: zoonotic origin of pandemic coronavirus. Acta Virol. **64** (2020)
11. Zc, A., Rdab, C., Jmga, D., Lr, D., Qcab, C.: SARS-CoV-2 cell entry and targeted antiviral development. *Acta pharmaceutica Sinica*. B.

12. Lan, J., et al.: Structure of the SARS-CoV-2 spike receptor-binding domain bound to the ACE2 receptor. Nature.
13. V'kovski, P., Kratzel, A., Steiner, S., Stalder, H., Thiel, V.: Coronavirus biology and replication: implications for SARS-CoV-2. Nat. Rev. Microbiol. **19**, 155–170 (2021). https://doi.org/10.1038/s41579-020-00468-6
14. Jackson, C.B., Farzan, M., Chen, B., Choe, H.: Mechanisms of SARS-CoV-2 entry into cells. Nat. Rev. Mol. Cell Biol. (2021). https://doi.org/10.1038/s41580-021-00418-x
15. Stephens, D.S., Mcelrath, M.J.: COVID-19 and the path to immunity. JAMA The Journal of the American Medical Association. (2020)
16. Marovich, M., Mascola, J., Cohen, M.S.: Monoclonal antibodies for prevention and treatment of COVID-19. JAMA. **324**, 131–132
17. Liu, L., Wang, P., Nair, M.S., Yu, J., Ho, D.D.: Potent neutralizing antibodies directed to multiple epitopes on SARS-CoV-2 spike. Nature. **584** (2020)
18. Rabaan, A.A., Al-Ahmed, S.H., Haque, S., Sah, R., Rodriguez-Morales, A.J.: SARS-CoV-2, SARS-CoV, and MERS-CoV: a comparative overview. Le infezioni in medicina: rivista periodica di eziologia, epidemiologia, diagnostica, clinica e terapia delle patologie infettive. **28**, 174–184 (2020)
19. Hatmal, M., Alshaer, W., Al-Hatamleh, M., Hatmal, M., Plebanski, M.: Comprehensive structural and molecular comparison of spike proteins of SARS-CoV-2, SARS-CoV and MERS-CoV, and their interactions with ACE2. Cell. **9**, 2638 (2020)
20. Yan, L., et al.: Coupling of N7-methyltransferase and 3′-5′ exoribonuclease with SARS-CoV-2 polymerase reveals mechanisms for capping and proofreading. Cell. **184**, 3474–3485 e3411 (2021). https://doi.org/10.1016/j.cell.2021.05.033
21. Smith, E.C., Denison, M.R.: Coronaviruses as DNA wannabes: a new model for the regulation of RNA virus replication fidelity. PLoS Pathog. **9**, e1003760 (2013). https://doi.org/10.1371/journal.ppat.1003760
22. Lin, S., et al.: Crystal structure of SARS-CoV-2 nsp10 bound to nsp14-ExoN domain reveals an exoribonuclease with both structural and functional integrity. Nucleic Acids Res. **49**, 5382–5392 (2021). https://doi.org/10.1093/nar/gkab320
23. Gribble, J., et al.: The coronavirus proofreading exoribonuclease mediates extensive viral recombination. PLoS Pathog. **17**, e1009226 (2021). https://doi.org/10.1371/journal.ppat.1009226
24. Joo, H.H., Kang, H.G., Lee, J.J.I.J.o. E.E.: Social Media, Consumer Behaviour and Information Disclosure: Evidence from the MERS Outbreak in South Korea. **01** (2022)
25. Wang, C., Xia, S., Zhang, P., Zhang, T., Liu, K.: Discovery of hydrocarbon-stapled short α-helical peptides as promising Middle East respiratory syndrome coronavirus (MERS-CoV) fusion inhibitors. J. Med. Chem. **61** (2018)
26. Krause, P.R., Fleming, T.R., Longini, I.M., Peto, R., Henao-Restrepo, A.M.: SARS-CoV-2 variants and vaccines. N. Engl. J. Med. (2021)
27. Hacisuleyman, E., Hale, C., Saito, Y., Blachere, N.E., Darnell, R.B.: Vaccine breakthrough infections with SARS-CoV-2 variants. N. Engl. J. Med. **384**, 2212–2218
28. Fr, A., Atba, B.: Is Omicron the Last SARS-CoV-2 Variant of Concern? (2022)
29. Cao, Y., et al.: BA.2.12.1, BA.4 and BA.5 escape antibodies elicited by omicron infection. Nature. (2022). https://doi.org/10.1038/s41586-022-04980-y
30. Tegally, H., et al.: Emergence of SARS-CoV-2 omicron lineages BA.4 and BA.5 in South Africa. Nat. Med. (2022). https://doi.org/10.1038/s41591-022-01911-2
31. William, C.G., Francesca, F., Keith, G., Peter, T.C., James, T.N.: Pseudotype-based neutralization assays for influenza: a systematic analysis. Front. Immunol. **6**, 161 (2015)
32. Salazar-García, M., et al.: Pseudotyped vesicular stomatitis virus-severe acute respiratory syndrome-Coronavirus-2 spike for the study of variants, vaccines, and therapeutics against coronavirus disease 2019. Front. Microbiol. **12** (2022). https://doi.org/10.3389/fmicb.2021.817200

33. Nie, J., et al.: Establishment and validation of a pseudovirus neutralization assay for SARS-CoV-2. Emerging Microbes and Infections. **9**, 680–686 (2020)
34. Xiong, R., et al.: Establishment of hDPP4 Transgenic and Knock-In Mouse Models and Comparison of Their Susceptibility to MERS-CoV
35. Fujioka, Y., et al.: A Method for the Generation of Pseudotyped Virus Particles Bearing SARS Coronavirus Spike Protein in High Yields. Cold Spring Harbor Laboratory (2021)
36. Zhang, L., Cui, Z., Li, Q., Yu, Y., Wang, Y.: Comparison of 10 Emerging SARS-CoV-2 Variants: Infectivity, Animal Tropism, and Antibody Neutralization (2021)
37. Establishment of replication-competent vesicular stomatitis virus-based recombinant viruses suitable for SARS-CoV-2 entry and neutralization assays. Emerging Microbes and Infections. **9**, 1–24 (2020)
38. Li, Q., Liu, Q., Huang, W., Li, X., Wang, Y.: Current status on the development of pseudoviruses for enveloped viruses. Rev. Med. Virol. **e1963** (2018)
39. Debnath, A.K.: Discovery of highly potent fusion inhibitors with potential Pan-coronavirus activity that effectively inhibit major COVID-19 variants of concern (VOCs) in Pseudovirus-based assays. Viruses. **14** (2021)
40. Hu, J., Gao, Q., He, C., Huang, A., Wang, K.: Development of cell-based pseudovirus entry assay to identify potential viral entry inhibitors and neutralizing antibodies against SARS-CoV-2. Genes & Diseases. (2020)
41. Zhao, G., et al.: A safe and convenient pseudovirus-based inhibition assay to detect neutralizing antibodies and screen for viral entry inhibitors against the novel human coronavirus MERS-CoV. Virol. J. **10**, 266 (2013). https://doi.org/10.1186/1743-422x-10-266
42. Cai, Y., et al.: A bivalent protein targeting glycans and HR1 domain in spike protein potently inhibited infection of SARS-CoV-2 and other human coronaviruses. Cell Biosci. **11**, 128 (2021). https://doi.org/10.1186/s13578-021-00638-w
43. Xia, S., et al.: A pan-coronavirus fusion inhibitor targeting the HR1 domain of human coronavirus spike. Sci. Adv. **5, eaav4580** (2019). https://doi.org/10.1126/sciadv.aav4580
44. Insights into neutralizing antibody responses in individuals exposed to SARS-CoV-2 in Chile. *Science*. Advances. **7**, eabe6855 (2021)
45. Huang, Y., et al.: Calibration of two validated SARS-CoV-2 pseudovirus neutralization assays for COVID-19 vaccine evaluation. Sci. Rep. **11**, 23921 (2021). https://doi.org/10.1038/s41598-021-03154-6
46. Giroglou, T., et al.: Retroviral vectors Pseudotyped with severe acute respiratory syndrome coronavirus S protein. J. Virol. **78**, 9007 (2004)
47. Millet, J.K., Whittaker, G.R.: Murine Leukemia virus (MLV)-based coronavirus spike-pseudotyped particle production and infection. Bio-protocol. **6** (2016)
48. Johnson, M.C., Lyddon, T.D., Suarez, R., Salcedo, B., Ritter, D.G.: Optimized Pseudotyping conditions for the SARS-COV-2 spike glycoprotein. J. Virol. **94** (2020)
49. Gorshkov, K., et al.: Quantum dot-conjugated SARS-CoV-2 spike pseudo-virions enable tracking of angiotensin converting enzyme 2 binding and endocytosis. ACS Nano. **14**, 12234–12247 (2020). https://doi.org/10.1021/acsnano.0c05975
50. Naskalska, A., et al.: Functional severe acute respiratory syndrome coronavirus 2 virus-like particles from insect cells. Front. Microbiol. **12** (2021). https://doi.org/10.3389/fmicb.2021.732998
51. Xie, X., et al.: An infectious cDNA clone of SARS-CoV-2. Cell Host Microbe. (2020)
52. Zhang, S., et al.: Bat and pangolin coronavirus spike glycoprotein structures provide insights into SARS-CoV-2 evolution. Nat. Commun. **12**, 1607 (2021). https://doi.org/10.1038/s41467-021-21767-3
53. Liu, K., Pan, X., Li, L., Yu, F., Wang, Q.: Binding and molecular basis of the bat coronavirus RaTG13 virus to ACE-2 in humans and other species. Cell. **184** (2021)
54. Hoffmann, M., Kleine-Weber, H., Schroeder, S., Krüger, N., Phlmann, S.: SARS-CoV-2 cell entry depends on ACE2 and TMPRSS2 and is blocked by a clinically proven protease inhibitor. Cell. **181** (2020)

55. Zhu, Y., Yu, D., Yan, H., Chong, H., He, Y.: Design of Potent Membrane Fusion Inhibitors against SARS-CoV-2, an Emerging Coronavirus with High Fusogenic Activity (2020)
56. Yang, Y., Du, L., Liu, C., Wang, L., Li, F.: Receptor usage and cell entry of bat coronavirus HKU4 provide insight into bat-to-human transmission of MERS coronavirus. Proc. Natl. Acad. Sci. U. S. A. **111**, 12516–12521 (2014)
57. Yi-Ning, et al.: Entry of Scotophilus Bat Coronavirus-512 and Severe Acute Respiratory Syndrome Coronavirus in Human and Multiple Animal Cells, vol. 8, Pathogens (Basel, Switzerland) (2019)
58. Liu, L., et al.: Natural mutations in the receptor binding domain of spike glycoprotein determine the reactivity of cross-neutralization between palm civet coronavirus and severe acute respiratory syndrome coronavirus. J. Virol. **81**, 4694–4700 (2007)
59. Lau, S., et al.: Receptor usage of a novel bat lineage C Betacoronavirus reveals evolution of Middle East respiratory syndrome-related coronavirus spike proteins for human dipeptidyl peptidase 4 binding. J. Infect. Dis. **2** (2018)
60. Li, Q., et al.: The impact of mutations in SARS-CoV-2 spike on viral infectivity and antigenicity. Cell. **182**, 1284–1294.e1289 (2020). https://doi.org/10.1016/j.cell.2020.07.012
61. Ma, Y., Mao, G., Wu, G., Chen, M., Zhang, X.E.: Dual-fluorescence Labeling Pseudovirus for real-time imaging of single SARS-CoV-2 entry in respiratory epithelial cells. ACS Appl. Mater. Interfaces. **13**, 24477–24486
62. Fukushi, M., et al.: Monitoring of S protein maturation in the endoplasmic reticulum by calnexin is important for the infectivity of severe acute respiratory syndrome coronavirus. J. Virol. **86**, 11745 (2012)
63. Li, Q., et al.: SARS-CoV-2 501Y.V2 variants lack higher infectivity but do have immune escape—ScienceDirect (2021)
64. Cele, S., Gazy, I., Jackson, L., Hwa, S.H., Sigal, A.: Escape of SARS-CoV-2 501Y.V2 variants from neutralization by convalescent plasma. Nature. (2021)
65. Wang, Z., et al.: mRNA vaccine-elicited antibodies to SARS-CoV-2 and circulating variants. Nature. **592**, 616–622 (2021). https://doi.org/10.1038/s41586-021-03324-6
66. Yuan, M., Huang, D., Lee, C., Wu, N.C., Wilson, I.A.: Structural and functional ramifications of antigenic drift in recent SARS-CoV-2 variants.. *bioRxiv : the preprint server for biology.* **2021**(2002), 430500 (2016)
67. Kemp, S.A., Collier, D.A., Datir, R.P., Ferreira, I., Gupta, R.K.: SARS-CoV-2 evolution during treatment of chronic infection. Nature.
68. Yi, C.E., Ba, L., Zhang, L., Ho, D.D., Chen, Z.: Single amino acid substitutions in the severe acute respiratory syndrome coronavirus spike glycoprotein determine viral entry and immunogenicity of a major neutralizing domain. J. Virol. **79**, 11638–11646 (2005)
69. Diego, C., Martin, M.N., Nigel, T.: The Role of Pseudotype Neutralisation Assays in Understanding SARS CoV-2. Oxford Open Immunology (2021)
70. Xiong, H.L., et al.: Robust neutralization assay based on SARS-CoV-2 S-protein-bearing vesicular stomatitis virus (VSV) pseudovirus and ACE2-overexpressing BHK21 cells. Emerg Microbes Infect. **9**, 2105–2113 (2020). https://doi.org/10.1080/22221751.2020.1815589
71. Neerukonda, S.N., Vassell, R., Herrup, R., Liu, S., Weiss, C.D.: Establishment of a well-characterized SARS-CoV-2 lentiviral pseudovirus neutralization assay using 293T cells with stable expression of ACE2 and TMPRSS2. PLoS One. **16**, e0248348 (2021)
72. Donofrio, G., et al.: A simplified SARS-CoV-2 Pseudovirus neutralization assay. Vaccines (Basel). **9** (2021). https://doi.org/10.3390/vaccines9040389
73. Bonjak, B., et al.: Low serum neutralizing anti-SARS-CoV-2 S antibody levels in mildly affected COVID-19 convalescent patients revealed by two different detection methods. Cell. Mol. Immunol.
74. Trinité, B., et al.: SARS-CoV-2 infection elicits a rapid neutralizing antibody response that correlates with disease severity. Sci. Rep. **11**, 2608 (2021). https://doi.org/10.1038/s41598-021-81862-9

75. Liu, L., et al.: High neutralizing antibody titer in intensive care unit patients with COVID-19. Emerging Microbes & Infections. **9**, 1–30 (2020). https://doi.org/10.1080/22221751.2020.1791738
76. Dispinseri, S., et al.: Neutralizing antibody responses to SARS-CoV-2 in symptomatic COVID-19 is persistent and critical for survival. Nat. Commun. **12**, 2670 (2021). https://doi.org/10.1038/s41467-021-22958-8
77. Corti, D., Passini, N., Lanzavecchia, A., Zambon, M.J.J., o. I. & Health, P.: Rapid generation of a human monoclonal antibody to combat Middle East respiratory syndrome. **9**, 231–235 (2016)
78. Modjarrad, Vaccine, K.J.: MERS-CoV vaccine candidates in development: The current landscape. **34**, 2982–2987 (2016)
79. Immunological mechanisms of vaccine-induced protection against COVID-19 in humans. Nat. Rev. Immunol.
80. Sadarangani, M., Marchant, A., Kollmann, T.R.: Immunological mechanisms of vaccine-induced protection against COVID-19 in humans. Nat. Rev. Immunol. **21**, 475–484 (2021). https://doi.org/10.1038/s41577-021-00578-z
81. Fong, Y., et al.: Immune correlates analysis of a single Ad26.COV2.S dose in the ENSEMBLE COVID-19 Vaccine efficacy clinical trial. medRxiv. (2022). https://doi.org/10.1101/2022.04.06.22272763

Chapter 8
Pseudotyped Viruses for Influenza

Joanne Marie M. Del Rosario, Kelly A. S. da Costa, and Nigel J. Temperton

Abstract We have developed an influenza hemagglutinin (HA) pseudotype (PV) library encompassing all influenza A (IAV) subtypes from HA1-HA18, influenza B (IBV) subtypes (both lineages), representative influenza C (ICV), and influenza D (IDV) viruses. These influenza HA (or hemagglutinin-esterase fusion (HEF) for ICV and IDV) pseudotypes have been used in a pseudotype microneutralization assay (pMN), an optimized luciferase reporter assay, that is highly sensitive and specific for detecting neutralizing antibodies against influenza viruses. This has been an invaluable tool in detecting the humoral immune response against specific hemagglutinin or hemagglutinin-esterase fusion proteins for IAV to IDV in serum samples and for screening antibodies for their neutralizing abilities. Additionally, we have also produced influenza neuraminidase (NA) pseudotypes for IAV N1-N9 subtypes and IBV lineages. We have utilized these NA-PV as surrogate antigens in in vitro assays to assess vaccine immunogenicity. These NA PV have been employed as the source of neuraminidase enzyme activity in a pseudotype enzyme-linked lectin assay (pELLA) that is able to measure neuraminidase inhibition (NI) titers of reference antisera, monoclonal antibodies, and postvaccination sera. Here we show the production of influenza HA, HEF, and NA PV and their employment as substitutes for wild-type viruses in influenza serological and neutralization assays. We also introduce AutoPlate, an easily accessible web app that can analyze data from pMN and pELLA quickly and efficiently, plotting inhibition curves and calculating half-maximal concentration (IC_{50}) neutralizing antibody titers. These serological techniques coupled with user-friendly analysis tools are faster, safer, inexpensive alternatives to classical influenza assays while also offering the reliability and reproducibility to advance influenza research and make it more accessible to laboratories around the world.

J. M. M. Del Rosario · K. A. S. da Costa · N. J. Temperton (✉)
Viral Pseudotype Unit, Medway School of Pharmacy, University of Kent and Greenwich at Medway, Chatham, UK
e-mail: N.Temperton@kent.ac.uk

Y. Wang (ed.), *Pseudotyped Viruses*, Advances in Experimental Medicine and Biology 1407, https://doi.org/10.1007/978-981-99-0113-5_8

Keywords Influenza · Hemagglutinin · Neuraminidase · Pseudotypes · Neutralization · ELLA · Influenza vaccine · Influenza immunity · AutoPlate

Abbreviations

HA	Hemagglutinin
HEF	Hemagglutinin-esterase fusion
HPAI	Highly pathogenic avian influenza
HRP	Horseradish peroxidase
IAV	Influenza A virus
IBV	Influenza B virus
IC_{50}	Half-maximal inhibitory concentration
ICV	Influenza C virus
ID_{50}	Half-maximal inhibitory dilution
IDV	Influenza D virus
IIV	Inactivated influenza vaccine
LAIV	Live attenuated influenza vaccine
LPA	Lectin peanut agglutinin
LPAI	Low pathogenic avian influenza
LV	Lentiviral vectors
NA	Neuraminidase
NI	Neuraminidase inhibition
OD	Optical density
pELLA	Pseudotype enzyme-linked lectin assay
pMN	Pseudotype microneutralization assay
PV	Pseudotype virus
RIV	Recombinant influenza vaccine
SA	Sialic acid

8.1 Introduction

Influenza viruses belong to the *Orthomyxoviridae* family and have segmented, negative sense, single-stranded, enveloped RNA as genetic material [1, 2]. Within this family, there are three influenza types that circulate in humans, influenza A, B, and C [3, 4]. Influenza D, on the other hand, mainly affects cattle and swine with the latter being a preferred reservoir [5]. Influenza A virus (IAV) is classified into subtypes, while influenza B virus (IBV) is divided into lineages based on host range and different combinations of glycoproteins on their surface, most notably HA and NA [6]. IAV and IBV are ever-present threats, causing seasonal epidemics worldwide. Influenza seasons in both the northern and southern hemispheres cause morbidity and mortality to afflicted populations as well as considerable burden to economies [7–9].

The IAV and IBV lipid membrane surfaces are dominated by influenza's two major membrane glycoproteins, hemagglutinin (HA) and neuraminidase (NA) [6, 10, 11] (Fig. 8.1). The surface glycoprotein HA is responsible for viral

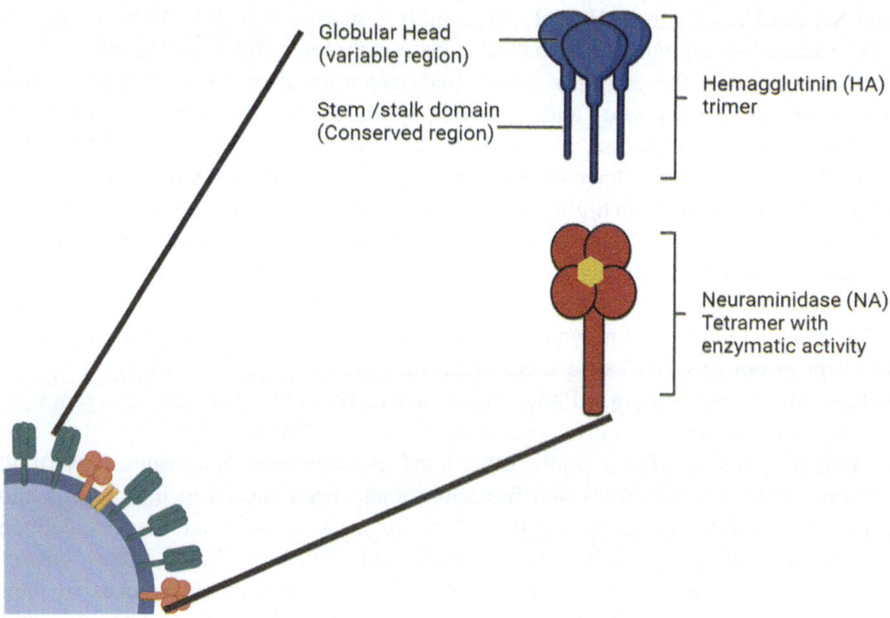

Fig. 8.1 The hemagglutinin (HA) and neuraminidase (NA) glycoproteins present on the surface of influenza A (IAV) and B (IBV) viruses. These proteins are responsible for viral entry and egress, respectively, and are the focus of studies to ascertain mechanisms of neutralization and inhibition and assess antiviral capacity of anti-influenza antibodies and drugs. (Image created using Biorender)

cell attachment and entry into host cells via sialic acid receptors [6, 10], while NA is known for its enzymatic activity that enables the release of virus progeny during the later stages of infection [11–13]. Alternatively, influenza C virus (ICV) and influenza D virus (IDV) have a single surface glycoprotein called hemagglutinin-esterase fusion (HEF) replacing HA and NA found in IAV and IBV [14]. ICV and IDV also express the enzyme esterase that has a similar enzymatic function to neuraminidase.

Influenza A viruses (IAV) are known to cause the most severe and common infections in its hosts compared to IBV and ICV as it is able to infect a wide host range from humans to other mammals, in addition to its known reservoir, the avian species [15, 16]. Influenza A viruses can be classified into combinations of 18 HA and 11 NA subtypes. These subtypes are clustered into two phylogenetic groups: Group 1 comprising H1, H2, H5, H6, H8, H9, H11, H12, H13, H16, H17, and H18, and Group 2 comprising H3, H4, H7, H10, H14, and H15 [17]. IAV zoonoses from wild birds to domestic poultry have been identified as the leading cause of its spread into mammals, specifically swine and humans [18]. These avian zoonotic viruses are not readily transmitted from their natural avian reservoir to humans, but nonetheless this has still occurred in the past, with a virus undergoing antigenic drift or, worse, antigenic shift with possible acquisition of new genes from influenza viruses from other species leading to the emergence of a novel virus that can start a pandemic [18–20]. While there are many genetically distinct subtypes from recombination of HA and NA in IAV, only three HA (H1, H2, and H3) and two NA (N1 and N2) subtypes have caused human pandemics that have been recorded in history [18, 20, 21].

In recent years, the spread of highly pathogenic avian influenza (HPAI) viruses such as H5 and H7 in poultry and wild bird populations has led to devastating effects with flocks being culled and, in the case of human spillovers, death [22–25]. HPAI viruses that are known to have the requisite polybasic insertional mutation at the HA cleavage site making them highly pathogenic in poultry are currently restricted to the H5 and H7 subtypes [26, 27]. Although these HPAI have caused sporadic deaths in humans, none of these strains was easily transmitted from person to person [26, 28–30], and their acquisition of genes from other circulating influenza viruses does not seem apparent [21]. It is known that these viruses require further adaptation in order to adapt to humans and cause widespread transmission [21, 25]. Infection by low pathogenic avian influenza (LPAI) viruses such as H9 or H10 has also been reported, but their effects are not as pronounced as H5 or H7 HPAI [31, 32].

Influenza B virus (IBV), on the other hand, is responsible for around 25% of all seasonal influenza infections and has additionally been shown to infect seals and pigs [33–35]. IBV co-circulates with IAV in humans as two distinct lineages, defined as the B/Yamagata/16/1988-like viruses and B/Victoria/2/1987-like viruses [36]. The two lineages are defined by genetic and antigenic differences of their major surface glycoprotein hemagglutinin (HA). Probably due to low reporting during the COVID pandemic and a host of other factors, only B/Victoria-like lineage infections have been diagnosed in the last 3 years [37].

Vaccination remains the cornerstone of influenza prevention and treatment. Commonly, human flu vaccines are designed to protect against four types of influenza virus that officially designated laboratories around the world have predicted to be in circulation during the upcoming season [33]. There are different types of flu vaccines available, and there are recommendations for use for each one of them. Available influenza vaccines include quadrivalent inactivated influenza vaccine [IIV4], recombinant influenza vaccine [RIV4], or live attenuated influenza vaccine (LAIV4) [33, 38, 39]. No preference is expressed for any influenza vaccine over another, but there are recommendations about who would be better suited for each type of vaccine accounting for factors such as age, location, and prior exposure.

Seasonal influenza vaccines must protect against and restrict virus transmission of H1N1, H3N2, and influenza B viral strains from both Victoria and Yamagata lineages currently circulating in humans globally. These candidate vaccine viruses are identified using painstaking, laborious, and very time-critical methods the season prior. In parallel, discovery of new vaccine antigens that could pave the way for longer protection that would last more than a single season and can cover a broad range of IAV and IBV viruses is now the holy grail of influenza vaccine research [40, 41].

The hemagglutination inhibition assay (HI) and the single radial hemolysis assay (SRH) have been the gold standards for assaying seroprotective titers and amounts of specific antigens (mostly hemagglutinin), respectively, in vaccines. However, these classical methods have many limitations [42, 43]. As such, innovative methods that can improve on these traditional assays and provide advantages such as ease of production and access, safety, reproducibility, and specificity are in the works. We have developed and fine-tuned an influenza pseudotyping platform technology that utilizes lentiviral vectors (LV) to act as surrogate viruses that can be used to screen vaccination sera, small molecule drugs, and monoclonal and polyclonal antibodies, to name a few, that can target the influenza hemagglutinin head and stem domains and neuraminidase [44–46]. These pseudotype viruses (PV) undergo abortive replication and do not give rise to replication-competent progeny [47–49]. While it is logistically possible to deal with low pathogenic strains of influenza, studies on novel strains and those that are not part of the seasonal influenza repertoire require higher containment facilities as well as highly trained and qualified personnel, hampering discovery and innovation.

Together with the HI assay, virus neutralization assays are standard in influenza research and development laboratories and are employed to detect immune responses against influenza elicited by vaccination or natural exposure in human and animal models. Using influenza PV in place of wild-type viruses, we have developed and optimized a pseudotype neutralization assay (pMN) that can quantify viral neutralization by serum neutralizing antibodies [44, 45, 50]. Unlike the HI which only detects anti-HA head antibodies that bind around the receptor binding site, the pMN can also detect HA stalk directed antibodies, making it more sensitive and versatile [51–56]. The pMN can be applied to many enveloped RNA viruses as it measures how much of the virus is able to transduce cells in the presence of an inhibitor which can be serum, antibodies, small molecules, drugs, or antivirals. In recent years, the pMN has also become crucial to characterizing functional immune responses during the ongoing SARS-CoV-2 pandemic [57, 58], indicating its remarkable applicability to new and emerging viruses, including those on the WHO Blueprint list.

In addition to HA neutralization, assays that measure neuraminidase inhibition are also important in future vaccine approaches. Neuraminidase has been a successful target for antiviral drugs and has been shown to control influenza infection in vivo [59–61]. Together with HA, NA is considered a prime candidate to be part of a "universal" or "cross-subtype" influenza vaccine [40, 41, 62]. Currently licensed inactivated influenza vaccines contain NA, but the quality and stability of NA in these preparations have not been fully investigated [63]. Anti-NA antibodies are not considered as classically neutralizing as they do not block infection but instead reduce viral spread [64]. A MUNANA substrate-based assay [65] has been utilized in the past to indirectly measure neuraminidase enzymatic activity; however, this assay employs hazardous chemicals making it unsuitable for day to day testing in research laboratories. Safer and more straightforward alternatives such as the enzyme-linked lectin assay (ELLA) [66–68] and fluorescence-linked MUNANA based assays (e.g., NA-Star/NA Fluor™) have since been developed to quantify NA enzymatic activity as an indirect measure of antibody mediated inhibition of viral egress from infected cells.

Neuraminidase PV have already been successfully used in place of reassortant virus or Triton X-treated wild-type virus in the ELLA assay for N1 and N2 subtypes [67]. Therefore, we have produced an NA-PV library encompassing IAV N1-N11 and IBV from the Victoria-like (B/Vic) and Yamagata-like (B/Yam) lineages for use in a pseudotype ELLA (pELLA) assay. Like their HA PV counterparts, these NA-PV can be effectively employed in pELLA to assess NA inhibition of reference anti-NA antisera, monoclonal antibodies (mAb), and anti-NA antibodies generated through vaccination and/or natural immunity without the requirement for containment above BSL 2.

After the pMN and pELLA assays have been run, the next challenge is to analyze the data and convert them into meaningful results showing neutralization potency and viral inhibition of immune responses. These are usually reported as half-maximal concentration (IC_{50}) in the case of antibodies or half-maximal dilution (ID_{50}) for antisera. This is obtained by measuring neutralization along a dilution series of antibody, serum, or drug in log scale, giving rise to a dose-response curve following a classic S shape which is well represented by the 4-parameter log-logistic regression curve [69]. IC_{50}/ID_{50} is then defined as the concentration/dilution which gives 50% neutralization or inhibition of the curve.

For the pMN, the main output is raw luminescence (or fluorescence) data that can be exported as CSV or Excel files for ease. Values are shown as relative luminescence units (RLU) for each well [45, 50, 51]. For pELLA, it is optical density (OD) collected at 450 nm. These raw outputs have to be copy-pasted from their source files to a chosen statistics or software package where reformatting is required and errors can accumulate. There are proprietary and open-source tools available for these analyses, but most of them require a bioinformatics or coding background to use, while some of the popular software come with expensive licenses.

As an alternative approach to data analysis that we have evaluated and tested with both the pMN and pELLA, we present AutoPlate [70]. AutoPlate employs a user-friendly interface which involves adding experimental metadata to the system with a few clicks bypassing the need for opening individual files and copying blocks of raw data, automatically reformatting the data into the standard pMN and pELLA plate layouts, and performing statistical analysis. AutoPlate produces publication-ready figures and also allows users to export data for further analysis into R or GraphPad prism. AutoPlate is open source and can be accessed through an online Shiny app or installed as an R package that is available at https://github.com/PhilPalmer/AutoPlate.

In this book chapter, we highlight the production and use of lentiviral vectors pseudotyped with influenza hemagglutinins in microneutralization assays (pMN) and influenza neuraminidase in enzyme lectin-linked assays (pELLA) as a safe and sensitive alternative to study specific antibody responses elicited by natural influenza infection or vaccination against certain antigens (Fig. 8.2). These versatile tools allow evaluation of broad and cross-subtype neutralizing antibody responses that may inform further preclinical studies involving antigen discovery, vaccination dosing regimens in animal models, and predictive antigen and vaccine designs for evolving viruses and future pandemics.

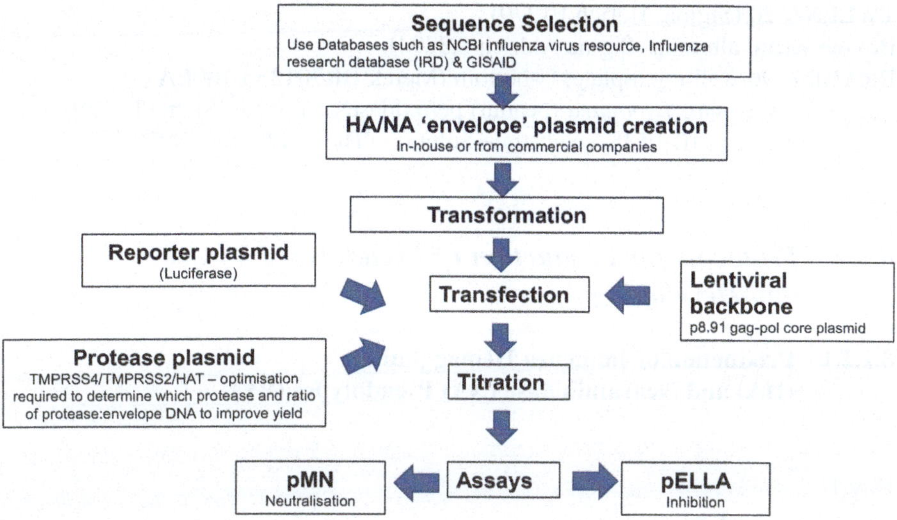

Fig. 8.2 Overall workflow for influenza pseudotype virus (PV) production and downstream assays. PV can be employed to assess anti-influenza neutralization (pMN) and inhibition (pELLA)

8.2 Production of Pseudotyped Viruses and Developing Assays Based on Pseudotyped Viruses

8.2.1 Materials

HEK293T/17 cells (ATCC: CRL-11268[a]).
MDCKII cells (ECACC: 00062107): required as target cells for H17/18 PV.
ST cells (ATCC: CRL-1746™): required as target cells for ICV/IDV HEF PV.
Dulbecco's Modified Essential Medium (DMEM) (PANBiotech P04-04510).
Heat-inactivated fetal bovine serum (PANBiotech P30-8500).
Penicillin-streptomycin (PenStrep) (Sigma Aldrich, Dorset, UK P4333).
Opti-MEM™ (Thermo Fisher Scientific 31985062).
FuGENE® HD Transfection Reagent (ProMega E2312).
Exogenous Neuraminidase (Sigma N2876).
Nunc F96 MicroWell white opaque polystyrene plates (Thermo Fisher Scientific 136101).
VACUSIP benchtop aspiration system (Integra 159010).
Bright-Glo™ luciferase assay system (Promega PR-E2650).
Phosphate buffered saline (Sigma D8662).
GloMax® Navigator (ProMega GM2000).
Nunc Maxisorp™ flat-bottom 96-well plates (Thermo Fisher 44-2404-21).
Fetuin (Sigma F3385).
KPL coating buffer (Sera Care 50-84-00).

TWEEN® 20 (Sigma Aldrich P1379).
Bovine serum albumin (Sigma Aldrich A3294).
BRAND® 96-well microplate, U-bottom (Merck BR701330-100EA).
Lectin from *Arachis hypogaea* (peanut) peroxidase conjugate (Sigma L7759).
1-Step™ Ultra TMB-ELISA Substrate Solution (Thermo Fisher 34029).

8.2.2 Protocols for Production of Pseudotyped Viruses and Developing Assays

8.2.2.1 Production of Influenza Hemagglutinin (HA) and Neuraminidase (NA) Pseudotypes (PV)

Day 1

1. Seed 4 × 10^5 HEK 293 T/17 cells per well of a 6-well plate in 2 mL complete DMEM (10% heat-inactivated FBS, 1% PenStrep).
2. Incubate cells at 37 °C, 5% CO_2 overnight.

Day 2

3. Replace media with fresh 2 mL complete DMEM per well. Incubate plates at 37 °C, 5% CO_2.
4. Prepare transfection mixture as per Table 8.1. Briefly, the following plasmids are needed depending on the desired PV: HA encoding plasmid, NA encoding plasmid, protease-encoding plasmid, luciferase reporter plasmid pCSFLW [71], and p8.91 gag-pol (Gag-Pol expression plasmid [72]). For ICV/IDV, a HEF encoding plasmid is required.

   Note 1: For transfection of low pathogenicity avian influenza (LPAI) and other subtypes with a monobasic cleavage site, an additional plasmid expressing

Table 8.1 Amounts of influenza HA and/or HA-NA transfection components needed per well of a 6-well plate

Solutions/plasmids	Amount	Type of PV
OptiMEM	100 μL	HA, HA-NA
p8.91	250 ng	HA, HA-NA
pCSFLW	375 ng	HA, HA-NA
HA (pEVAC)	10 ng (50 ng for HA18)	HA
HA (pI.18/phCMV1)	50–500 ng	HA
NA (pEVAC)	10 ng	HA-NA
NA (pI.18/phCMV1)	50–500 ng	HA-NA
Protease-encoding plasmid	2.5–500 ng	LPAI HA
FuGENE® HD	3 μL per μg total plasmid DNA	HA, HA-NA

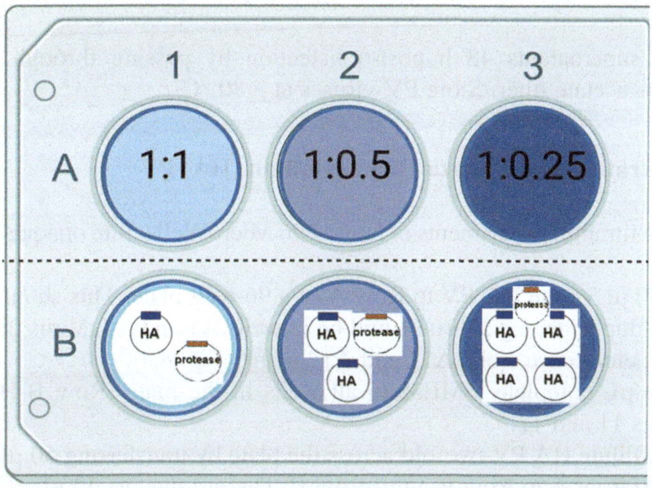

Fig. 8.3 PV yield optimization by changing the ratio of protease plasmid to HA plasmid DNA. PV titers can be increased by transfecting in a "checkerboard" approach with different proteases (e.g., HAT, TMPRSS4, and TMPRSS2). Protease plasmid is added at a ratio of 1:1, 1:0.5, and 1:0.25 to HA plasmid DNA for rapid optimization in a 6-well plate format. (Image created in BioRender)

either type II transmembrane protease serine 2 (TMPRSS2) [73], type IV transmembrane protease serine 4 (TMPRSS4) [74], or human airway trypsin-like protease (HAT) [73] is also needed. Optimization of amount of protease to be added may be required [44] (Fig. 8.3).

Note 2: For NA-PV, we found that co-transfection with 10 ng H11 encoding A/red shoveler/Chile/C14653/2016 (H11) in our plasmid of choice (pEVAC) with 10 ng NA encoding plasmid in pEVAC produces the best results when used in pELLA [46].

5. Combine components from Table 8.1 in OptiMEM as indicated and add FuGENE® HD dropwise followed by incubation for 15 min.
6. Add the plasmid DNA-OptiMEM mixture to the cells with constant swirling.

Note 3: HA and NA are presented (Table 8.1) in different plasmid vectors (i.e., pEVAC, pI.18, phCMV1) to demonstrate that the choice of plasmid will impact the amount of DNA required for transfection to obtain optimal PV production titers.

Day 3

7. At least 8 h post-transfection, add 1 unit of exogenous bacterial neuraminidase per well for HA PV only (with the exception of the H18 subtype which requires an N11 plasmid). There is no need to add this to HA-NA-PV transfection mixtures.

Day 4

8. Collect supernatants 48 h post-transfection by passing through a 0.45 μm cellulose acetate filter. Store PV viruses at −80 °C.

8.2.2.2 Titration of Influenza Hemagglutinin (HA) PV

1. Perform titration experiments in Nunc F96 MicroWell white opaque polystyrene plates.
2. Add 100 μL of the HA PV to Row A of a 96-well plate. This should ideally be done in duplicate; e.g., Virus 1 will be in wells A1 and A2, Virus 2 in wells A3 and A4, etc. Leave wells A11 and A12 blank (Fig. 8.4).
3. Add 50 μL complete DMEM to all wells in the plate (Row B-H) including Columns 11 and 12.
4. Serially dilute HA PV twofold across the plate by transferring 50 μL of the HA PV from Row A to Row B, until Row H. Discard the last 50 μL.
5. Add 100 μL 1 × 10^4 HEK 293 T cells to each well (for H17/H18, use MDCKII, and for ICV/IDV, use ST cells).
6. Incubate plates at 37 °C, 5% CO_2 for 48 h.
7. After 48 h, discard liquid on the plate using VACUSIP.
8. Thaw Bright-Glo® luciferase assay substrate and dilute 1:1 with PBS.
9. Add diluted 25 μL Bright-Glo® to each well. Incubate plates for 5 min at room temperature.
10. Read plates on the GloMax® Navigator (ProMega) using the Promega GloMax® Luminescence Quick-Read protocol.
11. Determine viral pseudotype titer in relative luminescence units/mL (RLU/mL).

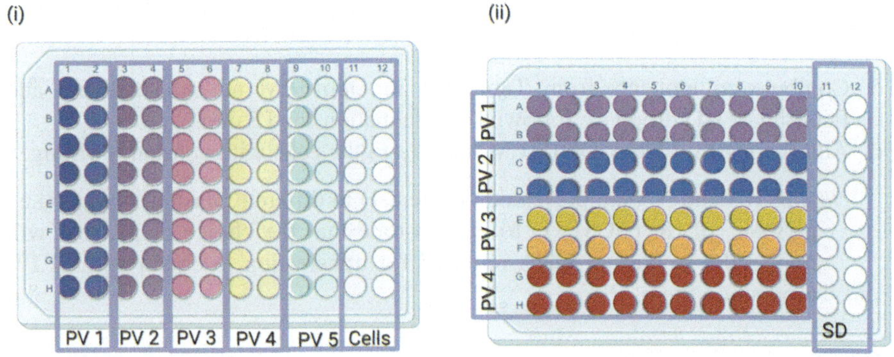

Fig. 8.4 Plate layout for titration of HA and NA-PV. (i) HA PV titration. It is recommended to titrate a maximum of 5 HA PV per plate with inclusion of cell-only controls. (ii) NA-PV titration. It is recommended to titrate a maximum of 4 NA-PV per plate with "blank" sample diluent (SD) only controls

Note 4: Luminescence output from the GloMax® Navigator is in RLU; therefore, dilution factors must be applied to raw data to calculate for RLU/mL. For example, for 50 μL of pseudotype virus added, conversion to RLU/mL requires multiplication of the raw values with 20; the next values from the twofold dilution require multiplication by 40, the next by 80, and so on.

8.2.2.3 Titration of H11_Neuraminidase (H11_NA) PV via Enzyme Linked Lectin Assay (pELLA)

1. Coat clear Nunc Maxisorp™ flat-bottom 96-well plates with 100 μL per well of 25 μg/mL fetuin in 1X KPL coating buffer (diluted from 10X stock). Incubate overnight at 4 °C.
2. The next day, wash plates three times with wash buffer (WB) (0.5% (v/v) Tween-20 in PBS).
3. Add 240 μL of H11_NA(X) (X indicating any N1-N9) PV to well A1 of a U-bottom 96-well mixing plate. Subsequent virus should be added to well B1, then to C1, and so forth.
4. Add 120 μL sample diluent (SD) (1% (w/v) bovine serum albumin (BSA), 0.5% (v/v) Tween-20 in PBS) to all other wells of the mixing plate.
5. Serially dilute H11_NA(X) PV twofold by transferring 120 μL of PV from well A1 to well A2, then from well A2 to well A3, and so forth, across the plate until Column 10. Do this for all other rows containing PV (Fig. 8.4).
6. Discard the last wash from the fetuin plate. Transfer 50 μL of the PV dilutions from the mixing plate to two rows of fetuin plate. Each row of PV from the mixing plate will account for duplicate rows in the fetuin plate.
7. Leave Columns 11 and 12 blank, as these should only contain SD (no PV control).
8. Add 50 μL SD to all wells. Total volume of liquid in each well should be 100 μL.
9. Incubate plates overnight at 37 °C, 5% CO_2.
10. The next day, wash plates six times with WB.
11. Add 100 μL of conjugate (1 μg/mL lectin from *Arachis hypogaea* (peanut) peroxidase conjugate in conjugate diluent (1% (w/v) BSA in PBS)) to all wells.
12. Incubate plates at room temperature for 2 h with shaking (225 rpm).
13. Wash plates three times with WB.
14. Add 100 μL 1-Step™ Ultra TMB Substrate and incubate in the dark with shaking (225 rpm) for 10 min.
15. Stop reaction by addition of 100 μL 0.1 M H_2SO_4 per well.
16. Read optical density at 450 nm (OD_{450}) using a microplate reader.
17. Normalize readings to 100% and 0% OD_{450}, and the dilution that results in 90% OD_{450} is selected as the PV dilution input for inhibition assays.

8.2.2.4 Pseudotype Microneutralization (pMN) Assay Using HA PV (Fig. 8.5)

1. Dilute samples to starting concentrations. Monoclonal antibody concentrations used are usually in the range of 0.5 ng/mL–1000 ng/mL, and serum and antiserum samples are recommended to be initially diluted 1:20 or 1:50 in 50 μL complete DMEM.
2. Similar to HA PV titration, 100 μL of the samples are serially diluted twofold across a Nunc F96 MicroWell white opaque polystyrene plate starting from well A2.
3. Leave Column 1 for controls, with A1 to E1 for HA PV-only controls and F1 to H1 for cell-only controls.
4. Add 50 μL of PV at a titer of 1.0×10^6 RLU/well as determined via titration to all wells except for F1-H1.

   Note 5: Adding the PV to the mAb or serum dilutions will double the final dilution of mAb or sera. For example, if you started with a 1:20 dilution, the final dilution that would be used for computations would be 1:40, to take into account addition of the PV.

5. Incubate plate for 1 h at 37 °C, 5% CO_2.
6. Afterward, add 50 μL of 1.5×10^4 HEK293T/17 cells to each well (for H17/H18, use MDCKII and for ICV/IDV, use ST cells).
7. Incubate plate for 48 h at 37 °C and 5% CO_2.
8. After 48 h, discard media and add 25 μL Bright-Glo® luciferase assay substrate to each well.

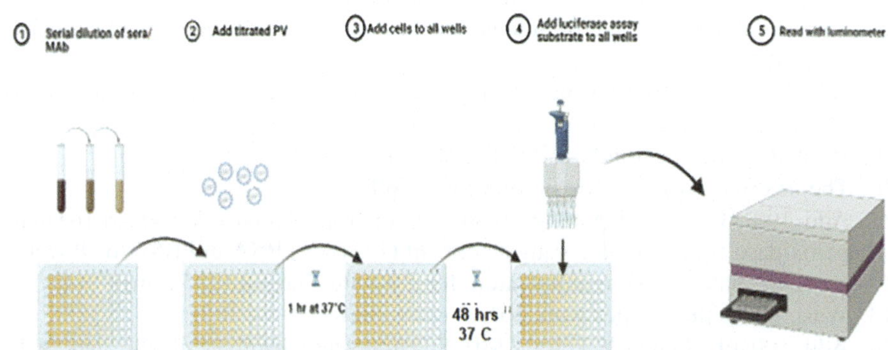

Fig. 8.5 Graphical representation of pMN assay. Sera or mAbs are serially diluted and added to 96-well plate. Titrated PV is diluted and added to all wells, except cell-only controls, and incubated for 1 h before addition of target cells. Plates are incubated for 48 h before addition of luciferase substrate. The final readout of luminescence production is obtained and the IC_{50} determined. (Image created in BioRender)

9. Read plate using the GloMax® Navigator (ProMega) using the Promega GloMax® Luminescence Quick-Read protocol.
10. Calculate half-maximal inhibitory dilution or concentration (IC_{50}). A detailed analysis is described in Ferrara, 2018 [50].

8.2.2.5 Inhibition of H11_NA(X) PV by Antisera and Monoclonal Antibodies via Enzyme-Linked Lectin Assay (pELLA) (Fig. 8.6)

1. Dilute samples to starting concentrations. Monoclonal antibody concentrations used are usually in the range of 0.5 ng/mL–32 μg/mL, and serum and antiserum samples are recommended to be initially diluted 1:10 or 1:20 in SD.
2. Similar to H11_NA(X) PV titration, 240 μL of the samples are serially diluted twofold across a U-bottom 96-well mixing plate with the highest concentrations in Column 1 and the most diluted in Column 10. Leave Columns 11 and 12. Column 11 will be the PV-only control (0% inhibition), and Column 12 will be the SD only control (100% inhibition).

   Note 6: Steps are similar to H11_NA(X) PV titration, but this time instead of PV being titrated, it is the sample (usually mAbs or antisera) that is diluted in the mixing plate and transferred to the fetuin plate.

3. Repeat steps 4–9 as described in Sect. 8.2.2.3.
4. Add 50 μL of the H11_NA(X) PV that resulted in 90% OD_{450} as determined in Sect. 8.2.2.3 to all wells of the fetuin-coated plate except for Column 12 which contains SD only (100% inhibition).
5. Add 50 μL and 100 μL of SD to Columns 11 and 12, respectively.
6. Repeat steps 9–17 as described in Sect. 8.2.2.3.
7. Calculate the IC_{50} as the inverse dilution of serum or antibody concentration that resulted in 50% inhibition of NA activity.

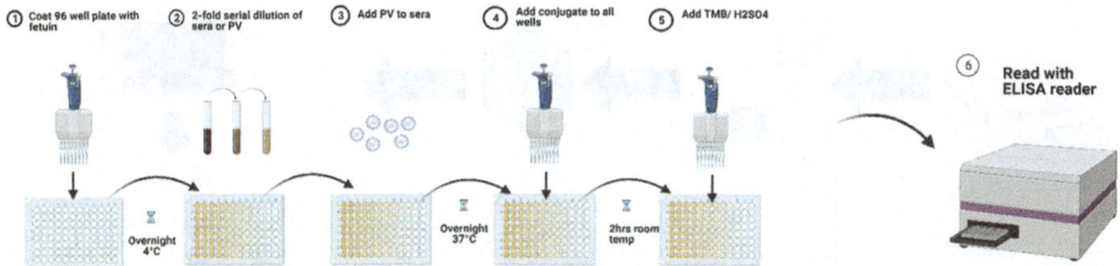

Fig. 8.6 Graphical representation of pELLA assay. Plates are coated with fetuin overnight, and then the sera/titered PV mixture is added to wells; detection of NA enzymatic activity assessed by breakdown of fetuin to galactose, detected by binding to HRP conjugated peanut lectin agglutinate. Final readout of color change following addition of TMB reported as OD_{450} and the IC_{50} determined. (Image created in BioRender)

8.2.2.6 AutoPlate Analysis

To facilitate high-throughput analysis of pMN and pELLA data, the AutoPlate web app was developed (Fig. 8.7). This enables researchers to produce publication-ready figures quickly. AutoPlate is a user-friendly application software which employs R based technology.

1. Access AutoPlate web app by following the link https://shinyapps.io/AutoPlate/.
2. Follow instructions on the quick start guide that can be accessed here: https://philpalmer.github.io/AutoPlate/index.html#web-application-quick-start.
3. Detailed step by step protocol can be accessed in Palmer et al., 2022 [70].

8.3 Commentary

8.3.1 Background

The entire process from PV production to neutralization and inhibition assays has been streamlined into a high-throughput system [44–46] which decreases time required from commencing experiments to obtaining data that produce results comparable to wild-type virus assays. Our system can be used at BSL-2, requiring setup in an MSC class II cabinet to retain sterility for the duration of the assay. This negates the need for BSL-3 facilities and trained staff while maximizing flexibility. This platform makes it possible to produce new/emerging viral strains, including those which are classified as BSL-3, i.e., possible pandemic and/or actual pandemic strains, relatively quickly and for IAV, IBV, ICV, and IDV.

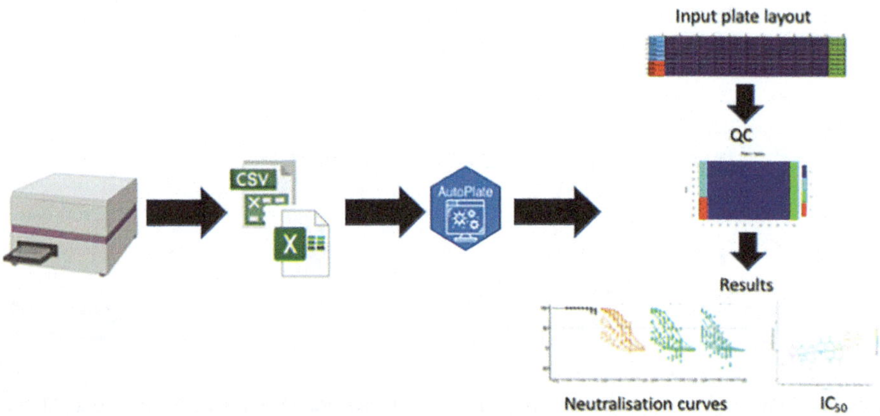

Fig. 8.7 Graphical representation of AutoPlate software workflow. Data is exported from the luminometer as csv files which are annotated and computationally processed within AutoPlate. The output of the software is IC_{50} titers and publication-ready inhibition curves

Transfection is designed to be carried out in a 6-well plate format which allows for optimization that can then be scaled up to tissue culture flasks. The pMN assay described herein can evaluate the neutralizing activity of both polyclonal (sera) and monoclonal anti-influenza antibodies. The system uses readily available cell lines (HEK293T for HA1–16 and IBV, MDCKII for HA17 and HA18 and ST cells for ICV and IDV) which do not require any modifications. Readout is luminescence-based, and the reporter gene is already incorporated into the lentiviral pseudotype produced. This can also be done using FACS analysis, by encoding a fluorescence-based reporter such as GFP. As the whole HA protein is presented in native conformation, antibodies can bind to both the HA head and the stalk as happens in vivo, an approach which is being assessed for novel vaccine and drug design [17, 75, 76]. The pELLA assay described in this chapter evaluates the ability of anti-influenza polyclonal and monoclonal antibodies as well as NA inhibitory drugs to restrict the enzymatic activity of influenza NA and therefore viral egress [46]. The PV used for the pELLA assay display both NA and HA on the surface, and therefore, the effect of steric hindrance of NA via non-neutralizing antibodies which bind to HA stalk exhibits more sensitivity than other methods [77]. As with our pMN assay, results are comparable to traditional assays, e.g., MUNANA substrate-based [46], but without the use of hazardous chemicals or expensive kits, and reassortant virus or Triton X-treated wild-type virus ELLA assays [67]. Additionally, pELLA requires an ELISA plate reader to obtain results, an instrument commonly found in standard laboratories. Both the pMN and pELLA assays utilize a 96-well plate format which can be used to carry out an initial screen of up to 40 samples in duplicate or a serial dilution of five samples per plate with cell-only and PV-only controls and reference antiserum (if required).

8.3.2 Critical Parameters and Troubleshooting

There are several factors which need to be considered to obtain optimal results for pseudotype production and both the pMN and pELLA assays. Firstly, the cells used for transfection and pMN should be kept at a consistent density and passaged three times a week. It is advisable to check the confluence and overall health of the cells before carrying out transfection or pMN and ensure that all reagents are within date and media recipes are followed exactly. Cells should be in a monolayer and at about 70% confluence during transfection. Generally, with experience, operators will be able to determine a maximum number of passages (this would be around passage 30) that will produce consistent results. Accurate counting of cells is critical, and where an automated cell counter is available, the viability of cells can be monitored. Additionally, FBS batch variation should be considered, and where possible, it is best practice to test batch variability on PV titers and purchase a large amount of a favorable lot from manufacturers.

With regard to transfection specifically, the plasmid utilized for the surface glycoprotein (s) “envelope,” i.e., HA/NA plasmid, can influence yield and will affect the amount of plasmid required for PV production. If different plasmids are used for different influenza subtypes, this should be considered. The selection of protease (to cleave HA) is important as this will affect PV yield. We have previously described the proteases which are optimal for each subtype [44]. It should also be noted that some PV will not require a protease plasmid (i.e., those with HPAI HA). The 6-well plate checkerboard approach enables operators to test combinations of the ratio of envelope (HA and/or NA) plasmid to protease plasmid. Titration of PV for neutralization assays is critical, and operators should aim for a minimum titer of 5×10^7 RLU/mL (with the GloMax) and shouldn't be concerned if this requires several attempts of varying the amount of envelope plasmid and amount and type of protease plasmid. When assessing NA enzymatic activity of PV for inhibition assays, this should be calculated as 90% OD_{450}, but we found that the dilution that gives a minimum OD_{450} of 2.5 should be employed for the assay to work consistently. All PV should be stored at −80 °C, and where possible, aliquots should not be freeze-thawed repeatedly especially the NA-PV prepared for the pELLA assay as enzymatic activity decreases with each freeze-thaw cycle.

Specifically, regarding the pELLA assay, it is critical that all reagents are stored correctly and all buffer components are measured accurately, adhering to the recipes stated herein and not kept for longer than a week. Plates need to be coated with fetuin at least 18 h before use although plates can be kept for up to 1 month at 4 °C. However, it is advisable to check that plates are covered sufficiently with a plate cover or tightly fitted lid and that the fetuin has not evaporated prior to use as this can affect results. This will be evident from a consistently low OD obtained from a column or row on the plate. It is also important, as with ELISA assays, that the plate is not allowed to dry out completely at any step as this will also result in lower OD. It is also essential to carry out all washing steps as stipulated in our protocol as this will limit the effects of nonspecific binding. The conjugate diluent should be prepared fresh on the day of the experiment. If using a plate washer, it is advised to make sure this is cleaned as per manufacturer's guidelines at the end of each day and any waste and remaining buffer are removed.

For both the pMN and pELLA, conditions for the assay such as cell density and health, media used, quality control of reagents, incubation times, and strict adherence to protocol, to name a few, should be consistent to allow comparison between timepoints and to assure assay to assay reproducibility. It is also advised that longitudinal samples should be assessed with PV from the same production batches where possible.

Additionally, when performing both assays, the controls detailed in this chapter should be included on every plate. This serves several functions: firstly, when analyzing data, the results will normalize the conditions on each plate, allowing for comparison between plates processed on the same day or different days. Secondly, these controls can highlight any operative or technical issues, especially if a standard antiserum is included, as the IC_{50} can be tracked over long periods of time. Additionally, for both assays, we recommend that a negative control serum is included in each study to assess if a background is observed as this can be corrected in final cutoff calculations. This may be important if using serum from nonhuman species; we have tested and used successfully sera from goat, mice, bat, and ferrets in these assays [44–46, 78]. Where possible, the same luminometer/plate reader should be used for related samples. Instruments should be serviced regularly to ensure they are working optimally. If all these considerations are adhered to and assays are performed systematically, highly robust results are obtained with this versatile high-throughput system.

8.3.3 Understanding Results

The final readout of the PV production titer presented herein for HA pseudotypes is relative luminescence units/ml. The pseudotype viruses have been encoded with firefly luciferase in their lentiviral vector genome, and after entry into target cells, the measured expression of luciferase is proportional to the number of cells transduced. For pMN, the levels of luminescence detected in test wells are compared to the signal from untransduced "cell-only" control wells (representing 100% neutralization) and "PV-only" control wells (representing 0% neutralization), where neutralizing antibodies/sera are not present. A lack of or reduction in luminescence is indicative of neutralizing antibody activity. This indirect measurement permits calculation of neutralizing antibody titers which are reported as IC_{50} dilution. This IC_{50} can then be compared between groups at single or multiple time points following vaccination, treatment, or infection.

The final readout of the pELLA presented herein is relative optical density at 450 nm (OD_{450}). This assay measures the enzymatic activity of NA by assessing the ability of NA to cleave the terminal sialic acid (SA) residue from the fetuin substrate that is coated on test plates. This gives rise to the production of galactose which in turn is recognized specifically by lectin peanut agglutinin (LPA). The LPA used here is conjugated to horseradish peroxidase (HRP) which in the presence of TMB produces a detectable color change which is measured as OD_{450}. The effect of antibodies or drugs which inhibit this enzymatic reaction can be indirectly measured following incubation with the NA-PV. The OD_{450} detected in test wells is compared to the signal from sample diluent-only control wells (representing 100% inhibition) and "PV-only" control wells (representing 0% inhibition), where neutralizing antibodies/sera are not present. A lack of change or reduction in OD_{450} is indicative of the presence of inhibiting antibodies or substances. This indirect measurement permits calculation of inhibition titers which are reported as IC_{50} dilution. As with the pMN assay, this IC_{50} can then be compared between groups at single or multiple time points following vaccination, treatment, or infection. We suggest that all samples are assessed in duplicate with control wells present on each plate in triplicate.

8.3.4 Time Considerations

The lead time for receipt of HA and NA plasmids varies depending on the manufacturer employed, but usually this is approximately 2 weeks. The transfection process requires the cells to be seeded overnight, and PV can be harvested in 48 h or extended to 72 h. The pMN assay is read 48 h after initial setup, and the pELLA assay takes 24 h to complete as it only requires overnight incubation. It is recommended that a plate washer is used for the pELLA assay, especially when more than five plates are assessed at one time to maximize efficiency.

References

1. Palese, P.: Influenza: old and new threats. Nat. Med. **10**, S82–S87 (2004)
2. Bouvier, N.M., Lowen, A.C.: Animal models for influenza virus pathogenesis and transmission. Viruses. **2**, 1530–1563 (2010)
3. Petrova, V.N., Russell, C.A.: The evolution of seasonal influenza viruses. Nat. Rev. Microbiol. **16**, 47–60 (2018)
4. World Health Organization: A checklist for pandemic influenza risk and impact management: building capacity for pandemic response. World Health Organization (2018)
5. Ferguson, L., et al.: Pathogenesis of influenza D virus in cattle. J. Virol. **90**, 5636–5642 (2016)
6. Webster, R.G., Rott, R.: Influenza virus a pathogenicity: the pivotal role of hemagglutinin. Cell. **50**, 665–666 (1987)
7. Neumann, G., Kawaoka, Y.: Transmission of influenza A viruses. Virology. **479–480**, 234–246 (2015)
8. Putri, W.C.W.S., Muscatello, D.J., Stockwell, M.S., Newall, A.T.: Economic burden of seasonal influenza in the United States. Vaccine. **36**, 3960–3966 (2018)
9. Dawood, F.S., et al.: Estimated global mortality associated with the first 12 months of 2009 pandemic influenza A H1N1 virus circulation: a modelling study. Lancet Infect. Dis. **12**, 687–695 (2012)
10. Wiley, D.C.: The structure and function of the hemagglutinin membrane glycoprotein of influenza virus. Annu. Rev. Biochem. **56**, 365–394 (1987)
11. Gottschalk, A.: The influenza virus neuraminidase. Nature. **181**, 377–378 (1958)
12. Gottschalk, A.: Neuraminidase: the specific enzyme of influenza virus and vibrio cholerae. Biochim. Biophys. Acta. **23**, 645–646 (1957)
13. McAuley, J.L., Gilbertson, B.P., Trifkovic, S., Brown, L.E., McKimm-Breschkin, J.L.: Influenza virus neuraminidase structure and functions. Front. Microbiol. **10**, 39 (2019)
14. Hause, B.M., et al.: Isolation of a novel swine influenza virus from Oklahoma in 2011 which is distantly related to human influenza C viruses. PLoS Pathog. **9**, e1003176 (2013)
15. Cauldwell, A.V., Long, J.S., Moncorgé, O., Barclay, W.S.: Viral determinants of influenza A virus host range. J. Gen. Virol. **95**, 1193–1210 (2014)
16. Long, J.S., Mistry, B., Haslam, S.M., Barclay, W.S.: Host and viral determinants of influenza A virus species specificity. Nat. Rev. Microbiol. **17**, 67–81 (2019)
17. Gamblin, S.J., Skehel, J.J.: Influenza hemagglutinin and neuraminidase membrane glycoproteins. J. Biol. Chem. **285**, 28403–28409 (2010)
18. Russell, C.J., Hu, M., Okda, F.A.: Influenza hemagglutinin protein stability, activation, and pandemic risk. Trends Microbiol. **26**, 841–853 (2018)
19. Fineberg, H.V.: Pandemic preparedness and response—lessons from the H1N1 influenza of 2009. N. Engl. J. Med. **370**, 1335–1342 (2014)

20. Rabozzi, G., et al.: Emerging zoonoses: the "one health approach". Saf. Health Work. **3**, 77–83 (2012)
21. Neumann, G., Kawaoka, Y.: Predicting the next influenza pandemics. J. Infect. Dis. **219**, S14–S20 (2019)
22. Zecchin, B., et al.: Evolutionary dynamics of H5 highly pathogenic avian influenza viruses (clade 2.3.4.4B) circulating in Bulgaria in 2019–2021. Viruses. **13**, 2086 (2021)
23. Gao, R., et al.: Human infection with a novel avian-origin influenza A (H7N9) virus. N. Engl. J. Med. **368**, 1888–1897 (2013)
24. Guo, Y., et al.: Analysis of hemagglutinin-mediated entry tropism of H5N1 avian influenza. Virol. J. **6**, 39 (2009)
25. Yu, D., et al.: The re-emergence of highly pathogenic avian influenza H7N9 viruses in humans in mainland China, 2019. Eur. Secur. **24** (2019)
26. Chen, L.-M., et al.: In vitro evolution of H5N1 avian influenza virus toward human-type receptor specificity. Virology. **422**, 105–113 (2012)
27. Stech, O., et al.: Acquisition of a Polybasic Hemagglutinin Cleavage Site by a low-pathogenic avian influenza virus is not sufficient for immediate transformation into a highly pathogenic strain. J. Virol. **83**, 5864–5868 (2009)
28. Chen, M., Chen, M., Tan, Y.: An avian influenza A (H7N9) virus with polybasic amino acid insertion was found in human infection in southern China, Guangxi, February 2017. Infect. Dis. **50**, 71–74 (2018)
29. Huo, X., et al.: Significantly elevated number of human infections with H7N9 virus in Jiangsu in eastern China, October 2016 to January 2017. Eur. Secur. **22** (2017)
30. Imai, M., Kawaoka, Y.: The role of receptor binding specificity in interspecies transmission of influenza viruses. Curr. Opin. Virol. **2**, 160–167 (2012)
31. Taubenberger, J.K., Kash, J.C.: Influenza virus evolution, host adaptation, and pandemic formation. Cell Host Microbe. **7**, 440–451 (2010)
32. Taubenberger, J.K., Morens, D.M.: Influenza: the once and future pandemic. Public Health Rep. **125**, 15–26 (2010)
33. Grohskopf, L.A., et al.: Prevention and control of seasonal influenza with vaccines: recommendations of the Advisory Committee on Immunization Practices—United States, 2018–19 influenza season. **67**, 24 (2018)
34. Iuliano, A.D., et al.: Estimates of global seasonal influenza-associated respiratory mortality: a modelling study. Lancet. **391**, 1285–1300 (2018)
35. Ran, Z., et al.: Domestic pigs are susceptible to infection with influenza B viruses. J. Virol. **89**, 4818–4826 (2015)
36. Hanson, A., et al.: Identification of stabilizing mutations in an H5 hemagglutinin influenza virus protein. J. Virol. **90**, 2981–2992 (2016)
37. Koutsakos, M., Wheatley, A.K., Laurie, K., Kent, S.J., Rockman, S.: Influenza lineage extinction during the COVID-19 pandemic? Nat. Rev. Microbiol. **19**, 741–742 (2021)
38. Cox, R.J., Brokstad, K.A., Ogra, P.: Influenza virus: immunity and vaccination strategies. Comparison of the immune response to inactivated and live, attenuated influenza vaccines. Scand. J. Immunol. **59**, 1–15 (2004)
39. Gianchecchi, E., et al.: How to assess the effectiveness of nasal influenza vaccines? Role and measurement of sIgA in mucosal secretions. Influenza Other Respir. Viruses. **13**, 429–437 (2019)
40. Paules, C.I., Marston, H.D., Eisinger, R.W., Baltimore, D., Fauci, A.S.: The pathway to a universal influenza vaccine. Immunity. **47**, 599–603 (2017)
41. Sautto, G.A., Kirchenbaum, G.A., Ross, T.M.: Towards a universal influenza vaccine: different approaches for one goal. Virol. J. **15**, 17 (2018)
42. Long, B.C., Goldberg, T.L., Swenson, S.L., Erickson, G., Scherba, G.: Adaptation and limitations of established hemagglutination inhibition assays for the detection of porcine anti—swine influenza virus H1N2 antibodies. J. Vet. Diagn. Investig. **16**, 264–270 (2004)

43. Kumar, A., Meldgaard, T.S., Bertholet, S.: Novel platforms for the development of a universal influenza vaccine. Front. Immunol. **9**, 600 (2018)
44. Del Rosario, J.M.M., et al.: Exploiting pan influenza A and pan influenza B Pseudotype libraries for efficient vaccine antigen selection. Vaccine. **9** (2021)
45. Ferrara, F., et al.: Development of lentiviral vectors Pseudotyped with influenza B hemagglutinins: application in vaccine immunogenicity, mAb potency, and Sero-surveillance studies. Front. Immunol. **12** (2021)
46. da Costa, K.A.S., et al.: Influenza A (N1-N9) and Influenza B (B/Victoria and B/Yamagata) Neuraminidase Pseudotypes as tools for pandemic preparedness and improved influenza vaccine. **10** (2022)
47. Garcia, J.-M., Lai, J.C.: Production of influenza pseudotyped lentiviral particles and their use in influenza research and diagnosis: an update. Expert Rev. Anti-Infect. Ther. **9**, 443–455 (2011)
48. Duvergé, A., Negroni, M.: Pseudotyping lentiviral vectors: when the clothes make the virus. Viruses. **12**, 1311 (2020)
49. Naldini, L., et al.: In vivo gene delivery and stable transduction of nondividing cells by a lentiviral vector. Science. **272**, 263–267 (1996)
50. Ferrara, F., Temperton, N.: Pseudotype neutralization assays: from laboratory bench to data analysis. Methods Protoc. **1**, 8 (2018)
51. Del Rosario, J.M.M., et al.: Protection from influenza by intramuscular gene vector delivery of a broadly neutralizing nanobody does not depend on antibody dependent cellular cytotoxicity. Front. Immunol. **11** (2020)
52. Ekiert, D.C., Wilson, I.A.: Broadly neutralizing antibodies against influenza virus and prospects for universal therapies. Curr. Opin. Virol. **2**, 134–141 (2012)
53. Ekiert, D.C., et al.: A highly conserved neutralizing epitope on group 2 influenza A viruses. Science. **333**, 843–850 (2011)
54. Dreyfus, C., et al.: Highly conserved protective epitopes on influenza B viruses. Science. **337**, 1343–1348 (2012)
55. Ekiert, D.C., et al.: Antibody recognition of a highly conserved influenza virus epitope. Science. **324**, 246–251 (2009)
56. Corti, D., et al.: A neutralizing antibody selected from plasma cells that binds to group 1 and group 2 influenza A hemagglutinins. Science. **333**, 850–856 (2011)
57. Nie, J., et al.: Quantification of SARS-CoV-2 neutralizing antibody by a pseudotyped virus-based assay. Nat. Protoc. **15**, 3699–3715 (2020)
58. Sampson, A.T., et al.: Coronavirus Pseudotypes for all circulating human coronaviruses for quantification of cross-neutralizing antibody responses. Viruses. **13** (2021)
59. Rijal, P., et al.: Broadly inhibiting Antineuraminidase monoclonal antibodies induced by trivalent influenza vaccine and H7N9 infection in humans. J. Virol. **94**, 17 (2020)
60. Madsen, A., et al.: Human antibodies targeting influenza B virus neuraminidase active site are broadly protective. Immunity. **53**, 852–863.e7 (2020)
61. Stadlbauer, D., et al.: Broadly protective human antibodies that target the active site of influenza virus neuraminidase. Science. **366**, 499–504 (2019)
62. Estrada, L.D., Schultz-Cherry, S.: Development of a universal influenza vaccine. J. Immunol. **202**, 392–398 (2019)
63. Krammer, F., et al.: NAction! How can neuraminidase-based immunity contribute to better influenza virus vaccines? mBio. **9** (2018)
64. Klasse, P.J., Sattentau, Q.J.: Occupancy and mechanism in antibody-mediated neutralization of animal viruses. J. Gen. Virol. **83**, 2091–2108 (2002)
65. Potier, M., Mameli, L., Bélisle, M., Dallaire, L., Melançon, S.B.: Fluorometric assay of neuraminidase with a sodium (4-methylumbelliferyl-alpha-D-N-acetylneuraminate) substrate. Anal. Biochem. **94**, 287–296 (1979)
66. Eichelberger, M.C., et al.: Comparability of neuraminidase inhibition antibody titers measured by enzyme-linked lectin assay (ELLA) for the analysis of influenza vaccine immunogenicity. Vaccine. **34**, 458–465 (2016)

67. Biuso, F., et al.: Use of lentiviral pseudotypes as an alternative to reassortant or triton X-100-treated wild-type influenza viruses in the neuraminidase inhibition enzyme-linked lectin assay. Influenza Other Respir. Viruses. **13**, 504–516 (2019)
68. Lambré, C.R., Terzidis, H., Greffard, A., Webster, R.G.: Measurement of anti-influenza neuraminidase antibody using a peroxidase-linked lectin and microtitre plates coated with natural substrates. J. Immunol. Methods. **135**, 49–57 (1990)
69. Ritz, C., Baty, F., Streibig, J.C., Gerhard, D.: Dose-response analysis using R. PLoS One. **10**, e0146021 (2016)
70. Palmer, P., et al.: AutoPlate: rapid dose-response curve analysis for biological assays. Front. Immunol. **12** (2022)
71. Temperton, N.J., et al.: A sensitive retroviral pseudotype assay for influenza H5N1-neutralizing antibodies. Influenza Other Respir. Viruses. **1**, 105–112 (2007)
72. Zufferey, R., Nagy, D., Mandel, R.J., Naldini, L., Trono, D.: Multiply attenuated lentiviral vector achieves efficient gene delivery in vivo. Nat. Biotechnol. **15**, 871–875 (1997)
73. Böttcher, E., et al.: Protcolytic activation of influenza viruses by serine proteases TMPRSS2 and HAT from human airway epithelium. J. Virol. **80**, 9896–9898 (2006)
74. Matrosovich, M.N., Matrosovich, T.Y., Gray, T., Roberts, N.A., Klenk, H.-D.: Human and avian influenza viruses target different cell types in cultures of human airway epithelium. Proc. Natl. Acad. Sci. **101**, 4620–4624 (2004)
75. Kirkpatrick, E., Qiu, X., Wilson, P.C., Bahl, J., Krammer, F.: The influenza virus hemagglutinin head evolves faster than the stalk domain. Sci. Rep. **8**, 10432 (2018)
76. Nachbagauer, R., et al.: A chimeric hemagglutinin-based universal influenza virus vaccine approach induces broad and long-lasting immunity in a randomized, placebo-controlled phase I trial. Nat. Med. **27**, 106–114 (2021)
77. Chen, Y.-Q., Lan, L.Y.-L., Huang, M., Henry, C., Wilson, P.C.: Hemagglutinin stalk-reactive antibodies interfere with influenza virus neuraminidase activity by steric hindrance. J. Virol. **93** (2019)
78. Carnell, G.W., Trombetta, C.M., Ferrara, F., Montomoli, E., Temperton, N.J.: Correlation of influenza B haemagglutination Inhibiton, single-radial haemolysis and Pseudotype-based microneutralisation assays for immunogenicity testing of seasonal vaccines. Vaccine. **9** (2021)

Chapter 9
Pseudotyped Virus for *Henipavirus*

Tao Li, Ziteng Liang, Weijin Huang, and Youchun Wang

Abstract The genus *Henipavirus* (HNV) includes two virulent infectious viruses, Nipah virus (NiV) and Hendra virus (HeV), which are the focus of considerable public health research efforts and have been classified as priority infectious diseases by the World Health Organization. Both viruses are high risk and should be handled in biosafety level 4 laboratories. Pseudotyped viruses containing the envelope proteins of HNV viruses have the same envelope protein structure as the authentic viruses; thus, they can mimic the receptor-binding and membrane fusion processes of authentic viruses with host cells and can be handled in biosafety level 2 laboratories. These characteristics enable pseudotyped viruses to be widely used in studies of viral infection mechanisms (packaging, budding, virus attachment, membrane fusion, viral entry, and glycosylation), inhibitory drug screening assays, and monoclonal antibody neutralization characteristics. This review will provide an overview of the progress of research concerning pseudotyped virus packaging systems for NiV and HeV.

Keywords *Henipavirus* · Pseudotyped virus · Packaging system · Membrane fusion · Receptor

T. Li · Z. Liang · W. Huang
Division of HIV/AIDS and Sex-transmitted Virus Vaccines, National Institutes for Food and Drug Control (NIFDC) and WHO Collaborating Center for Standardization and Evaluation of Biologicals, Beijing, China

Y. Wang (✉)
Chinese Academy of Medical Sciences & Peking Union Medical College, Beijing, China

Institute of Medical Biology, Chinese Academy of Medicine Sciences & Peking Union Medical College, Kunming, China
e-mail: wangyc@nifdc.org.cn; wangyc@imbcams.com.cn

Y. Wang (ed.), *Pseudotyped Viruses*, Advances in Experimental Medicine and Biology 1407, https://doi.org/10.1007/978-981-99-0113-5_9

Abbreviations

BSL-2	Biosafety level laboratory-2
BSL-4	Biosafety level laboratory-2
EFNB2	Ephrin B2
EFNB3	Ephrin B3
F	Fusion protein
G	Glycoprotein
GFP	Green fluorescent protein
HeV	Hendra virus
HeVpp	HeV envelope pseudotyped reporter virus
HIV-1	The human immunodeficiency virus type 1
HNV	Henipavirus
L	Polymerase
M	Matrix protein
MOI	Multiplicity of infection
MuLV	Moloney murine leukemia virus
N	Nucleoprotein
NiV	Nipah virus
NiVpp	NiV envelope pseudotyped reporter virus
P	Phosphoprotein
SEAP	Secretory alkaline phosphatase
VLP	Viruslike particle
VSV	Vesicular stomatitis virus
WHO	World Health Organization
βla	β-Lactamase gene

9.1 General Information About Henipaviruses

Nipah virus (NiV) and Hendra virus (HeV) are animal-derived viruses with multiple hosts and high lethality [1, 2]. Based on the classification of complete L protein amino acid sequences of *Paramyxoviridae* by the Committee on Classification of Viruses, NiV and HeV belong to the *Henipavirus* (HNV) genus of *Paramyxovirinae*. The receptors for entry of NiV and HeV into host cells are the ephrin B2 (EFNB2) and EFNB3 proteins [2–4]. NiV and HeV infections cause severe symptoms in animals and humans; the main manifestations are respiratory disease (approximately 5% mortality in pigs) and febrile encephalitis (40%–75% mortality in humans) [5–8]. The World Health Organization (WHO) has classified both viruses as global health problems because of their high pathogenicity and mortality in humans, as well as their zoonotic characteristics with potential for human-to-human transmission. NiV- and HeV-related experimental manipulations must be conducted in biosafety level 4 (BSL-4) controlled facilities. There are currently no commercially available vaccines or therapeutic drugs for either virus. NiV and other henipaviral diseases are listed by the WHO as priority infectious diseases for research, and there is an urgent need to accelerate studies of these viruses [9, 10]. Current diagnostic tests to detect serum and antibodies to NiV and HeV mainly use authentic viruses; thus, safer and effective alternative evaluation methods are needed.

9.1.1 Transmission of HNV Viruses

HNV viruses have high infectivity and mortality. Sequence analysis has shown that NiV has evolved into at least two branches: Bangladeshi and Malaysian [11, 12]. The first outbreak of NiV occurred in Malaysia in 1998–1999. During this outbreak, the virus was transmitted from fruit bats to pigs and from pigs to humans; a subsequent outbreak in Bangladesh involved direct viral transmission from fruit bats to humans, as well as human-to-human transmission [13], as shown in Fig. 9.1. HeV was identified during an outbreak in Australia in 1994; the virus was named after Hendra in the suburb of Brisbane, where several horses and breeders died from severe respiratory disease and human-to-human transmission occurred [14]. HeV is also transmitted by fruit bats; only a few outbreaks have occurred in Australia, probably through horse-to-human and human-to-human transmission after horses have eaten in pastures contaminated by fruit bats [15, 16], as shown in Fig. 9.1. Although the geographic distributions of HeV and NiV transmission have differed, there is considerable sequence similarity between the two viruses; thus, cross-neutralization reactions and similarly high infectivity have been observed [17, 18].

9.1.2 Structure of HNV

HNV has the longest genome in the *Paramyxoviridae* family (approximately 18 kb). The genome encodes six main proteins (Fig. 9.2A), in the following order from the 3′ end of the genome to the 5′ end of the genome: nucleoprotein (N), phosphoprotein (P), matrix protein (M), fusion protein (F), glycoprotein (G)/receptor-binding protein (RBP), and polymerase (L) (3’-N-P-M-F-G (RBP)-L-5′; Fig. 9.2B) [19, 20]. Similar to other paramyxoviruses, HNV has two glycosylated proteins on its surface—the G protein, which is primarily responsible for binding to receptors on the cell surface, and the F protein, which induces fusion between the viral capsid and the cell membrane; both are required for viral infection [21]. Membrane fusion is the first step in HNV virulence, a process that is initiated in a receptor-dependent manner and requires the mediation of the G and F proteins on the envelope surface [22]. The G protein binds to a specific receptor on the host cell surface and undergoes a conformational change that causes the F protein to mediate the virus-cell and cell-cell fusion reactions; this facilitates viral invasion of host cells and leads to

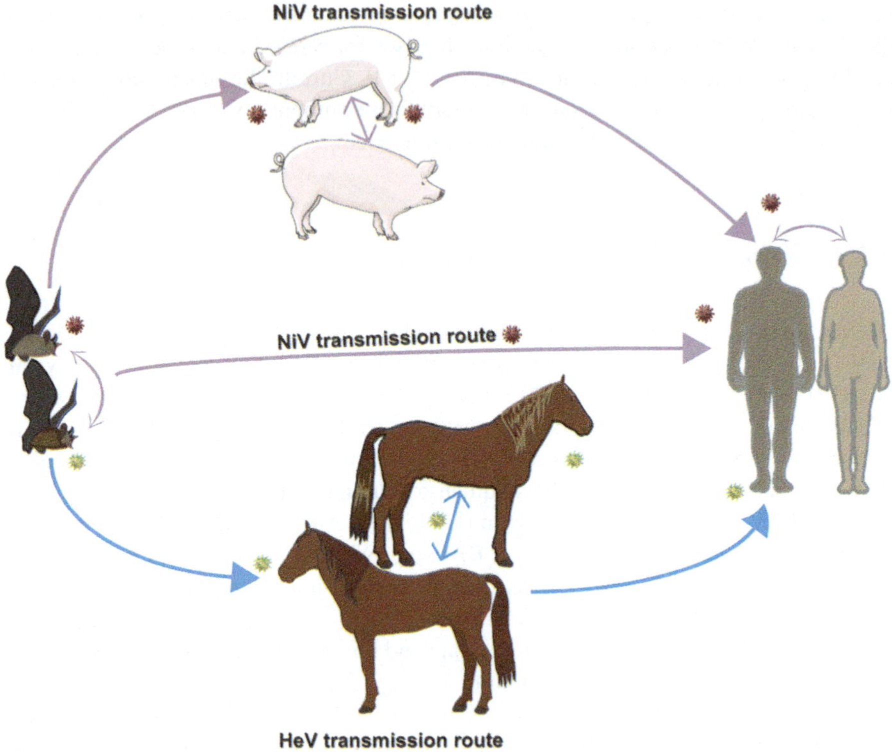

Fig. 9.1 Schematic representation of the transmission routes of HeV and NiV. Top: fruit bats (the hosts of HeV) may contaminate pastures through placentas and excrement, thus spreading HeV to horses; human contact with horses and their excrement leads to infection. Middle and bottom: there are two NiV transmission pathways. First, NiV virus in Malaysia is transmitted from fruit bats to pigs and from pigs to humans. Second, NiV virus in Bangladesh is directly transmitted from fruit bats to humans, followed by human-to-human transmission

syncytium formation. Thus, the F and G glycoproteins are commonly used to construct HNV pseudotyped viruses.

Next, the F and G proteins of HNV will be described in greater detail. The HNV-F protein is a type I transmembrane glycoprotein. The precursor of the HNV-F protein (F0) can be cleaved by cathepsin L into F1 and F2 fragments of appropriate lengths [23, 24]. The N-terminus of the F1 fragment contains a hydrophobic region that is responsible for the fusion of the viral capsid and cell membrane. The two α-helical regions of the HNV-F protein are essential for mediating the fusion of viral capsids with cellular membranes; when synthesized by artificial methods, the peptide sequences of these two α-helical regions can efficiently inhibit virus-mediated cell fusion [25, 26]. Therefore, this region has great potential as a target for antiviral drugs. HNV-G is a type II glycoprotein encoded by the G gene in HNV; it consists of a hydrophilic cytoplasmic tail region, a hydrophobic transmembrane region, a stem domain, and a globular head domain, all of which are connected on the viral surface

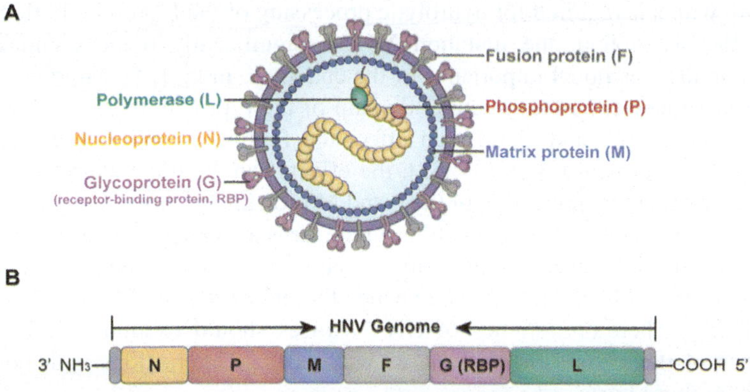

Fig. 9.2 Schematic representation of the major structural proteins of HNV. (**A**) Locations of the six structural proteins. (**B**) The genome encodes nucleoprotein (N), phosphoprotein (P), matrix protein (M), fusion protein (F), glycoprotein (G)/receptor-binding protein (RBP), and polymerase (L) in the order 3'-N-P-M-F-G(RBP)-L-5′. There are untranslated regions between genes, not shown in the figure

by disulfide bonds and stem domain bundles in a tetrameric form [27]. In contrast to other paramyxovirus adhesion proteins, HNV-G has neither neuraminidase activity nor blood-adsorption activity. It has two main functions: mediating the adsorption of viral particles to specific receptors on the surface of target cells and activating HNV-F to initiate cell fusion.

Furthermore, HNV entry occurs through a pH-independent membrane fusion mechanism; viral attachment is mediated by the interaction of the globular head domain of the viral G glycoprotein with the EFNB2 or EFNB3 receptors (functional receptor proteins for HNV) on the surface of target cells [25, 28, 29].

9.1.3 Diversity of HNV

Similar to mutations in the Spike protein of severe acute respiratory syndrome coronavirus 2, dozens of mutations have been reported in HNV-F and HNV-G proteins. Analyses of mutations have demonstrated that the presence of mutations in the G protein receptor-binding domain and F protein transmembrane domain can cause changes in viral receptor-binding and viral fusion ability, respectively; these findings highlight the importance of these two specific regions. For example, analysis of the HeV-G chimeric protein showed that residue 507 in the extracellular domain is associated with efficient binding to EFNB3. Mutations in nearby residues 504 and 505 (i.e., W504A and E505A) were also identified that reduce EFNB3-dependent binding and viral entry without affecting the use of the EFNB2 receptor; another mutation (E533Q) abolishes binding interactions with EFNB2 and EFNB3 [30]. The cytoplasmic tail of the HeV-F protein contains an endocytic shared motif,

YXXPhi, which is required for hydrolytic processing of the F protein. In this region of the HeV-F protein, the mutation Y525A significantly reduces endocytosis, confirming the functional importance of the endocytic motif [21]. Moreover, mutations in the transmembrane structural domain of the F protein (e.g., residues S490, M491, L492, Y498, and G508) can affect the folding, stability, and membrane fusion of the F protein trimer [31, 32]; the effects can disrupt endocytic transport, thereby impeding the interaction of the F and M proteins for viruslike particle (VLP) assembly [33]. Notably, while mutations in the virus affect viral infection, mutations in the receptor influence the intensity of viral entry. For example, the L124A mutation in the EFNB2 protein can enhance the infectivity of NiV pseudotypes by 30-fold [28]. Therefore, studies of HNV viruses should examine the biological properties of the viruses themselves, as well as their interactions with hosts and impacts on those hosts.

9.2 Construction of Pseudotyped Viruses

Because of the high-risk nature of HNV viruses, the use of authentic viruses for research has safety and generalization disadvantages. Accordingly, HNV pseudotyped viruses are in development. Pseudotyped viruses available for studies of HNV have distinct functional applications and advantages because they use distinct pseudoviral backbone systems; this section describes their construction in detail.

9.2.1 Pseudotyped Viruses Using the Moloney Murine Leukemia Virus (MuLV) Packaging System

Pseudotyped NiV viruses and closely related HeV viruses expressing envelope attachment (G) and fusion (F) glycoproteins are developed using the MuLV packaging system (a retrovirus packaging system) (Fig. 9.3A).

1. To generate pseudotyped NiV (Malaysian strain, M), the expression plasmids pCAGIG-NiV(M)-G and pCAGIG-NiV(M)-F are co-transfected at a 5:2 ratio into the MuLV env-negative packaging cell line TELCeB6 using a transfection reagent.
2. After 48 h, the cell culture supernatants are collected and cell debris are removed by centrifugation at 1500 g for 5 min.
3. Subsequently, the supernatant containing pseudotyped viral particles is fractionated and stored at −80 °C.

This procedure is also used for the production of pseudotyped NiV (Bangladesh strain, B) and HeV viruses [17].

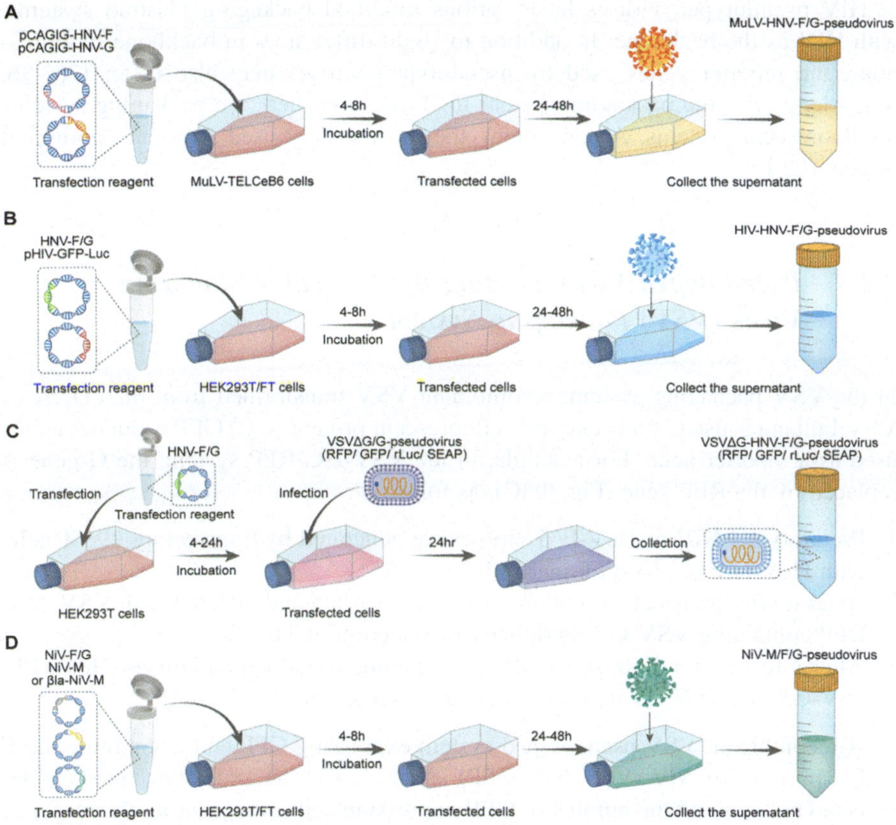

Fig. 9.3 Packaging systems for HNV pseudotyped viruses. Packaging processes are shown for (**A**) MuLV, (**B**) HIV, (**C**) VSV/ΔG, and (**D**) VLP pseudotyped viruses

9.2.2 *Pseudotyped Viruses Using the Human Immunodeficiency Virus Type 1 (HIV-1) Packaging System*

The HIV-1 packaging system with luciferase or other fluorescence genes is used to construct pseudotyped HNV viruses (Fig. 9.3B). The general process of pseudotyped virus construction using the HIV packaging system is as follows [34]:

1. Single-cycle pseudotyped NiV particles expressing luciferase are generated in HEK293T cells by co-transfection of HIV-Lai-luc Δenv with NiV F/G expression plasmids.
2. Cell supernatants containing pseudotyped viruses are collected 2 days after transfection.
3. Cell debris are removed by centrifugation (300 g, 5 min).
4. Supernatants are filtered through 0.45-μm pores and stored at −80 °C [35].

HIV pseudotyped viruses have various modified packaging plasmid systems, with HIV as the backbone. In addition to slight differences in backbone modifications, the reporter genes used by pseudotyped viruses may also differ [28, 36, 37]. Along with the packaging system for luciferase, there are packaging systems for fluorescent proteins. Among them, the most widely used is green fluorescent protein (GFP).

9.2.3 Pseudotyped Viruses Using the Vesicular Stomatitis Virus (VSV) Packaging System

In the VSV packaging system, recombinant VSV transformed from the cDNA of VSV Indiana is used as a backbone; a fluorescent protein (e.g., GFP or luciferase) is used as a reporter gene. For example, in the VSV-ΔG-RFP system, the G gene is replaced by the RFP gene (Fig. 9.3C), as follows:

1. Pseudotyped NiV or HeV-F/G viruses are generated by transfecting 293 T cells with HeV-G/F or NiV-G/F plasmids.
2. At 24 h after transfection, culture dishes are washed and infected with VSV-ΔG-RFP containing VSV-G (multiplicity of infection of 1).
3. After infection for 18 h, supernatants containing pseudotyped viruses (HeV-G/F, NiV-G/F, or VSV-G) are collected and stored at −80 °C [38–40].

A recombinant VSV pseudotyped system expressing GFP and assembling the F and G proteins of NiV (VSV-NiV-GFP) is also in use; neutralization titers can be detected by counting the number of GFP-expressing cells or measuring fluorescence intensity [41].

In addition to fluorescent proteins, other reporter genes can be used to generate pseudotyped viral libraries, such as the recombinant VSV-ΔG-rLuc pseudotyped viral system. In this VSV system derived from a full-length cDNA clone of the VSV Indiana serotype, the G envelope protein is replaced by Renilla luciferase (Fig. 3C) [30, 42, 43].

9.2.4 Pseudotyped Viruses Using the VSV-NiV-SEAP Novel Packaging System

Both fluorescent reporter proteins and luciferase reporter proteins require increased instrumental capabilities, and the costs of the corresponding assays are higher. However, a new assay has the unique ability to obtain neutralization titers by measuring secretory alkaline phosphatase (SEAP) activity in the supernatant using a common enzyme-linked immunosorbent assay (ELISA) plate reader. VSV pseudotyped NiV viruses expressing SEAP can be generated as described by

Fukushi et al. (Fig. 9.3C) [41, 44]. The titration of VSV-NiV-SEAP novel pseudotyped virus is performed as follows [45]:

1. VSV-NiV-SEAP is serially diluted tenfold in DMEM-2% fetal calf serum medium and inoculated on a monolayer of Vero cells in a 96-well culture plate.
2. After 20–24 h of incubation, 40 μL of supernatant are transferred to a new 96-well culture dish.
3. Then, 200 μL of substrate solution (SIGMAFAST p-Nitrophenyl Phosphate Tablets, Sigma) are added and incubated for 2 h at 37 °C.
4. SEAP activity is measured at OD_{405} using an ELISA plate reader. Thereafter, serum neutralization assays can be performed using supernatant dilutions with OD_{405} values of 1.0–1.5.

9.2.5 Pseudotyped Viruses Using the Self-Assembling NiV-M-VLP Packaging System

Many viruses form VLPs by expressing only their matrix proteins (e.g., Sendai virus) or by binding only to envelope proteins (e.g., measles virus) [46, 47]. Paramyxovirus matrix proteins direct VLPs to bud from the surface of infected cells and interact with the inner domains of envelope proteins, ultimately assisting in viral assembly [46, 48]. Notably, the matrix protein of NiV—alone or in combination with its fusion protein and receptor-binding protein (NiV-G)—can efficiently bud and form VLPs [49, 50]. Moreover, mutations in the transmembrane structural domain or cytoplasmic tail of the F protein disrupt its endocytic transport in host cells, impeding the ability of the F protein to undergo VLP assembly together with the M protein [33]. Therefore, VLPs based on the NiV matrix M protein are the simplest VLPs and may more closely reflect the biological properties of the virus during cell entry.

The M protein is essential for the release of HNV virus particles from infected cells, as reported by Ciancanelli et al. [50]. During initial characterization of the assembly process of the highly lethal, emerging paramyxovirus NiV, analysis of NiV-M budding showed that the expression of the NiV-M protein was sufficient to produce budding VLPs that were physically and morphologically similar to NiV[49]. The specific method for packaging NiV-M-VLPs is as follows (Fig. 3D):

1. NiV-M expression plasmids (3 μg) are transfected into HEK293T cells using Lipofectamine 2000 (Invitrogen, Carlsbad, CA, USA).
2. Forty-eight hours after transfection, the transfected cell medium is collected and clarified at 200 g for 5 min.
3. The culture supernatant is passed through a 20% sucrose cushion by centrifugation at 160,000 g at 4 °C for 2 h.
4. After centrifugation, the supernatant is aspirated and the VLPs are resuspended in NTE buffer [50].

The C-terminus of many M proteins is involved in complex functions, such as budding and viral assembly [51, 52]. Wolf et al. performed mammalian codon optimization of the β-lactamase gene (βla) fused to the N-terminus of NiV-M; this yielded a fully codon optimized βla-M gene for efficient expression in mammalian cells. βla-M constructs can bud and form VLPs in cells (Fig. 9.3D) [53].

9.3 Applications of HNV Pseudotyped Viruses

9.3.1 Studies of Potential Virus Receptors

Pseudotyped HNV viruses have a G glycoprotein that can bind to receptors EFNB2 and/or EFNB3 [54, 55]. Pseudotyped virus experiments using the VSV-ΔG-LUC system have shown that the NiV and HeV-G have distinct receptor-binding affinities; NiV envelope pseudotyped (NiVpp) reporter viruses enter EFNB3-expressing cells more efficiently, compared with HeVpp, while soluble EFNB3 effectively inhibits the entry of NiVpp but not HeVpp. These data suggest that NiV binds EFNB3 receptors more efficiently, compared with HeV[29]. The use of recombinant soluble EFNB2 and EFNB3 receptors blocks pseudotyped HNV particles from binding and infecting target cells; HeV- and NiV-F/G-bearing particles preincubated with soluble EFNB2 or -B3 are unable to infect host cells [30, 55]. Additionally, competition experiments with NiV and HeV pseudotyped viruses have shown that heparin can limit pseudotyped virus binding to EFNB2 and EFNB3 [6].

The key amino acid residues on receptors can also be studied by using NiV pseudotyped viruses. After mutations of Tyr (to Leu) and Met (to Trp) at specific positions in EFNB1, EFNB1 was able to function as a receptor for G protein; this finding demonstrated that Leu and Trp are required for EFNB3 and EFNB2 to function as cellular receptors for NiV[53]. The binding of the G protein to the host cell receptor induces target cells to initiate a fusion mechanism that leads to a change in the internal conformation of the F0 protein, thereby promoting fusion.

9.3.2 Viral Infectivity Studies

Because HNV viruses are BSL-4 pathogens, the NiV/HeV pseudotyped viral system that allows BSL-2 handling is useful for studying the infectivity of these high-risk viruses. For example, relationships with cell-cell fusion and cell infectivity were confirmed using pseudotyped virus infectivity experiments. In that study, the cholesterol level was positively correlated with the NiV viral infectivity level through the virus-cell membrane fusion pathway [7]. In contrast, Bradel-Tretheway et al. found that viral infection and cell-cell fusion mechanisms were not always related to each other, based on work with the VSV pseudotyped viral system; combinations with a high fusion phenotype did not necessarily have a higher level of infection with pseudotyped viruses, compared with wild-type combinations [43]. Pseudotyped viral systems can also be used to mimic viral replication cycles. For example, the VSV-ΔG-RFP system mimics viral replication cycles by using pseudotyped VSV-ΔG-RFP encapsulated by HeV-G and HeV-F glycoproteins to infect cells that can stably express HeV glycoproteins [39]. Confocal microscopy shows RFP at 24, 48, and 72 h; this reveals the infectious changes of the pseudotyped virus.

The use of the pseudotyped virus system also facilitates investigations of whether mutations in viral glycoproteins or receptor proteins affect viral infection. NiV pseudotyped viruses can enter CHO cells expressing EFNB2 or EFNB3. Pseudotyped viruses formed by introducing mutations of W504A, E505A, and V507S into the NiV-G protein exhibit infectivity similar to wild-type NiV viruses; however, compared with wild-type NiV pseudotyped viruses, the triple mutant pseudotyped viruses exhibited lower EFNB3 receptor-dependent infectivity [30]. Furthermore, G-E533Q mutant pseudotyped virus particles were largely unable to complete EFNB2- or EFNB3-mediated viral infection and exhibited considerably lower infectivity, suggesting that amino acid residue 533 is essential for viral infectivity [30]. Additionally, Yuan et al. used NiV pseudotyped virus particles to explore differences in the ability of pseudotyped viruses to bind and enter cells with >10 different EFNB2 mutants; they found that the L124A mutant of EFNB2 permitted a significantly greater level of viral entry [28].

Additionally, glycosylation can affect viral infectivity [56]. Stone et al. constructed six pseudotyped virus variants with mutations at glycosylation sites; they found that the deletion of O-glycosylation influenced the infectivity and F/G protein-mediated cellular entry of HNV pseudotyped viruses [57].

9.3.3 Evaluation of Neutralization Detection Systems and Potential Antibody Candidates

Because of the risks involved in using authentic viruses to evaluate HNV serum neutralizing antibodies, there is a need to develop safer and more reliable alternative methods using pseudotyped viruses. The NiV and HeV pseudotyped viruses can be differentially neutralized with monoclonal antibody 7G9, demonstrating the specificity and safety of neutralization assays; these pseudotyped viruses provide a validated tool for assaying antibody effectiveness [17]. Additionally, the human monoclonal antibody m102.4 neutralizes pseudotyped HNV viruses by specifically binding to the HNV-G glycoprotein and blocking the regions that bind to EFNB2 and EFNB3 receptors; this completely blocks infection by pseudotyped HNV viruses [34].

Generally, pseudotyped virus systems with demonstrated infectivity will express luciferase or green fluorescence for instrumental readout, which can be used to determine the neutralization results. However, this type of assessment often requires more expensive substrates or specific equipment. The new generation of SEAP pseudotyped viruses can be used to determine neutralization results with common ELISA readouts, thus lowering the threshold for neutralization experiments. Kaku et al. used the VSV-NiV-SEAP neutralization assay system to test the neutralizing antibody statuses (positive/negative) of NiV- and HeV-specific sera from different species, in comparison with the findings obtained by conventional NiV biopsy [45]. When 66 fruit bat sera were screened at the same dilution, the VSV-NiV-SEAP test produced results similar to the live NiV test [45], indicating that the new VSV-NiV-SEAP test is safe and has high sensitivity and specificity.

9.3.4 Screening Studies of Inhibitory Drugs

Considering the safety of pseudotyped viral particle formation and the inexpensive nature of the experiment, pseudotyped viruses can be used for high-throughput in vitro screening of antiviral compounds with inhibitory effects on viral infections. Compounds that specifically inhibit target viral infection can be selected by pseudotyped virus particle inhibition assays of target viral glycoproteins. Elshabrawy et al. screened a library of 5000 small molecules and found that the small molecule 5705213 and its derivative 7402683 inhibited endogenous processing of NiV and HeV fusion glycoproteins, as well as the entry of pseudotyped viruses [37]; the two small molecules inhibited the entry of HNV-F0 pseudotyped viruses into 293FT cells by 80%–100% [37]. Luciferase reporter screening with pseudotyped viruses revealed that the HNV-specific peptide fusion inhibitor NiV-FC2 (a 36-amino-acid peptide), corresponding to heptapeptide repeat region 2 of HNV-F glycoprotein, completely blocked the entry of pseudotyped HNV [25, 34, 58]. Further comparisons of the effects of various inhibitors can be conducted using pseudotyped viral systems. To compare the inhibitory effects of NiV heptad repeat regions (HRC) and human parainfluenza virus 3 HRC peptide, a plaque expansion test was performed using NiV-derived peptide and HPIV3-derived peptide against HeV and NiV pseudotyped viruses [59]. Based on the IC_{50} of both peptides to inhibit plaque expansion of pseudotyped HNV viruses, the HPIV3-derived peptide achieved up to 50-fold greater inhibition of HNV, compared with the NiV-derived peptide. Specifically, the HPIV3-derived peptide blocked HeV at 0.45 nm and NiV at 0.25 nm, while the NiV-derived FC2 peptide blocked HeV at 4.2 nm and NiV at 11.4 nm[40].

Additionally, pseudotyped viruses are useful when screening for discrimination-specific inhibitors. For example, brilliant green and gliotoxin showed similar IC_{50} values against VSV wild-type pseudotyped viruses, suggesting that their effects are related to the backbone of the VSV pseudotyped virus system or the presence of general nonspecific antiviral activity, rather than to the presence of virus specific inhibitors [38]. Moreover, gentian violet exhibited significant selective inhibition of HNV, according to the NiV/HeV-G-VSV pseudotype assay [39]. These findings demonstrate the superiority of pseudotyped viruses in screening for specific antiviral substances. The use of pseudotyped viruses allows for high-throughput screening of suitable inhibitors; it also allows further optimization and modification of inhibitors after the initial screening, which greatly improves the ease of screening for inhibitory drugs.

9.4 Conclusions

Generally, pseudotyped viruses are advantageous in terms of their safety and sensitivity; they have diverse applications in studies of emerging and virulent infectious diseases, HNV pseudotyped viruses have made extensive progress in evaluations of virus receptors, viral infectivity, vaccine and antibody neutralization responses, and ease of inhibitory drug screening. HNV pseudotyped viruses can reduce the risk of laboratory handling and offer viable replacements for authentic viruses in terms of evaluating and developing prevention and control products. These attributes may help to reduce dependence on BSL-4 laboratory requirements and the use of authentic viruses, thus reducing the biosafety considerations required for most laboratories to conduct research on these highly virulent pathogens. The development of HNV pseudotyped virus technology can also verify the consistency of pseudotyped viruses with authentic viruses or traditional standard analysis methods.

In conclusion, pseudotyped HNV viruses constitute important tools for biological analyses of these pathogens.

Acknowledgments This work was funded by the Beijing Municipal Science and Technology Project (Z211100002521018), National Natural Science Foundation of China (82073621, 82172244, and 32070678), National Key Research and Development Program of China (2021YFC0863300), and the Bill and Melinda Gates Foundation (INV-006379). We thank Ryan Chastain-Gross, Ph.D., from Liwen Bianji (Edanz) (www.liwenbianji.cn/) for editing the English text of a draft of this manuscript.

References

1. Field, H., et al.: The natural history of Hendra and Nipah viruses. Microbes Infect. **3**, 307–314 (2001)
2. Negrete, O.A., et al.: EphrinB2 is the entry receptor for Nipah virus, an emergent deadly paramyxovirus. Nature. **436**, 401–405 (2005)
3. Thibault, P.A., Watkinson, R.E., Moreira-Soto, A., Drexler, J.F., Lee, B.: Zoonotic potential of emerging paramyxoviruses: knowns and unknowns. Adv. Virus Res. **98** (2017). https://doi.org/10.1016/bs.aivir.2016.12.001
4. Xu, K., Broder, C.C., Nikolov, D.B.: Ephrin-B2 and ephrin-B3 as functional henipavirus receptors. Semin. Cell Dev. Biol. **23**, 116–123 (2012). https://doi.org/10.1016/j.semcdb.2011.12.005
5. Lo Presti, A., et al.: Origin and evolution of Nipah virus. J. Med. Virol. **88**, 380–388 (2016). https://doi.org/10.1002/jmv.24345

6. Mathieu, C., et al.: Heparan sulfate-dependent enhancement of henipavirus infection. *mBio*. **6, e02427** (2015). https://doi.org/10.1128/mBio.02427-14
7. Contreras, E.M., et al.: Roles of cholesterol in early and late steps of the Nipah virus membrane fusion Cascade. J. Virol. **95** (2021). https://doi.org/10.1128/JVI.02323-20
8. Chua, K.B., et al.: Nipah Virus: a Recently Emergent Deadly Paramyxovirus. *Science*, vol. 288, pp. 1432–1435, New York, N.Y (2000)
9. Gómez Román, R., et al.: Medical countermeasures against henipaviruses: a review and public health perspective. Lancet Infect. Dis. **22**, e13–e27 (2022). https://doi.org/10.1016/s1473-3099(21)00400-x
10. Sun, B., Jia, L., Liang, B., Chen, Q., Liu, D.: Phylogeography, transmission, and viral proteins of Nipah virus. Virol. Sin. **33**, 385–393 (2018). https://doi.org/10.1007/s12250-018-0050-1
11. Lo, M.K., et al.: Characterization of Nipah virus from outbreaks in Bangladesh, 2008-2010. Emerg. Infect. Dis. **18**, 248–255 (2012). https://doi.org/10.3201/eid1802.111492
12. Chadha, M.S., et al.: Nipah virus-associated encephalitis outbreak, Siliguri, India. *Emerg Infect Dis*. **12**, 235–240 (2006)
13. Diederich, S., Maisner, A.: Molecular characteristics of the Nipah virus glycoproteins. Ann. N. Y. Acad. Sci. **1102**, 39–50 (2007)
14. Hooper, P.T., Gould, A.R., Russell, G.M., Kattenbelt, J.A., Mitchell, G.: The retrospective diagnosis of a second outbreak of equine morbillivirus infection. Aust. Vet. J. **74**, 244–245 (1996)
15. Annand, E.J., et al.: Novel Hendra virus variant detected by sentinel surveillance of horses in Australia. Emerg. Infect. Dis. **28**, 693–704 (2022). https://doi.org/10.3201/eid2803.211245
16. Skowron, K., et al.: Nipah virus-another threat from the world of zoonotic viruses. Front. Microbiol. **12**, 811157 (2021). https://doi.org/10.3389/fmicb.2021.811157
17. Bae, S.E., et al.: Construction of the safe neutralizing assay system using pseudotyped Nipah virus and G protein-specific monoclonal antibody. Biochem. Biophys. Res. Commun. **513**, 781–786 (2019). https://doi.org/10.1016/j.bbrc.2019.03.212
18. Rahman, M.A., et al.: Date palm sap linked to Nipah virus outbreak in Bangladesh, 2008. Vector-Borne and Zoonotic Diseases. **12**, 65–72 (2012). https://doi.org/10.1089/vbz.2011.0656
19. Rima, B., et al.: ICTV virus taxonomy profile: Paramyxoviridae. J. Gen. Virol. **100**, 1593–1594 (2019). https://doi.org/10.1099/jgv.0.001328
20. Rockx, B., Winegar, R., Freiberg, A.N.: Recent progress in henipavirus research: molecular biology, genetic diversity, animal models. Antivir. Res. **95**, 135–149 (2012). https://doi.org/10.1016/j.antiviral.2012.05.008
21. Meulendyke, K.A., Wurth, M.A., McCann, R.O., Dutch, R.E.: Endocytosis plays a critical role in proteolytic processing of the Hendra virus fusion protein. J. Virol. **79**, 12643–12649 (2005). https://doi.org/10.1128/JVI.79.20.12643-12649.2005
22. Iorio, R.M., Melanson, V.R., Mahon, P.J.: Glycoprotein interactions in paramyxovirus fusion. Future Virol. **4**, 335–351 (2009). https://doi.org/10.2217/fvl.09.17
23. Pager, C.T., Craft Jr., W.W., Patch, J., Dutch, R.E.: A mature and fusogenic form of the Nipah virus fusion protein requires proteolytic processing by cathepsin L. Virology. **346**, 251–257 (2006). https://doi.org/10.1016/j.virol.2006.01.007
24. Pager, C.T., Dutch, R.E.: Cathepsin L is involved in proteolytic processing of the Hendra virus fusion protein. J. Virol. **79**, 12714–12720 (2005)
25. Bossart, K.N., et al.: Inhibition of Henipavirus fusion and infection by heptad-derived peptides of the Nipah virus fusion glycoprotein. Virol. J. **2**, 57 (2005)
26. Bossart, K.N., et al.: Receptor binding, fusion inhibition, and induction of cross-reactive neutralizing antibodies by a soluble G glycoprotein of Hendra virus. J. Virol. **79**, 6690–6702 (2005). https://doi.org/10.1128/JVI.79.11.6690-6702.2005
27. Harcourt, B.H., et al.: Molecular characterization of Nipah virus, a newly emergent paramyxovirus. Virology. **271**, 334–349 (2000). https://doi.org/10.1006/viro.2000.0340
28. Yuan, J., et al.: Mutations in the G-H loop region of ephrin-B2 can enhance Nipah virus binding and infection. J. Gen. Virol. **92**, 2142–2152 (2011). https://doi.org/10.1099/vir.0.033787-0

29. Pryce, R., et al.: A key region of molecular specificity orchestrates unique ephrin-B1 utilization by cedar virus. Life Sci Alliance. **3** (2020). https://doi.org/10.26508/lsa.201900578
30. Negrete, O.A., Chu, D., Aguilar, H.C., Lee, B.: Single amino acid changes in the Nipah and Hendra virus attachment glycoproteins distinguish ephrinB2 from ephrinB3 usage. J. Virol. **81**, 10804–10814 (2007)
31. Smith, E.C., et al.: Trimeric transmembrane domain interactions in paramyxovirus fusion proteins: roles in protein folding, stability, and function. J. Biol. Chem. **288**, 35726–35735 (2013). https://doi.org/10.1074/jbc.M113.514554
32. Barrett, C.T., et al.: Analysis of Hendra virus fusion protein N-terminal transmembrane residues. Viruses. **13** (2021). https://doi.org/10.3390/v13122353
33. Cifuentes-Muñoz, N., et al.: Mutations in the transmembrane domain and cytoplasmic tail of Hendra virus fusion protein disrupt virus-like-particle assembly. J. Virol. **91** (2017). https://doi.org/10.1128/JVI.00152-17
34. Khetawat, D., Broder, C.C.: A functional henipavirus envelope glycoprotein pseudotyped lentivirus assay system. Virol. J. **7**, 312 (2010). https://doi.org/10.1186/1743-422X-7-312
35. Akiyama, H., et al.: Virus particle release from glycosphingolipid-enriched microdomains is essential for dendritic cell-mediated capture and transfer of HIV-1 and henipavirus. J. Virol. **88**, 8813–8825 (2014). https://doi.org/10.1128/JVI.00992-14
36. Tiffen, J.C., Bailey, C.G., Ng, C., Rasko, J.E., Holst, J.: Luciferase expression and bioluminescence does not affect tumor cell growth in vitro or in vivo. Mol. Cancer. **9**, 299 (2010). https://doi.org/10.1186/1476-4598-9-299
37. Elshabrawy, H.A., et al.: Identification of a broad-spectrum antiviral small molecule against severe acute respiratory syndrome coronavirus and Ebola, Hendra, and Nipah viruses by using a novel high-throughput screening assay. J. Virol. **88**, 4353–4365 (2014). https://doi.org/10.1128/JVI.03050-13
38. Aljofan, M., et al.: Antiviral activity of gliotoxin, gentian violet and brilliant green against Nipah and Hendra virus in vitro. Virol. J. **6**, 187 (2009). https://doi.org/10.1186/1743-422X-6-187
39. Porotto, M., et al.: Simulating henipavirus multicycle replication in a screening assay leads to identification of a promising candidate for therapy. J. Virol. **83**, 5148–5155 (2009). https://doi.org/10.1128/JVI.00164-09
40. Porotto, M., et al.: Molecular determinants of antiviral potency of paramyxovirus entry inhibitors. J. Virol. **81**, 10567–10574 (2007). https://doi.org/10.1128/JVI.01181-07
41. Kaku, Y., et al.: A neutralization test for specific detection of Nipah virus antibodies using pseudotyped vesicular stomatitis virus expressing green fluorescent protein. J. Virol. Methods. **160**, 7–13 (2009). https://doi.org/10.1016/j.jviromet.2009.04.037
42. Bradel-Tretheway, B.G., Liu, Q., Stone, J.A., McInally, S., Aguilar, H.C.: Novel functions of Hendra virus G N-glycans and comparisons to Nipah virus. J. Virol. **89**, 7235–7247 (2015). https://doi.org/10.1128/JVI.00773-15
43. Bradel-Tretheway, B.G., et al.: Nipah and Hendra virus glycoproteins induce comparable homologous but distinct heterologous fusion phenotypes. J. Virol. **93** (2019). https://doi.org/10.1128/JVI.00577-19
44. Fukushi, S., Watanabe, R., Taguchi, F.: Pseudotyped vesicular stomatitis virus for analysis of virus entry mediated by SARS coronavirus spike proteins. Methods Mol. Biol. **454**, 331–338 (2008). https://doi.org/10.1007/978-1-59745-181-9_23
45. Kaku, Y., et al.: Second generation of pseudotype-based serum neutralization assay for Nipah virus antibodies: sensitive and high-throughput analysis utilizing secreted alkaline phosphatase. J. Virol. Methods. **179**, 226–232 (2012). https://doi.org/10.1016/j.jviromet.2011.11.003
46. Takimoto, T., Portner, A.: Molecular mechanism of paramyxovirus budding. Virus Res. **106**, 133–145 (2004)

47. Sugahara, F., et al.: Paramyxovirus Sendai virus-like particle formation by expression of multiple viral proteins and acceleration of its release by C protein. Virology. **325**(1), 1–10 (2004)
48. Lamb, R. A. & Kolakofsky, D. J. F. V. Paramyxoviridae: the Viruses and their Replication. (1996)
49. Patch, J.R., Crameri, G., Wang, L.F., Eaton, B.T., Broder, C.C.: Quantitative analysis of Nipah virus proteins released as virus-like particles reveals central role for the matrix protein. Virol. J. **4**, 1 (2007). https://doi.org/10.1186/1743-422X-4-1
50. Ciancanelli, M.J., Basler, C.F.: Mutation of YMYL in the Nipah virus matrix protein abrogates budding and alters subcellular localization. J. Virol. **80**, 12070–12078 (2006). https://doi.org/10.1128/JVI.01743-06
51. Schmitt, A.P., Leser, G.P., Morita, E., Sundquist, W.I., Lamb, R.A.: Evidence for a new viral late-domain core sequence, FPIV, necessary for budding of a paramyxovirus. J. Virol. **79**, 2988–2997 (2005)
52. Irie, T., Licata, J.M., McGettigan, J.P., Schnell, M.J., Harty, R.N.: Budding of PPxY-containing rhabdoviruses is not dependent on host proteins TGS101 and VPS4A. J. Virol. **78**, 2657–2665 (2004)
53. Wolf, M.C., et al.: A catalytically and genetically optimized beta-lactamase-matrix based assay for sensitive, specific, and higher throughput analysis of native henipavirus entry characteristics. Virol. J. **6**, 119 (2009). https://doi.org/10.1186/1743-422X-6-119
54. Negrete, O.A., et al.: Two key residues in ephrinB3 are critical for its use as an alternative receptor for Nipah virus. PLoS Pathog. **2**, e7 (2006)
55. Bonaparte, M.I., et al.: Ephrin-B2 ligand is a functional receptor for Hendra virus and Nipah virus. Proc. Natl. Acad. Sci. U. S. A. **102**, 10652–10657 (2005)
56. Vigerust, D.J., Shepherd, V.L.: Virus glycosylation: role in virulence and immune interactions. Trends Microbiol. **15**, 211–218 (2007)
57. Stone, J.A., Nicola, A.V., Baum, L.G., Aguilar, H.C.: Multiple novel functions of Henipavirus O-glycans: the first O-glycan functions identified in the paramyxovirus family. PLoS Pathog. **12**, e1005445 (2016). https://doi.org/10.1371/journal.ppat.1005445
58. Bossart, K.N., Wang, L.F., Eaton, B.T., Broder, C.C.: Functional expression and membrane fusion tropism of the envelope glycoproteins of Hendra virus. Virology. **290**, 121–135 (2001)
59. Porotto, M., et al.: Inhibition of hendra virus fusion. J. Virol. **80**, 9837–9849 (2006)

Chapter 10
Pseudotyped Viruses for Lyssavirus

Wenbo Wang, Caifeng Long, Lan Wang, and Youchun Wang

Abstract Lyssaviruses, which belong to the family *Rhabdoviridae*, are enveloped and bullet-shaped ssRNA viruses with genetic diversity. All members of *Lyssavirus* genus are known to infect warm-blooded animals and cause the fatal disease rabies. The rabies virus (RABV) in lyssavirus is the major pathogen to cause fatal rabies. The pseudotyped RABV is constructed to study the biological functions of G protein and evaluation of anti-RABV products including vaccine-induced antisera, rabies immunoglobulins (RIG), neutralizing mAbs, and other antiviral inhibitors. In this chapter, we focus on RABV as a representative and describe the construction of RABV G protein bearing pseudotyped virus and its applications. Other non-RABV lyssaviruses are also included.

Keywords Lyssavirus · RABV · Glycoprotein · Pseudovirus

Abbreviations

ABLV	Australian bat lyssavirus
ARAV	Aravan virus
BBLV	Bokeloh bat lyssavirus
CNS	Central nervous system
DUVV	Duvenhage virus

W. Wang · C. Long · L. Wang (✉)
Division of Monoclonal Antibody Products, National Institutes for Food and Drug Control (NIFDC) and WHO Collaborating Center for Standardization and Evaluation of Biologicals, Beijing, China
e-mail: wanglan@nifdc.org.cn

Y. Wang (✉)
Chinese Academy of Medical Sciences & Peking Union Medical College, Beijing, China

Institute of Medical Biology, Chinese Academy of Medicine Sciences & Peking Union Medical College, Kunming, China
e-mail: wangyc@nifdc.org.cn; wangyc@imbcams.com.cn

Y. Wang (ed.), *Pseudotyped Viruses*, Advances in Experimental Medicine and Biology 1407, https://doi.org/10.1007/978-981-99-0113-5_10

EBLV-1	European bat-1 lyssavirus
EBLV-2	European bat-2 lyssavirus
FAVN	Fluorescent antibody virus neutralization
G	Glycoprotein
GBLV	Gannoruwa bat lyssavirus
HIV	Human immunodeficiency virus
IKOV	Ikoma lyssavirus
IRKV	Irkut virus
KBLV	Kotalahti bat lyssavirus
KHUV	Khujand virus
LBV	Lagos bat virus
LLEBV	Lleida bat lyssavirus
MLV	Murine leukemia virus
MOKV	Mokola virus
PEP	Post-exposure prophylaxis
PNGS	Potential N-linked glycosylation site
RABV	Rabies virus
RFFIT	Rapid fluorescent focus inhibition test
RNP	Ribonucleoprotein complex
SHIBV	Shimoni bat virus
TWBLV	Taiwan bat lyssavirus
VSV	Vesicular stomatitis virus
WCBV	West Caucasian bat virus

10.1 General Information About Lyssavirus

The *Lyssavirus* genus belongs to family *Rhabdoviridae* of the order *Mononegavirales* and comprises 17 proposed species (Fig. 10.1). All species in lyssavirus are highly neurotropic and known to cause fatal encephalitis. Lyssaviruses are membrane-enveloped, bullet-shaped viruses with ssRNA that have genetic diversity. The most extensively characterized virus within *Lyssavirus* genus is RABV that have been studied in great details for many decades and serve as model systems for elucidating the molecular mechanisms of lyssavirus life cycles in host cells.

Within the *Lyssavirus* genus, viruses have been classified according to both genetic and antigenic data into phylogroups [1–3]. Phylogroup I includes the classic RABV and Aravan virus (ARAV), Australian bat lyssavirus (ABLV), Bokeloh bat lyssavirus (BBLV), Duvenhage virus (DUVV), European bat-1 and European bat-2 lyssaviruses (EBLV-1 and EBLV-2), Gannoruwa bat lyssavirus (GBLV), Irkut virus (IRKV), and Khujand virus (KHUV). Phylogroup II includes Lagos bat virus (LBV), Mokola virus (MOKV), and Shimoni bat virus (SHIBV), and phylogroup III includes West Caucasian bat virus (WCBV), Ikoma lyssavirus (IKOV), and

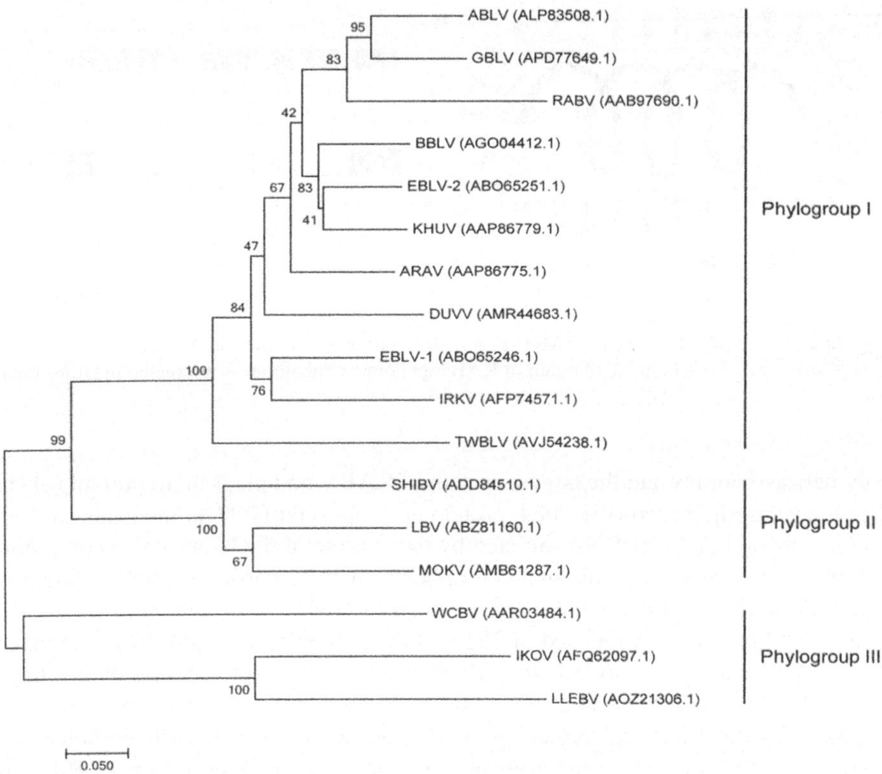

Fig. 10.1 Phylogenetic tree of *Lyssavirus* genus with complete G protein sequence

Lleida bat lyssavirus (LLEBV). Data indicated that the classical RABV vaccines generally protect the hosts against phylogroup I lyssaviruses [4–7], while little or no protection against more genetic and antigenic divergent phylogroup II and III lyssaviruses [6, 8, 9].

Because the pseudotyped RABV has been widely applied in studying the functions of G protein and evaluation of neutralizing antibodies, cell tropism, and screening of antiviral inhibitors, we introduce the pseudotyped viruses for RABV as a representative.

10.2 General Information About RABV

RABV is a single-stranded negative-stranded RNA virus with a typical bullet shape (Fig. 10.2). The viral genome is about 12 kb and encodes five proteins: the nucleoprotein (N), phosphoprotein (P), matrix protein (M), glycoprotein (G), and RNA-dependent RNA polymerase (L). The L protein is a component of the

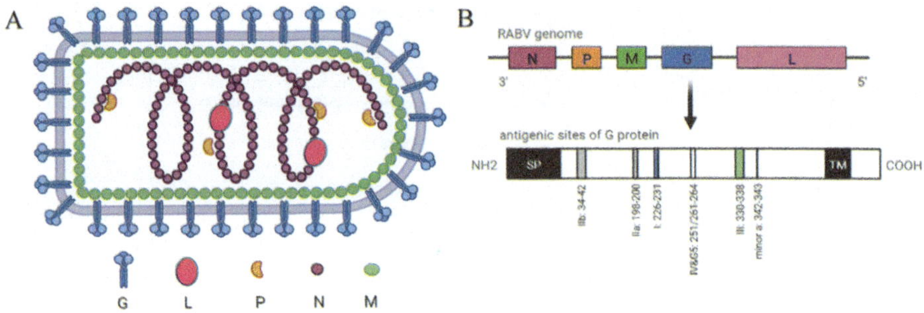

Fig. 10.2 Schematic diagram of RABV virion structure and genome. (**a**) Schematic diagram of RABV structure. (**b**) Schematic diagram of RABV genome. (This figure was created in BioRender.com)

polymerase complex and the largest protein of RABV and plays an important role in the viral transcription process; the P protein is composed of 297 amino acids and can promote the synthesis of RNA initiated by the L protein; the N protein is composed of 450 amino acids and is the main component of virus particles. The N protein is highly conserved and is often used in diagnosis, classification, and epidemiological research; M protein is composed of 202 amino acids and has a molecular weight of 20–25 kDa, which is an important component of virus particles. Some studies have shown that M protein is also related to virus pathogenicity and host tropism [10–12]; G protein is the only viral protein presented on the surface of virus particles and involved in cell infection, host tropism, virulence, and immune response. Ribonucleoprotein complex (RNP) of RABV aggregates to form endosomes in virus-infected neurons, which is the main feature of pathological damage after RABV infection. G protein is the only viral surface protein used for pseudotyping.

10.2.1 Glycoprotein Structure and Its Biological Role

G protein is a type 1 transmembrane protein that forms spikes on the surface of viral particles in the form of homotrimers, and the surface-rich hydrophobic amino acids are important for stabilizing the conformation of the trimers. The RABV G protein gene encodes a total of 524 amino acids to form the G protein precursor, and the N-terminal 19 amino acids are the signal peptide (SP), which mediates the anchoring of the G protein to the endoplasmic reticulum and the Golgi for post-translational modifications such as glycosylation, forming a mature G protein consisting of 505 amino acids. Mature G protein consists of three parts, including extracellular domain (ED), transmembrane domain (TM), and intracellular domain (CD) [9]. The extracellular domain ED is composed of amino acids at position 1–439, which is mainly involved in the binding to cell receptors and membrane fusion, and is the main region that contains antigenic epitopes; TM is composed of amino acids

440–461 and related to the fixation of the G protein on the viral phospholipid bilayer membrane; the intracellular domain CD is composed of amino acids 462–505 at the C-terminus, located on the inner surface of the viral envelope, and can interact with the M protein to assemble viral particles. As a surface glycoprotein, N-linked glycosylation is presented on RABV G protein and the number of N-linked glycosylation sites varies among different strains. Two glycosylation sites, N37 and N319, are highly conserved in street RABV isolates [13]. Other N-linked glycosylation sites such as N158, N204, and N247 are also presented in the fixed RABV whose pathogenicity is attenuated by tissue or cell culture. There are few functional studies of N-glycosylation on G proteins, and some studies suggest that N-glycosylation may be related to G protein folding and virulence [14–16].

RABV is a neurotropic virus, and could not invade blood or induce toxemia. RABV enters the body through the wound, and can directly infect the nerve cells around the wound, or firstly infect other non-nerve cells and then invade into the central nervous system, leading to the occurrence of clinical symptoms of rabies. The process of RABV infection initiate with the binding of G protein to cell surface receptors; however, the receptor molecules necessary for RABV infection are still unclear. Studies have shown that the possible RABV receptors on cell surface mainly include nicotinic acetylcholine receptor (nAchR) [17, 18], neurotrophin receptor (p75 neurotrophin receptor, p75NTR) [19], neuronal cell adhesion molecule (NCAM) [20], and metabotropic glutamate receptor subtype 2 (mGluR2) [21]. Once the G protein binds to cell receptors, RABV enters the cell through endocytosis and fuses with the endosomal membrane to release RNP into the cytoplasm. A low pH environment (5.8–6.0) in endosomes induces the initiation of RABV G protein-mediated membrane fusion, whereas fusion stops when pH is greater than 6.3 [22]. The region in G protein that interacts with the endosomal membrane, that is, the fusion peptide, is not clear, and some studies infer that it may be located at amino acids 103–179 [23].

G proteins directly react with neutralizing antibodies and contain multiple antigenic epitopes. Through a series of anti-G protein mAbs test, conformational and linear epitopes have been identified (Fig. 10.2), including the antigenic site I, antigenic site II (IIa and IIb), antigenic site III, antigenic site IV, G5, and minor a, etc. [24–29]. Antigenic site I is located at amino acids 226–231; antigenic site II is composed of two non-consecutive epitopes IIa and IIb where IIa is located at amino acids 198–200 and IIb is located at amino acids 34–42; antigenic site III is a conformational epitope, located at amino acids 330–338; antigenic site IV overlaps with G5, mainly located at amino acids 251 and 261–264; antigenic site minor a is located at amino acids 342–343. In addition to the above antigenic sites, new epitopes may exist [30]. Among these antigenic sites, the most important ones are antigenic site II and III [24]. Most of the neutralizing mAbs bind to these two sites, among which 70% bind antigenic site II and ~ 20% bind antigenic site III [31, 32].

RABV G protein also contributes to the viral pathogenicity. G protein affects the pathogenicity through the following functions: (i) G protein can bind to cell surface receptors to mediate RABV entry into cells; (ii) G protein can interact with viral RNP-M complex and lead to effective virus budding; (iii) the expression level of G protein affects the pathogenicity of the virus, and the expression of G protein must be controlled at a minimum level to prevent neuronal cell apoptosis after infection.

10.2.2 Variation and Mutation on RABV G Protein

Due to the lack of proofreading ability of the polymerases of RNA viruses, their genomes are prone to errors during viral replication. The resulting high mutation rate allows the virus to evade immune attack by the host; this is an important part of viral evolution and a major challenge in vaccine and antiviral drug development [33–35]. As an RNA virus, RABV has a high mutation rate due to low polymerase fidelity [36], which leads to the diversity of RABV. The mutation frequency of the RABV G protein gene is approximately 3.2 to 3.9×10^{-4} nucleotides/site/year [37, 38]. Although the ratio of non-synonymous/synonymous mutations is low, it still causes the sequence and antigenic variations. Studies indicate that in some regions, RABV G protein sequence of street isolates deviates from the vaccine strains. For example, the sequence of RABV street isolates in Iran is different from the vaccine strains PM and Flury-LEP used in this region; this difference is also found in India that the sequence similarity of street isolates to vaccine strains is 85–87%. The RABV street isolates in Poland also has similar results [39–41]. In terms of viral antigenicity, some studies have reported differences in antigenicity among RABV street isolates and vaccine strains. RABV street isolates from Nigerian rabid dogs was evaluated by 40 anti-G protein monoclonal antibodies, and data showed that there were antigenic differences between street isolates and the Flury-LEP vaccine strain [42, 43]. Similarly, RABV street isolates from rabid dogs in Plateau State, Nigeria, were evaluated by 44 anti-G protein monoclonal antibodies, and the antigenicity was significantly different from the vaccine strains Flury-LEP and PM [44]. The antigenicity of 35 RABV street isolates in Brazil was evaluated by anti-G protein monoclonal antibodies, and their antigenicity was different from that of PM, Flury-LEP, PV, and other vaccine strains [45]. Wiktor et al. evaluated eight American RABV street isolates and found differences in antigenicity between street virus and vaccine strains [46]. The differences in antigenicity could also affect the protective effectiveness of the RABV vaccine. Wiktor found that the vaccine strains have varied protective effects on street isolates with different antigenicity [45], while Pierre et al. found that vaccine strain showed good protection against RABV street isolates originated in European and African countries, but only partial protection against RABV isolates from Madagascar and Thailand [47]. Amino acid mutations have also been identified on the G protein with neutralizing mAbs [48, 49].

Genetic analysis has been widely used to predict the cross-protection of rabies vaccines. However, due to the lack of data on the effect of single amino acid on viral antigenicity, genetic analysis alone is limited to a certain extent. In fact, more than 30% of the antigenic differences cannot be determined by genetic analysis [1]. The pseudotyped virus system is a powerful tool to study the impact of mutations/variations on antigenicity and also the antigenic differences between street RABV variants, because the pseudotyped viruses bearing the mutated or naturally evolved G proteins are easily constructed. The combination of the genetic analysis and the pseudotyped virus-based antigenicity evaluation may provide a more accurate prediction of the cross-protection of rabies vaccines to street RABV variants and also the non-RABV lyssaviruses.

10.3 Construction of RABV G Protein Pseudotyped Virus

A variety of pseudotyped lyssaviruses have been constructed and widely applied. Within *Lyssavirus* genus, almost all species have been pseudotyped, and among them, RABV is the most pseudotyped lyssavirus (Table 10.1).

The packaging systems for preparing RABV-G bearing pseudoviruses mainly include lentivirus packaging system, VSV packaging system, and murine leukemia virus (MLV) packaging system [64]. These packaging systems are used in viral packaging and preparation and have their own advantages and characteristics. The lentivirus vector system is highly efficient in packaging envelope-pseudotyped viruses, and vectors are mainly derived from human immunodeficiency virus (HIV-1), simian immunodeficiency virus (SIV), or feline immunodeficiency virus (FIV). These vectors maintain all viral gene sequences except for gene encoding the envelope proteins. HIV-1 packaging system is the most widely used pseudotyped virus packaging system and also used for RABV and other rhabdovirus pseudotyping. To minimize the viral gene recombination and reduce the possibility of reversion to the wild-type virus, two to four plasmids that contain HIV genes are usually used for packaging. HIV-1 vector pseudotyped viruses are usually produced by plasmid co-transfection to suitable cell lines such as HEK293T and harvest the supernatant to obtain the pseudovirus stocks. Briefly, the RABV-G expressing plasmids and HIV-1 backbone plasmids were co-transfected into HEK293T cells by transfection reagents such as Lipofectamine-2000, and the supernatant containing

Table 10.1 Pseudovirus packaging systems for RABV and non-RABV lyssaviruses

Packaging vector	Pseudotyped lyssavirus	Reporting Gene	Application	Reference
HIV	RABV (street isolates and mutants), ARAV, ABLV, BBLV, DUVV, EBLV-1, EBLV-2, IRKV, KHUV, LBV, MOKV, SHIBV, WCBV, IKOV	Luciferase, GFP, β-galactosidase	Serosurveillance, cross-neutralization, antigenicity evaluation, mAb screen and epitope mapping, antiviral drug screen, gene therapy	[50–59]
MLV	RABV, EBLV-1, EBLV-2	Luciferase, GFP	Neutralization	[53, 60]
VSV	RABV, ABLV, EBLV-1, EBLV-2, MOKV	Luciferase, GFP	Neutralization, cell tropism, antiviral drug screen	[61–63]

pseudotyped RABV was harvested 48 h after transfection. Through 0.45-μm pore size filtration, the pseudovirus stocks were obtained and frozen at −70 °C for further use.

As a rhabdovirus with bullet shape, the VSV system is also a common used packaging vector for RABV. VSV packaging systems typically utilize replication-defective VSV viruses to infect cells and bearing RABV G protein by transfection of G protein expressing plasmid, to package pseudotyped RABV. Briefly, HEK293T cells were transfected with RABV-G expressing plasmids and then infected with G*ΔG-VSV 24 h after transfection. At 1–2 h after infection, HEK293T cells were washed with PBS to remove the residual G*ΔG-VSV and cultured with new complete medium for 24 h. Supernatants containing pseudotyped RABV were harvested, filtered, and stored at −70 °C in aliquots for further use [61]. Since some VSV virus will remain in the pseudotyped virus preparations which may interfere the detection results, the residue of VSV in the pseudotyped virus should be minimized, and corresponding controls should be added in subsequent tests.

The MLV system, also called the retroviral system, is used to make pseudotyped viruses due to its high efficiency and may be a better choice than HIV-1 packaging system in some cases. Pseudotyped RABV were generated by co-transfection of HEK293T cells with plasmids that encoding RABV G protein, MLV gag-pol, and a retroviral vector encoding the reporter genes such as GFP and luciferase. Supernatants were harvested 48 or 72 h post-transfection and titrated on HEK293T and N2A cell lines [53, 60]. Diana et al. titered the neutralizing antibodies using a MLV-based pseudotyped virus bearing the RABV-G protein [60].

Pseudotyped virus systems for RABV usually introduce reporter genes on their backbone plasmids, which are easier to operate and quantify accurately than live virus experiments. The reporter genes used for RABV pseudotyped virus usually encode enzymes or fluorescent proteins, including luciferase, GFP, and β-galactosidase. Wright et al. established a lentiviral pseudotyped RABV using luciferase, GFP, and β-galactosidase as reporter gene, respectively [52, 53]. These two types of reporter genes have their own advantages and disadvantages. Reporter genes such as luciferase have high sensitivity and low background signal, but the detection and data analysis are relatively time-consuming and costly; while fluorescent protein reporter genes such as green fluorescent protein have low experimental costs and can be used in vivo or in vitro and easy to operate and detect in experiments, the sensitivity is low and the background signal is relatively high [65, 66].

A high yield pseudotyped virus packaging system is always needed for practical applications. Factors such as the efficiency of envelope expression, the packaging system, and the production cell lines can critically influence the yield and titer of pseudoviruses. Thus, to obtain a high titer of RABV pseudotyped viruses, such factors need to be optimized. Our lab constructed a high titer of RABV-G bearing pseudotyped virus system based on HIV-1 vector, through a serial of optimization on G protein expressing plasmid, HIV-1 backbone expressing plasmid, transfection ratio of these two plasmids, cell type, and number for pseudotyped virus production [50]. Using our optimized pSG3$^{\Delta env}$ two-plasmid system by inserting a firefly luciferase reporter gene and adding CMV promoter, the yield of pseudotyped virus was greatly improved. Utilizing the optimized pseudotyped virus system, our lab constructed pseudotyped viruses of various viruses, including Lassa fever virus, Chikungunya virus, Nipah virus, Marburg virus, H7N9 virus, etc. [50, 67–70]. Wright et al. also optimized the package viral backbone (HIV-1, MLV) and several cell lines for titration and establish a high yield lentiviral packaging system with RABV [53]. For some RABV strains (CVS-B2c) with low titers packaged by HIV-1 system, the packaging efficiency could be enhanced by constructing a chimeric G protein where the cytoplasmic domain of the RABV G protein was switched for that of VSV-G [71].

10.4 Application of Pseudotyped RABV

Due to its safety and easy to prepare and operate, the single round infectious pseudotyped RABVs have been widely applied in neutralizing antibody detection, anti-RABV mAb screening and epitope mapping, evaluation of viral infection, cell tropism and antigenicity, screening of antiviral drugs, and also gene therapy.

10.4.1 Application of Pseudotyped RABV in Neutralizing Antibody Detection

Although the mortality rate of rabies is high, timely and correct administration of post-exposure prophylaxis (PEP) after exposure to infection can effectively prevent rabies. The PEP is composed of vaccination and passive immune with RIG that could neutralize the viruses around the wound. The neutralizing antibody is considered as a key factor in protecting against rabies. According to WHO guideline, a titer of neutralizing antibody above 0.5 IU/ml is indicative of adequate minimum humoral response after the vaccination of humans against rabies. Thus, a robust and accurate assay for determination of anti-RABV neutralizing antibody is important. The rapid fluorescent focus inhibition test (RFFIT) [72] and fluorescent antibody virus neutralization (FAVN) test [73] for RVNA detection and titration are routinely used and recommended by the WHO and OIE as a gold standard for RVNA detection. Both tests can effectively determine the rabies vaccine induced neutralizing antibodies in humans and animals [72–74]. However, FAVN and RFFIT testing based on the live RABV, and an expensive fluorescent dye conjugated anti-rabies antibody and biosafety level 2 (BSL-2) are also required.

Pseudotyped virus-based neutralization assay offer great advantages over the live virus-based assay because they are safer to handle and versatile. Nie et al. developed a highly efficient RABV pseudotyped lentivirus production platform and established a robust neutralization assays both in vitro and in vivo for evaluation of rabies vaccines [50]. The pseudotyped virus-based assay showed a good linear correlation to RFFIT test ($R^2 = 0.946$, $p < 0.001$) and demonstrated better reproducibility. Wright also developed a lentiviral pseudotyped neutralization assay that has 100% specific and equal or greater sensitivity compared with live virus-based assay [52, 53], and a strong correlation between FAVN and the pseudotyped virus-based assay was also described. Sarah et al. developed a neutralization assay using VSV pseudotyped with RABV-G protein, and the testing results of RABV reference serum or immune sera from vaccinated mice showed a strong correlation with the RFFIT. Using this assay, they also found that neutralization of pseudotyped viruses bearing the other lyssaviruses G protein by reference serum were significantly reduced or even not neutralized [62]. Diana et al. adapted the RFFIT by using a MLV-based pseudotyped virus bearing the RABV-G protein and carrying a GFP marker gene and named this assay as pseudotyped micro-neutralization rapid fluorescent focus inhibition test (pmRFFIT). This pmRFFIT required much lesser serum volume and is safer to handle and has moderate sensitivity (78.79%) and high specificity (84.62%) [60].

A EuroSciCon conference organized by the Viral Pseudotype Unit discussed the recent advances of pseudotype technology in important human and animal pathogens, including RABV/lyssaviruses and other viruses [75]. Dr. Wright (University of Westminster, UK) reported that his group constructed RABV pseudotyped viruses and test the neutralizing response; they found those results similar or more sensitive than live virus assays. Dr. Giada Mattiuzzo (NIBSC, UK) has been working to validate an alternative pseudotyped virus-based neutralization assay to test the biological potency of RIG. Dr. Mattiuzzo reported that they evaluated five commercial RIGs using pseudotyped virus-based neutralization assay, and the potency results were similar to those obtained by RFFIT. Dr. Mattiuzzo reported the pseudotyped virus-based neutralization assay is a promising alternative system for potency testing of RIG, and he will present this assay for inclusion into the European Pharmacopoeia monograph for RIG, to increase safety and accessibility of potency testing [75].

The pseudotyped virus-based neutralization assay is also used in serosurveillance and cross-neutralization assessment between RABV and non-RABV lyssaviruses. Wright et al. analyzed 304 serum samples from Tanzanian dogs for detection of the anti-RABV antibodies by a lentiviral pseudotyped assay using *lacZ* gene as a reporter element. This method provided greater sensitivity compared to live virus assay and suggested the proposed pseudotyped assay is a suitable option for undertaking lyssavirus serosurveillance in areas most affected by these infections [52]. Wright et al. also made a serosurveillance for lyssaviruses LBV, MOKV, and WCBV using a HIV-1 pseudotyped virus system. They found 56% of screened bat serum neutralized LBV-G pseudotyped virus, 27% for MOKV-G pseudotyped virus, and 1% for WCBV-G pseudotyped virus. The high seroprevalence suggest that the infection rate of LBV in *Eidolon helvum* remains high [54].

All viruses within *Lyssavirus* genus are highly neurotropic and are capable of causing fatal rabies. Safe and effective RABV vaccines have been available for decades, and current vaccines are generally based on inactivated preparations of classical RABV strains. While current vaccines are highly effective against RABV, the ability of RABV vaccines to protect against non-RABV lyssaviruses is variable or undefined. Wright et al. compare the cross-neutralization of 60 sera from vaccinated humans, canines, and felines against RABV and two European bat lyssaviruses (EBLV-1 and EBLV-2) using lentiviral pseudotyped virus. They found the antibody-mediated neutralization in humans and cats were higher against RABV CVS-11, followed by EBLV-2 and finally EBLV-1, and consist with the hypothesis that the higher the degree of G protein similarity, the more comparable the neutralization titers [53]. Evans et al. assess immunogenic antigenic sites within the lyssavirus glycoprotein by swapping sites between phylogroup I and II lyssaviruses using a HIV-1 vector pseudotyped virus. They demonstrated intra- but limited inter-phylogroup cross-neutralization and indicated the immunodominance of antigenic site II [76].

10.4.2 Application of Pseudotyped RABV in mAb Screening and Epitope Mapping

Due to the global shortage of HRIG and the relatively high price, and certain immunogenicity of horse-derived immunoglobulins, the WHO recommends the use of mAb products to replace the current immunoglobulins. Because of the high fatality rate after the onset of rabies, the WHO also recommends the use of at least two mAbs with non-overlapping epitopes to form a "cocktail," which can broadly neutralize street RABV variants to the greatest extent. Recombinant mAb products have certain advantages in terms of availability, safety, purity, and price. At present, Rabishield [77, 78], an anti-RABV mAb produced by the Serum Research Institute of India, has been approved for marketing in India, mAb NM57 has been approved in China recently, and there are several mAbs in clinical trials or R&D stage [48, 49, 55, 79, 80].

For the fatal rabies, besides the potent neutralization activity, a mAb candidate need to have a broad neutralization coverage to diverse RABV circulating isolates to offer an efficient protection. Pseudotyped viruses with artificial amino acid mutations or RABV-G of circulating variants are easy to construct, and the broad neutralization properties of anti-RABV G mAbs could be well characterized. Paola et al. screened sera from 90 RABV vaccinees by lentiviral pseudotyped lyssavirus isolates of phylogroup I, II, and III viruses and further screened 500 mAbs from 4 vaccinees by testing the neutralization activity to pseudotyped CVS-11 strain [55]. Finally, mAb RVC20 and RVC58 were isolated with potent and broad spectrum activity to 35 RABV strains and 25 non-RABV lyssaviruses. The epitopes of these two mAbs were also mapped by testing against chimeric and mutant RABV and LBV pseudotyped viruses, and data indicated the epitopes of these two mAbs are non-overlapping that RVC20 bind to antigenic I and RVC58 bind to antigenic III, respectively. The combination of the RVC20 and RVC58 represents a treatment with breadth and potency for the development of a low-risk and affordable product to replace RIGs in rabies PEP. In a recent study, the RVC20 and RVC58 mAb cocktail cures symptomatic rabid mice [81]. Boruah et al. screened a llama-derived antibody library and identified two single-domain antibodies 26424 and 26434 with potent neutralizing activity against RABV CVS-11 G protein pseudotyped virus [56].

10.4.3 Application of Pseudotyped RABV in Evaluation of Viral Infection, Cell Tropism, and Antigenicity

Dawn et al. evaluated the host cell tropism of Australian bat lyssavirus (ABLV) using maxGFP-encoding recombinant vesicular stomatitis viruses that express ABLV G glycoproteins and found ABLV receptor is conserved but not ubiquitous among mammalian cell lines and ABLV variants can utilize alternate receptors for entry. Data indicated that the proposed RABV receptors were not sufficient to permit ABLV entry into resistant cells, suggesting that ABLV utilizes an unknown alternative receptor [63].

The G protein is a direct target to neutralizing antibodies and is also a key component of vaccine to induce the neutralizing antibodies, thus, closely related to the body's immune response and vaccine protection [26, 82–85]. Research on the antigenicity of viral envelope proteins, including antigenic sites, antigenic variation, etc., can provide a basis for the development and clinical usage of vaccines and antiviral products. In our previous study, we systematically analyzed 2890 RABV G sequences worldwide and analyzed the effect of single amino acid mutation on the antigenicity of G protein using a HIV-1 pseudotyped system [51]. A total of 99 natural occurring mutants for their reactivities to well-characterized neutralizing mAbs and vaccine-induced antisera were investigated. We found that the sequence similarity between RABV street virus and vaccine strains showed a decreasing trend, and some single amino acid mutations significantly affect the antigenicity of RABV, especially mutations in antigenic site III (R333P, I338T, etc.) that significantly reduced the neutralization sensitivity to mAbs, vaccine-induced antisera, and HRIG. The antigenic variants containing G protein mutations were observed in a wide range of animal hosts and geographic locations, with most of them emerging since 2010. As the number of antigenic variants has increased in recent years, we suggest that close monitoring on variation of street isolates should be strengthened. The pseudotyped virus-based neutralization assay was also used to compare the antigenicity between RABV and non-RABV lyssaviruses, to better understand the antigenicity differences and provide a guidance for development of vaccines and mAbs [53, 55, 76].

Glycosylation of viral G protein plays an important role in viral infectivity and evasion of antibody-mediated neutralization [86–89]. Previously, using a HIV-1 pseudotyped virus system, we systematically studied the effect of potential N-linked glycosylation site (PNGS) on HIV-1 infection and antibody-mediated neutralization, and we found specific individual PNGS, and their combination had significant impact on viral infection and neutralization [90, 91]. RABV G protein has less glycosylation level compared to HIV and SARS-CoV-2 and naturally presents two PNGS at N37 and N319 in most street RABV isolates. Recently, we constructed 11 mutants at 7 PNGS sites on CVS-N2c and evaluated their effect on infectivity, cell tropism, and antibody-mediated neutralization using HIV-1 vector pseudotyped assay. We found PNGS have certain impact on both viral infection and neutralization (data unpublished).

10.4.4 Application of Pseudotyped RABV in Screening of Antiviral Drugs

RABV-G pseudotyped virus is also an efficient and high-throughput tool for screen inhibitors for virus entry, and several molecules were identified with anti-RABV activity in vitro/vivo.

Our lab screened a library of 767 approved drugs using a high-throughput screening assay based on RABV-G pseudotyped HIV vectors [57]. We identified 11 candidates with anti-RABV activity, and clofazimine (CFZ), an anti-leprosy drug, displayed most profound anti-RABV effect that against RABV with EC_{50} of 2.28 μM and SI over 967. Investigations of mode of action revealed that CFZ inhibit RABV-G-mediated membrane fusion, and CFZ also exhibited elevated survival rates in vivo. Utilizing a single-cycle RABV reporter strain, Venice identified a first-in-class anti-RABV inhibitor, GRP-60367, with a specificity SI of >100,000 to a subset of RABV strains [92]. GRP-60367 directly act with RABV G protein and block RABV entry; this inhibitory effect was also confirmed in a RABV-G pseudotyped VSV-based assay. Dynamin-specific inhibitor dynasore, chlorpromazine, a drug that blocks clathrin-mediated endocytosis, was reported that could significantly inhibit ABLV-G protein pseudotyped VSV entry [93, 94].

10.5 Summary

Pseudotyped viruses have been widely used in lyssavirus research, and a variety of pseudotyped lyssaviruses have been successfully packaged and applied to study the functions of glycoprotein and evaluation of antiviral products, including neutralizing antibody detection, cell tropism, receptor identification, and screening of entry inhibitors. Clearly, given the evolutionary nature and lethality of lyssaviruses, we should continue our research efforts on these neglected viruses, and pseudotyped virus could be a powerful tool for this purpose.

References

1. Horton, D.L., et al.: Quantifying antigenic relationships among the lyssaviruses. J. Virol. **84**, 11841–11848 (2010)
2. Kuzmin, I.V., et al.: Possible emergence of West Caucasian bat virus in Africa. Emerg. Infect. Dis. **14**, 1887–1889 (2008)
3. Fooks, A.: The challenge of new and emerging lyssaviruses. Expert Rev. Vaccines. **3**, 333–336 (2004)
4. Brookes, S.M., Healy, D.M., Fooks, A.R.: Ability of rabies vaccine strains to elicit cross-neutralising antibodies. Dev. Biol. **125**, 185–193 (2006)
5. Nolden, T., et al.: Comparative studies on the genetic, antigenic and pathogenic characteristics of Bokeloh bat lyssavirus. J. Gen. Virol. **95**, 1647–1653 (2014)
6. Hanlon, C.A., et al.: Efficacy of rabies biologics against new lyssaviruses from Eurasia. Virus Res. **111**, 44–54 (2005)
7. Brookes, S.M., Parsons, G., Johnson, N., McElhinney, L.M., Fooks, A.R.: Rabies human diploid cell vaccine elicits cross-neutralising and cross-protecting immune responses against European and Australian bat lyssaviruses. Vaccine. **23**, 4101–4109 (2005)
8. Horton, D.L., et al.: Antigenic and genetic characterization of a divergent African virus, Ikoma lyssavirus. J. Gen. Virol. **95**, 1025–1032 (2014)
9. Badrane, H., Bahloul, C., Perrin, P., Tordo, N.: Evidence of two lyssavirus phylogroups with distinct pathogenicity and immunogenicity. J. Virol. **75**, 3268–3276 (2001)
10. Faber, M., et al.: Identification of viral genomic elements responsible for rabies virus neuroinvasiveness. Proc. Natl. Acad. Sci. U. S. A. **101**, 16328–16332 (2004)
11. Finke, S., Granzow, H., Hurst, J., Pollin, R., Mettenleiter, T.C.: Intergenotypic replacement of lyssavirus matrix proteins demonstrates the role of lyssavirus M proteins in intracellular virus accumulation. J. Virol. **84**, 1816–1827 (2010)
12. Pulmanausahakul, R., Li, J., Schnell, M.J., Dietzschold, B.: The glycoprotein and the matrix protein of rabies virus affect pathogenicity by regulating viral replication and facilitating cell-to-cell spread. J. Virol. **82**, 2330–2338 (2008)
13. Yamada, K., et al.: Serial passage of a street rabies virus in mouse neuroblastoma cells resulted in attenuation: potential role of the additional N-glycosylation of a viral glycoprotein in the reduced pathogenicity of street rabies virus. Virus Res. **165**, 34–45 (2012)
14. Hamamoto, N., et al.: Association between RABV G proteins transported from the perinuclear space to the cell surface membrane and N-glycosylation of the Sequon Asn(204). Jpn. J. Infect. Dis. **68**, 387–393 (2015)
15. Yamada, K., et al.: Addition of a single N-glycan to street rabies virus glycoprotein enhances virus production. J. Gen. Virol. **94**, 270–275 (2013)
16. Yamada, K., Noguchi, K., Nishizono, A.: Characterization of street rabies virus variants with an additional N-glycan at position 247 in the glycoprotein. Arch. Virol. **159**, 207–216 (2014)
17. Lentz, T.L., Burrage, T.G., Smith, A.L., Crick, J., Tignor, G.H.: Is the acetylcholine receptor a rabies virus receptor? Science. **215**, 182–184 (1982)
18. Gastka, M., Horvath, J., Lentz, T.L.: Rabies virus binding to the nicotinic acetylcholine receptor alpha subunit demonstrated by virus overlay protein binding assay. J. Gen. Virol. **77**(Pt 10), 2437–2440 (1996)
19. Tuffereau, C., Benejean, J., Blondel, D., Kieffer, B., Flamand, A.: Low-affinity nerve-growth factor receptor (P75NTR) can serve as a receptor for rabies virus. EMBO J. **17**, 7250–7259 (1998)
20. Thoulouze, M.I., et al.: The neural cell adhesion molecule is a receptor for rabies virus. J. Virol. **72**, 7181–7190 (1998)
21. Wang, J., et al.: Metabotropic glutamate receptor subtype 2 is a cellular receptor for rabies virus. PLoS Pathog. **14**, e1007189 (2018)
22. Gaudin, Y.: Rabies virus-induced membrane fusion pathway. J. Cell Biol. **150**, 601–612 (2000)

23. Durrer, P., Gaudin, Y., Ruigrok, R.W., Graf, R., Brunner, J.: Photolabeling identifies a putative fusion domain in the envelope glycoprotein of rabies and vesicular stomatitis viruses. J. Biol. Chem. **270**, 17575–17581 (1995)
24. Benmansour, A., et al.: Antigenicity of rabies virus glycoprotein. J. Virol. **65**, 4198–4203 (1991)
25. Bunschoten, H., et al.: Characterization of a new virus-neutralizing epitope that denotes a sequential determinant on the rabies virus glycoprotein. J. Gen. Virol. **70**(Pt 2), 291–298 (1989)
26. Dietzschold, B., et al.: Characterization of an antigenic determinant of the glycoprotein that correlates with pathogenicity of rabies virus. Proc. Natl. Acad. Sci. U. S. A. **80**, 70–74 (1983)
27. Kuzmina, N.A., Kuzmin, I.V., Ellison, J.A., Rupprecht, C.E.: Conservation of binding epitopes for monoclonal antibodies on the rabies virus glycoprotein. J. Antivir. Antiretrovir. **5**, 037–043 (2013)
28. Luo, T.R., et al.: A virus-neutralizing epitope on the glycoprotein of rabies virus that contains Trp251 is a linear epitope. Virus Res. **51**, 35–41 (1997)
29. Seif, I., Coulon, P., Rollin, P.E., Flamand, A.: Rabies virulence: effect on pathogenicity and sequence characterization of rabies virus mutations affecting antigenic site III of the glycoprotein. J. Virol. **53**, 926–934 (1985)
30. Matsumoto, T., et al.: Isolation and characterization of novel human monoclonal antibodies possessing neutralizing ability against rabies virus. Microbiol. Immunol. **54**, 673–683 (2010)
31. Lafon, M., Wiktor, T.J., Macfarlan, R.I.: Antigenic sites on the CVS rabies virus glycoprotein: analysis with monoclonal antibodies. J. Gen. Virol. **64**(Pt 4), 843–851 (1983)
32. Muller, T., et al.: Development of a mouse monoclonal antibody cocktail for post-exposure rabies prophylaxis in humans. PLoS Negl. Trop. Dis. **3**, e542 (2009)
33. Gao, G.F.: From "a"IV to "Z"IKV: attacks from emerging and re-emerging pathogens. Cell. **172**, 1157–1159 (2018)
34. Mumford, J.A.: Vaccines and viral antigenic diversity. Rev. Sci. Tech. **26**, 69–90 (2007)
35. Smith, D.J., et al.: Mapping the antigenic and genetic evolution of influenza virus. Science. **305**, 371–376 (2004)
36. Hanada, K., Suzuki, Y., Gojobori, T.: A large variation in the rates of synonymous substitution for RNA viruses and its relationship to a diversity of viral infection and transmission modes. Mol. Biol. Evol. **21**, 1074–1080 (2004)
37. Bourhy, H., et al.: The origin and phylogeography of dog rabies virus. J. Gen. Virol. **89**, 2673–2681 (2008)
38. Troupin, C., et al.: Large-scale phylogenomic analysis reveals the complex evolutionary history of rabies virus in multiple carnivore hosts. PLoS Pathog. **12**, e1006041 (2016)
39. Ajorloo, M., et al.: Evaluation and phylogenetic analysis of regular rabies virus vaccine strains. Arch. Iran. Med. **21**, 101–110 (2018)
40. Patel, A.C., et al.: Molecular and immunogenic characterization of BHK-21 cell line adapted CVS-11 strain of rabies virus and future prospect in vaccination strategy. Virus. **26**, 288–296 (2015)
41. Orlowska, A., Zmudzinski, J.F.: Molecular epidemiology of rabies virus in Poland. Arch. Virol. **159**, 2043–2050 (2014)
42. Aghomo, H.O., Rupprecht, C.E.: Antigenic characterisation of virus isolates from vaccinated dogs dying of rabies. Trop. Anim. Health Prod. **22**, 275–280 (1990)
43. Umoh, J.U., Cox, J.H., Schneider, L.G.: Antigenic characterization of street rabies virus isolates from Nigeria using monoclonal antibodies. Zentralbl. Veterinarmed. B. **37**, 222–228 (1990)
44. Okoh, A.E.: Antigenic characterization of rabies virus isolates from vaccinated dogs in plateau state, Nigeria. Vet. Res. Commun. **24**, 203–211 (2000)
45. Zanluca, C., et al.: Novel monoclonal antibodies that bind to wild and fixed rabies virus strains. J. Virol. Methods. **175**, 66–73 (2011)
46. Wiktor, T.J., Koprowski, H.: Antigenic variants of rabies virus. J. Exp. Med. **152**, 99–112 (1980)

47. Sureau, P., Rollin, P., Wiktor, T.J.: Epidemiologic analysis of antigenic variations of street rabies virus: detection by monoclonal antibodies. Am. J. Epidemiol. **117**, 605–609 (1983)
48. Bakker, A.B., et al.: Novel human monoclonal antibody combination effectively neutralizing natural rabies virus variants and individual in vitro escape mutants. J. Virol. **79**, 9062–9068 (2005)
49. Marissen, W.E., et al.: Novel rabies virus-neutralizing epitope recognized by human monoclonal antibody: fine mapping and escape mutant analysis. J. Virol. **79**, 4672–4678 (2005)
50. Nie, J., et al.: Development of in vitro and in vivo rabies virus neutralization assays based on a high-titer pseudovirus system. Sci. Rep. **7**, 42769 (2017)
51. Wang, W., et al.: Antigenic variations of recent street rabies virus. Emerg. Microbes Infect. **8**, 1584–1592 (2019)
52. Wright, E., et al.: A robust lentiviral pseudotype neutralisation assay for in-field serosurveillance of rabies and lyssaviruses in Africa. Vaccine. **27**, 7178–7186 (2009)
53. Wright, E., et al.: Investigating antibody neutralization of lyssaviruses using lentiviral pseudotypes: a cross-species comparison. J. Gen. Virol. **89**, 2204–2213 (2008)
54. Wright, E., et al.: Virus neutralising activity of African fruit bat (Eidolon helvum) sera against emerging lyssaviruses. Virology. **408**, 183–189 (2010)
55. De Benedictis, P., et al.: Development of broad-spectrum human monoclonal antibodies for rabies post-exposure prophylaxis. EMBO Mol. Med. **8**, 407–421 (2016)
56. Boruah, B.M., et al.: Single domain antibody multimers confer protection against rabies infection. PLoS One. **8**, e71383 (2013)
57. Wu, J., et al.: Clofazimine: a promising inhibitor of rabies virus. Front. Pharmacol. **12**, 598241 (2021)
58. Federici, T., et al.: Comparative analysis of HIV-1-based lentiviral vectors bearing lyssavirus glycoproteins for neuronal gene transfer. Genet. Vaccines Ther. **7**, 1 (2009)
59. Watson, D.J., Kobinger, G.P., Passini, M.A., Wilson, J.M., Wolfe, J.H.: Targeted transduction patterns in the mouse brain by lentivirus vectors pseudotyped with VSV, Ebola, Mokola, LCMV, or MuLV envelope proteins. Mol. Ther. **5**, 528–537 (2002)
60. Meza, D.K., et al.: Predicting the presence and titre of rabies virus-neutralizing antibodies from low-volume serum samples in low-containment facilities. Transbound. Emerg. Dis. **68**, 1564–1576 (2021)
61. Cai, M., et al.: Analysis of the evolution, infectivity and antigenicity of circulating rabies virus strains. Emerg. Microbes Infect. **11**, 1474–1487 (2022)
62. Moeschler, S., Locher, S., Conzelmann, K.K., Kramer, B., Zimmer, G.: Quantification of lyssavirus-neutralizing antibodies using vesicular stomatitis virus Pseudotype particles. Viruses. **8**, 254 (2016)
63. Weir, D.L., Smith, I.L., Bossart, K.N., Wang, L.F., Broder, C.C.: Host cell tropism mediated by Australian bat lyssavirus envelope glycoproteins. Virology. **444**, 21–30 (2013)
64. Li, Q., Liu, Q., Huang, W., Li, X., Wang, Y.: Current status on the development of pseudoviruses for enveloped viruses. Rev. Med. Virol. **28**, e1963 (2018)
65. Nie, J., Liu, Y., Huang, W., Wang, Y.: Development of a triple-color Pseudovirion-based assay to detect neutralizing antibodies against human papillomavirus. Viruses. **8**, 107 (2016)
66. Sakuma, T., et al.: Characterization of retroviral and lentiviral vectors pseudotyped with xenotropic murine leukemia virus-related virus envelope glycoprotein. Hum. Gene Ther. **21**, 1665–1673 (2010)
67. Li, Q., et al.: An LASV GPC pseudotyped virus based reporter system enables evaluation of vaccines in mice under non-BSL-4 conditions. Vaccine. **35**, 5172–5178 (2017)
68. Nie, J., et al.: Nipah pseudovirus system enables evaluation of vaccines in vitro and in vivo using non-BSL-4 facilities. Emerg. Microbes Infect. **8**, 272–281 (2019)
69. Ning, T., et al.: Antigenic drift of influenza a(H7N9) virus hemagglutinin. J. Infect. Dis. **219**, 19–25 (2019)

70. Wu, J., Zhao, C., Liu, Q., Huang, W., Wang, Y.: Development and application of a bioluminescent imaging mouse model for Chikungunya virus based on pseudovirus system. Vaccine. **35**, 6387–6394 (2017)
71. Carpentier, D.C., et al.: Enhanced pseudotyping efficiency of HIV-1 lentiviral vectors by a rabies/vesicular stomatitis virus chimeric envelope glycoprotein. Gene Ther. **19**, 761–774 (2012)
72. Khalil, W.M., Aboshanab, K.M., Aboulwafa, M.M.: Evaluation and correlation of rabies vaccine potency using the National Institute of health, rapid focus fluorescent inhibition, and passive hemagglutination tests. Viral Immunol. **35**, 159–169 (2022)
73. De Benedictis, P., Mancin, M., Cattoli, G., Capua, I., Terregino, C.: Serological methods used for rabies post vaccination surveys: an analysis. Vaccine. **30**, 5611–5615 (2012)
74. Liu, Y., Zhang, S., Zhang, F., Hu, R.: A semi-quantitative serological method to assess the potency of inactivated rabies vaccine for veterinary use. Virol. Sin. **27**, 259–264 (2012)
75. Bentley, E.M., Mather, S.T., Temperton, N.J.: The use of pseudotypes to study viruses, virus sero-epidemiology and vaccination. Vaccine. **33**, 2955–2962 (2015)
76. Evans, J.S., et al.: Antigenic site changes in the rabies virus glycoprotein dictates functionality and neutralizing capability against divergent lyssaviruses. J. Gen. Virol. **99**, 169–180 (2018)
77. Gogtay, N., et al.: Safety and pharmacokinetics of a human monoclonal antibody to rabies virus: a randomized, dose-escalation phase 1 study in adults. Vaccine. **30**, 7315–7320 (2012)
78. Sloan, S.E., et al.: Identification and characterization of a human monoclonal antibody that potently neutralizes a broad panel of rabies virus isolates. Vaccine. **25**, 2800–2810 (2007)
79. Chao, T.Y., et al.: SYN023, a novel humanized monoclonal antibody cocktail, for post-exposure prophylaxis of rabies. PLoS Negl. Trop. Dis. **11**, e0006133 (2017)
80. Franka, R., et al.: In vivo efficacy of a cocktail of human monoclonal antibodies (CL184) against diverse North American bat rabies virus variants. Trop. Med. Infect. Dis. **2**, 48 (2017)
81. de Melo, G.D., et al.: A combination of two human monoclonal antibodies cures symptomatic rabies. EMBO Mol. Med. **12**, e12628 (2020)
82. Dietzschold, B., et al.: Differences in cell-to-cell spread of pathogenic and apathogenic rabies virus in vivo and in vitro. J. Virol. **56**, 12–18 (1985)
83. Lodmell, D.L., et al.: DNA immunization protects nonhuman primates against rabies virus. Nat. Med. **4**, 949–952 (1998)
84. Prehaud, C., Takehara, K., Flamand, A., Bishop, D.H.: Immunogenic and protective properties of rabies virus glycoprotein expressed by baculovirus vectors. Virology. **173**, 390–399 (1989)
85. Morimoto, K., Hooper, D.C., Spitsin, S., Koprowski, H., Dietzschold, B.: Pathogenicity of different rabies virus variants inversely correlates with apoptosis and rabies virus glycoprotein expression in infected primary neuron cultures. J. Virol. **73**, 510–518 (1999)
86. Bonhomme, C.J., Knopp, K.A., Bederka, L.H., Angelini, M.M., Buchmeier, M.J.: LCMV glycosylation modulates viral fitness and cell tropism. PLoS One. **8**, e53273 (2013)
87. Ge, P., Ross, T.M.: Evolution of a(H1N1) pdm09 influenza virus masking by glycosylation. Expert Rev. Vaccines. **20**, 519–526 (2021)
88. Kim, P., et al.: Glycosylation of hemagglutinin and neuraminidase of influenza a virus as signature for ecological Spillover and adaptation among influenza reservoirs. Viruses. **10**, 183 (2018)
89. Yen, P.J., et al.: Loss of a conserved N-linked glycosylation site in the simian immunodeficiency virus envelope glycoprotein V2 region enhances macrophage tropism by increasing CD4-independent cell-to-cell transmission. J. Virol. **88**, 5014–5028 (2014)
90. Wang, W., et al.: A systematic study of the N-glycosylation sites of HIV-1 envelope protein on infectivity and antibody-mediated neutralization. Retrovirology. **10**, 14 (2013)

91. Wang, W., et al.: N463 glycosylation site on V5 loop of a mutant gp120 regulates the sensitivity of HIV-1 to neutralizing monoclonal antibodies VRC01/03. J. Acquir. Immune Defic. Syndr. **69**, 270–277 (2015)
92. Du Pont, V., et al.: Identification and characterization of a small-molecule rabies virus entry inhibitor. J. Virol. **94** (2020)
93. Mues, M.B., Cheshenko, N., Wilson, D.W., Gunther-Cummins, L., Herold, B.C.: Dynasore disrupts trafficking of herpes simplex virus proteins. J. Virol. **89**, 6673–6684 (2015)
94. Weir, D.L., Laing, E.D., Smith, I.L., Wang, L.F., Broder, C.C.: Host cell virus entry mediated by Australian bat lyssavirus G envelope glycoprotein occurs through a clathrin-mediated endocytic pathway that requires actin and Rab5. Virol. J. **11**, 40 (2014)

Chapter 11
Pseudotyped Viruses for Enterovirus

Xing Wu, Lisha Cui, Yu Bai, Lianlian Bian, and Zhenglun Liang

Abstract Using a non-pathogenic pseudotyped virus as a surrogate for a wide-type virus in scientific research complies with the recent requirements for biosafety. Enterovirus (EV) contains many species of viruses, which are a type of nonenveloped virus. The preparation of its corresponding pseudotyped virus often needs customized construction compared to some enveloped viruses. This article describes the procedures and challenges in the construction of pseudotyped virus for enterovirus (pseudotyped enterovirus, EVpv) and also introduces the application of EVpv in basic virological research, serological monitoring, and the detection of neutralizing antibody (NtAb).

Keywords Enterovirus · Pseudotyped virus · Neutralizing antibody

Abbreviations

CPE	Cytopathic effect
CMV	Cytomegalovirus
CVA	Coxsackievirus A
CVB	Coxsackievirus B
DNA	Deoxyribonucleic acid
EM	Electron microscopy
ECHO	Enteric cytopathic human orphan virus
EV	Enterovirus
EVpv	Pseudotyped enterovirus
EGFP	Enhanced green fluorescent protein

X. Wu · Y. Bai · L. Bian · Z. Liang (✉)
Division of Hepatitis Virus & Enterovirus Vaccines, Institute for Biological Products, National Institutes for Food and Drug Control, Beijing, China

WHO Collaborating Center for Standardization and Evaluation of Biologicals, Beijing, China

L. Cui
Minhai biotechnology Co. Ltd, Beijing, China

Y. Wang (ed.), *Pseudotyped Viruses*, Advances in Experimental Medicine and Biology 1407, https://doi.org/10.1007/978-981-99-0113-5_11

Fluc	Firefly luciferase
GMC	Geometric mean concentration
NtAb	Neutralizing antibody
PCR	Polymerase chain reaction
PV	Poliovirus
pvNAs	Pseudotyped virus neutralization assay
RLUs	Relative light units
RNA	Ribonucleic acid
RV	Rhinovirus
SCARB2	Scavenger receptor class B member 2
UTR	Untranslated region
VP	Structural viral proteins

11.1 Introduction of Enterovirus

The viruses in *Enterovirus* genus mainly replicate in the gastrointestinal tract or upper respiratory tract but are found in nerve and muscle tissues in rare cases. Their infection could result in a wide variety of illnesses ranging from the common cold to acute neurological disorders, including subclinical infection, herpangina, hand-foot-and-mouth disease, pneumonia, epidemic myalgia, hemorrhagic conjunctivitis, poliomyelitis, pericarditis, aseptic meningitis, encephalitis, acute flaccid paralysis, pharyngitis, tympanitis, rhinosinusitis, and pancreatitis [1–5] (Table 11.1). They could be transmitted from person to person through an air, a fecal-oral route, and/or contaminated objects.

Table 11.1 Common diseases and their pathogens

Affected system	Disease	Serotype	Reference (s)
Nervous system	Poliomyelitis, meningitis, encephalitis, flaccid paralysis	EV-A71, PV-1-3, EV-D68, CVA9, E11	[6–9]
Respiratory system	Pneumonia, herpangina, *acute respiratory viral infections (ARVI)*	CVA, CVB, EV-D68, RV	[10–12]
Circulatory system	Myocarditis, myopericarditis, myocardial infarction	EV-D70, CVA, CVB, ECHO	[13–16]
Locomotor system	Dermatomyositis, polymyositis, *epidemic myalgia pleurodynia*	EV-D68, RV, CVA, CVB, ECHO	[17]
Hormonal system	Pancreatitis, diabetes	CVA, EV-A71	[18]
Others	Hand, foot, and mouth disease; *acute hemorrhagic conjunctivitis*	EV-A71, CVA6, CVA16, EV-D70, CVA24	[19–22]

11.2 Biological Characteristics of Enterovirus

11.2.1 Classification of Enterovirus

Belonging to the *Picornaviridae* family, this genus *Enterovirus* (EV) consists of 15 species, including nine enterovirus species (Enterovirus A, B, C, D, E, F, G, H, and J) and three rhinovirus species (Rhinovirus A, B, and C). Among them, Enterovirus A to D and Rhinovirus species A to C are associated with human diseases, including Enterovirus A, 25 types; Enterovirus B, 63 types; Enterovirus C, 23 types; Enterovirus D, 5 types; Rhinovirus A, 80 types; Rhinovirus B, 32 types; and Rhinovirus C, 57 types [23] (Table 11.2).

The virion of EV consists of 60 protomers which could be assembled into an icosahedral capsid. Available cryo-electron microscopy (EM) data indicates that the virion particles are 30–32 nm in diameter with no envelope.

Enteroviruses have relatively high mutation and recombination rates which may lead to the emergence of new pathogenic strains. The molecular classification of enterovirus is based on VP1 sequences which correlate well with the known serotypes [24]. Intraserotypic divergence is about 25% nucleotide sequence difference or 12% amino acid sequence difference by the comparison of complete VP1 sequences. By far, more than 300 enterovirus strains have been characterized genetically by phylogenetic clustering [25–27].

Table 11.2 Classification of enteroviruses associated with human

Species	No. of types	Virus abbrev.
Enterovirus A	25	CVA (2~8, 10, 12, 14, 16); EV-A (71, 76, 89~92, 114, 119~125)
Enterovirus B	63	CVB (1~6), CVA9, E-(1~7, 9, 11~21, 24~27, 29~33), EV-B (69, 73~75, 77~88, 93, 97, 98, 100, 101, 106, 107, 110~114)
Enterovirus C	23	PV (1~3), CVA (1, 11, 13, 17, 19~22, 24), EV-C (95~96, 99, 102, 104~105, 109, 113, 116~118)
Enterovirus D	5	EV-D (68, 70, 94, 111, 120)
Rhinovirus A	80	RV-A (1, 1B, 2, 7~13, 15, 16, 18~25, 28~34, 36, 38~41, 43, 45~47, 49~51, 53~68, 71, 73~78, 80~82, 85, 88~90, 94, 96, 100~108
Rhinovirus B	32	RV-B (3~6, 14, 17, 26, 27, 35, 37, 42, 48, 52, 69, 70, 72, 79, 83, 84, 84, 91~93, 97, 99~104)
Rhinovirus C	57	RV-C (1~51, 54~57)

CV, coxsackievirus; EV, enterovirus; E, echovirus; PV, poliovirus; RV, rhinovirus

11.2.2 Genome and Life Cycle of the Enterovirus

The enterovirus genome is a single positive-stranded RNA of about 7500 nucleotides. The genome consists of a polyprotein coding region which is flanked by 5' and 3'-untranslated regions (UTRs). The 5′ end of the viral genome is covalently linked to the viral genome-linked protein (VPg/3B). The UTRs include secondary structural elements which control genome replication and translation. 5′-UTR contains the internal ribosome entry site that could initiate the translation of the mRNA by recruiting ribosomes, and 3′-UTR contains a secondary structure that controls viral RNA synthesis [28].

Different enteroviruses enter cells by recognizing distinct receptors. After entering into the host cell, the viral genomic RNA is translated into a single polyprotein containing the structural protein P1, the non-structural protein P2, and P3. This polyprotein then undergoes protease 2A-mediated self-cleavage to release capsid protein precursor P1 [29]. P1 is subsequently cleaved into 3 capsid proteins, VP0, VP1, and VP3, by either 3Cpro or 3CDpro, which are further assembled into protomers and pentamers. The antigenicity of enterovirus is determined by the viral structural proteins VP1, VP2, VP3, and VP4 (Fig. 11.1). In the meantime, non-structural proteins 2A–2C and 3A–3D are also cleaved and folded into mature nonstructural viral proteins, and RNA polymerase 3Dpol initiates the viral replication and produces negative-stranded intermediate RNA which in turn is used as a template for synthesis of positive-stranded viral genomic RNA. Positive-stranded viral genomic RNA can then either be used as template for translation or be packaged into capsids to produce infectious virus particles. Newly synthesized viral RNA and capsid protein pentamers together form the provirion. Finally, VP0 is cleaved into VP2 and VP4 to form the mature virions [30], which exit the host cell via the non-lytic release of extracellular vesicles or via cell lysis at the late stage of infection. In addition to viral polyproteins, viral protease 2Apro and 3Cpro also use the host cell proteins as substrates to optimize the translation, replication, and transmission of virus and inhibit cellular antiviral responses [31].

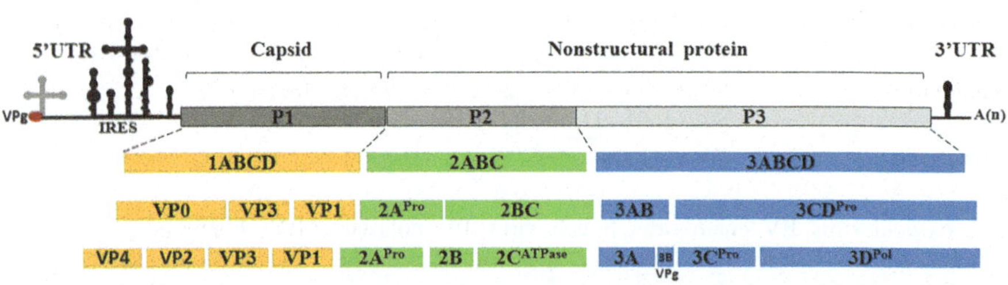

Fig. 11.1 Genome structure and polyprotein processing of enterovirus. Diagram of the enterovirus RNA genome as well as the proteolytic processing of enteroviral polyprotein. 5'UTR, 5' untranslated region; IRES, internal ribosome entry site; 3'UTR A(n), poly(A) tail

11.3 Construction of Pseudotyped Enterovirus

The first step of constructing a pseudotyped enterovirus is to get the full-length infectious cDNA clone either through molecular cloning from enterovirus infected samples or through de novo DNA synthesis based on the gene sequence from GenBank. Then, capsid expresser and replicon containing non-structural genes will be further constructed from the cDNA clones, respectively.

11.3.1 Plasmid Construction

Capsid expresser: Using full-length enteroviral infectious cDNA clone as a template, P1 containing all capsid proteins can be obtained through PCR. By adding EGFP to the 5' end of P1 while adding a 2A protease self-cleavage site between EGFP and P1, the target fragment was cloned into the eukaryotic expression vector to get the EGFP-2A-capsid expresser, which expresses all capsid proteins of the enteroviruses as well as the EGFP fluorescent reporter protein (Fig. 11.2Ba). The EGFP reporter gene can be used to monitor the expression efficiency of capsid protein under the fluorescence microscope.

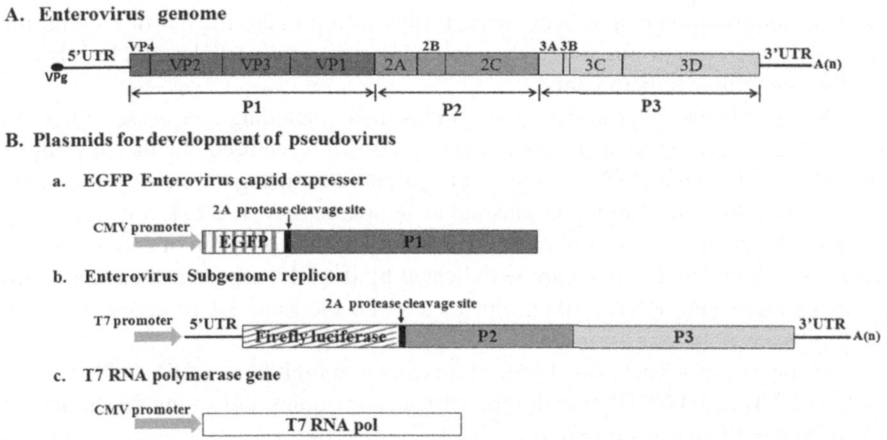

Fig. 11.2 Construction of pseudotyped enterovirus. (**A**) Organization of the enterovirus genome; (**B**) expression vector (a), enterovirus replicon (b), and T7 RNA polymerase plasmid used for the trans-encapsidation; (**Ba**) capsid expresser was used to express all the structural capsid genes in trans, an EGFP gene was inserted upstream of the enterovirus capsid gene and all other viral genome were deleted, and the EGFP was separated by 2A self-cleavage site from the structural genes; (**Bb**) enterovirus subgenome replicon was produced by replacing the capsid coding region with a firefly luciferase reporter gene in the full-length enterovirus genome and a T7 promoter was placed at the 5' end for transcription in vitro; (**Bc**) the T7 RNA polymerase plasmid contains the T7 RNA polymerase under the control of CMV promoter

Subgenomic replicon: by using molecular cloning, a capsid protein segment within the full-length enteroviral infectious clone is replaced with a firefly luciferase(Fluc) reporter gene. The 2A protease cleavage site is inserted between the reporter gene and P2 to get the replicon fragment (5′UTR-Fluc-P2-P3-3′UTR). The subgenomic replicon plasmid construct with a reporter gene is obtained by inserting this fragment into plasmid vector with a T7 promoter (Fig. 11.2Bb).

11.3.2 Preparation of Pseudotyped Enterovirus

There are two pseudotyped enterovirus packaging approaches based on in vitro or in vivo transcription of replicon, i.e., the one-step procedure and the two-step procedure.

For the one-step procedure, 293T cells are co-transfected with the three plasmids in Fig. 11.2 including expresser plasmid (Fig. 11.2Ba), replicon plasmid (Fig. 11.2Bb), and commercialized T7 RNA polymerase plasmid (Fig. 11.2Bc). After transfected into cells, enteroviral structural protein P1 and T7 RNA polymerase are expressed under the CMV promoter. T7 RNA pol will bind to the T7 promoter to drive the transcription from 5’ to 3’ direction to get the subgenomic RNA with the Fluc reporter, which can be subsequently translated into non-structural proteins. The structural P1 protein can be further catalytically cleaved by non-structural proteins into capsid proteins VP1-4. The capsid proteins then assemble into protomers and pentamers, further into pseudovirus particles with the newly synthesized genomic RNA. Pseudotyped enterovirus particles will be released into the cell culture supernatant.

For the two-step procedure, two plasmids including expresser plasmid (Fig. 11.2Ba) and replicon plasmid (Fig. 11.2Bb) are required for the procedure, while the subgenomic RNA are separately prepared by using T7 in vitro transcription kit and a linearized replicon plasmid as template: Step 1, 293T cells are firstly transfected with an expresser plasmid which allows the expression of enterovirus P1 protein to a high level which can be indicated by the GFP reporter. Step 2, in vitro purified subgenomic RNA is then transfected into the same P1 protein expressing 293T cells.

Both the two aforementioned procedure can yield high titers of EV-A71, CVA6, CVA16, CVB3, and CVB5 pseudotype virus in our studies, indicating no significant differences in the encapsidation efficiency.

11.3.3 Pseudotyped Enterovirus Shows High Physical, Chemical, and Antigenic Similarities with Wild-Type Virus

As the pseudotyped enterovirus shares the same capsid proteins as the wild-type virus, it is speculated that the pseudotype enterovirus retains the similar physical, chemical, and antigenic properties as wild-type virus. Compared with the wild-type enterovirus, the pseudotyped enteroviruses show the similar morphological characteristics in the transmission EM analysis [32, 33], similar sediment coefficient when analyzed using ultracentrifugation. Most importantly, the pseudotyped enteroviruses have the same cell tropism as well as antigenicity and immunogenicity compared with the wild-type virus [33]. All these demonstrate that the pseudotyped enterovirus could be used as a surrogate for the wild-type virus. To our knowledge, several pseudotyped enteroviruses have already been successfully constructed, including EV-A71, CVA16, CVA6, CVB3, CVB5, Polio1, Polio12, Polio13, CVA10, and EV-D68 (Table 11.3). These pseudotyped enteroviruses not only serve as an excellent tool for studying the wild-type virus entry process in fundamental research by avoiding any ambiguity possibly caused by post entry events but also can be used in other clinical applications, such as epidemiological surveillance, vaccine evaluation, and antiviral drug screening.

Table 11.3 Successfully constructed pseudoviruses and their packaging approaches

Species	Virus	Origins of plasmid genes	Packaging approach of pseudovirus	Reference
A	EV-A71	Live virus template and reverse genetics	Two-step procedure and one-step procedure	[34, 35]
	CVA16	Live virus template and reverse genetics	Two-step procedure and one-step procedure	[33, 36]
	CVA6	Live virus template	Two-step procedure	[37]
B	CVB3, CVB5	Live virus template	Two-step procedure	[38, 39]
C	Polio1, 2, 3	Live virus template and reverse genetics	Two-step procedure	[32, 40]
	CVA10	Live virus template	One-step procedure	[41]
D	EV-D68	Live virus template	One-step procedure	[42]

11.3.4 Investigation of Encapsidation Efficiency of Pseudotyped Enterovirus

11.3.4.1 Sequence Accuracy Is the Key to the Successful Package of Pseudotyped Enterovirus

Though the enterovirus has been known for a high mutation and recombination rate, there are conserved sites either at nucleotide or amino acid level, though have not been fully deciphered by far, critical to the life cycle of enterovirus, like polyprotein processing, RNA transcription, and viral entry.

Upon translation, the single polyprotein of enterovirus has to go through at least 10 rounds of cleavage by virus-coded proteases such as 2A, 3C, and 3CD before folding into mature structural and nonstructural viral proteins (Fig. 11.3). Both 2A and 3C proteases are cysteine proteases coded by enteroviruses. They play an important role in the viral life cycle including the processing of viral polyprotein precursors and the inhibition of host cell translation. Research on polioviruses by Krausslich HG et al. suggests that cleavage sites of 2A and 3C proteases be located in the tyrosine-glycine (Y-G) pairs and glutamine-glycine (Q-G) pairs [43]. However, not all of the Y-G and Q-G dipeptides in the polyprotein are cleaved. 3C protease can only cleave 8 among the 13 Q-G pairs within polyproteins [44]. Substrate recognition and cleavage require at least more than two adjacent particular amino acids. The topology and accessibility of potential cleavage sites certainly are further determinants [45]. Research on amino acid composition of rhinovirus HRV2 2A substrate TRPIITTA(P1)*G(P1')PSDMYVH(P8-P8') shows that a minimum of nine amino acids (comprising residues P8 to P1′) are necessary for cleavage to occur. The proteolysis of substituted peptides is highly tolerant toward changes at P1, P2′, and P3′ with glycine P1′ and a high preference for threonine P2 as prerequisites [46]. Pallai P. V. et al. provide evidence for strong preference of proline related to twisted or crooked protein structures by 3C proteins through synthesized 3C polypeptide substrates, among which the most sensitive polypeptide substrates contain the Gln-Gly-Pro sequence [47]. All of the above indicate the key importance of conformational specificity relevant to primary sequences of amino acids in recognizing 3C protein substrates. The homology of amino acid sequences among the same group of enteroviruses is higher. The homology of 2Apro between EV-A71, CVA6, and CVA16 viruses in group A is 79–96%. The sequence homology of CVB3 and CVB5 amino acid sequences in group B was 89%, and the 2A homology of polio 3 genotypes in group C was about 93%. In contrast, the amino acid

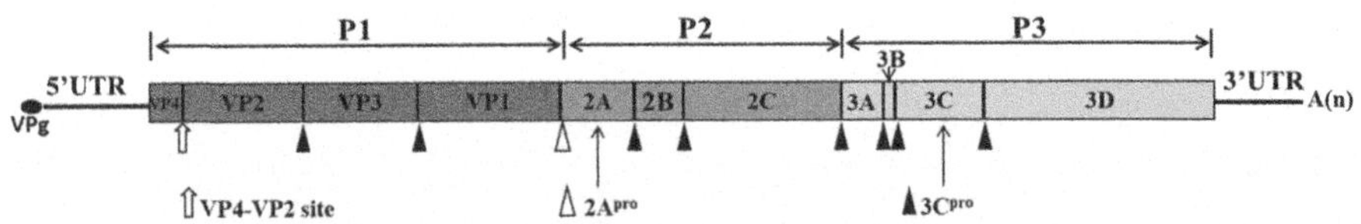

Fig. 11.3 Cleavage site in enterovirus polyprotein

homology of 2Apro between different groups of enteroviruses is mostly at a relatively low level of 40–60% (Table 11.4). The cleavage efficiency of different enteroviral proteases depends on the primary sequence and spatial structure of their substrates. Any mutations that would result in inefficient cleavage of polyprotein could be detrimental to the life cycle of enteroviruses.

Studies discover that genetic mutations 2B-A71E, 2B-I73E, and 2B-E40K lead to changes in hydrophobic domain of CVB3 virus 2B protein, which in turn could cause a more drastic reduction in RNA synthesis [48, 49].

Mutations could lead to the changes in antigenicity, infectivity, and cell adaptability of enteroviruses [50–56]. For example, a T60K mutation in VP3 has led to significant differences in antigenicity between type one Sabin strain and the parent Mahoney strain [57, 58].

Thus, for the process of retrieving a full-length viral cDNA clone from clinical samples or viral culture, it is crucial to get the functional cDNA clone out of the many viral variants known as viral quasispecies [59]. Besides, it is also important to use high-fidelity DNA polymerase to avoid any error induced to the viral cDNA clone during the PCR amplification process.

11.3.4.2 Compatibility of Trans-packaging

Research by Porter D.C et al. shows serial passage of the Enterovirus C poliovirus type 1 replicons with poliovirus type 1, 2, or 3 result in the trans-encapsidation into the type 1, 2, or 3 capsids, respectively, but they exhibit different efficiency of encapsidation for capsid proteins. In contrast, serial passage with bovine enterovirus, Coxsackievirus A21 or B3, or enterovirus 70 does not result in trans-encapsidation [60]. Wu et al. find out that capsid expresser from CV-A16 could package the EV-A71 replicon into infectious CV-A16 pseudovirus, and both EV-A71 and CV-A16 replicons could package CV-A6 capsid expresser into infectious CV-A6 pseudoviruses, and capsid expresser from CV-B5 could package the CV-B3 replicon RNA into infectious CV-B5 pseudovirus. However, the capsid expresser from CV-A6 could not package CV-B3 and CV-B5 replicon RNA into infectious CV-A6 pseudovirus. These results confirm the existence of compatibility of capsid proteins within the same species of enteroviruses. It is suggested that similar packaging mechanisms exist between EV-A71 and CV-A16 from enterovirus species A, and between CV-B3 and CV-B5 from enterovirus species B, which are different from each other [39] (Fig. 11.4).

Blom N. et al. use graphical visualization techniques and neural network algorithms to investigate the sequence specificity of the two proteases 2Apro and 3Cpro. Their study indicates that the prediction model of protease substrate for each virus performs poorly if a combination of enzyme substrate datasets from enterovirus, rhinovirus, and aphthovirus is used. However, using neural network originated from cleavage sites from a specific virus could accurately predict all the cleavage sites of

Table 11.4 Amino acid sequence homology of enterovirus 2A protein

Accession No.	Species	Amino acid sequence homology of 2Apro(%)											
		EV-A71	CVA6	CVA16	CVA10	CVB3	CVB5	PV1	PV2	PV3	EV-D68	RVA2	E1
U22521	EV-A71	100	83	79	80	13	12	55	57	56	42	39	67
AY421764	CVA6	83	100	97	96	59	61	57	57	58	47	33	76
AY421767	CVA10	80	96	97	100	58	62	57	58	59	47	34	76
U05876	CVA16	79	97	100	97	59	61	57	59	59	47	35	77
M88483	CVB3	13	59	59	58	100	89	50	51	50	46	39	78
AF114383	CVB5	12	61	61	62	89	100	48	50	49	47	35	80
AY184219	PV1	55	57	57	57	50	48	100	94	94	54	34	58
AY184220	PV2	57	57	59	58	51	50	94	100	93	54	35	59
AY184221	PV3	56	58	59	59	50	49	94	93	100	52	34	59
AY426531	EV-D68	42	47	47	47	46	47	54	54	52	100	39	52
X02316	RVA2	39	33	35	34	39	35	34	35	34	39	100	39
AF029859	E1	67	76	77	76	78	80	58	59	59	52	39	100

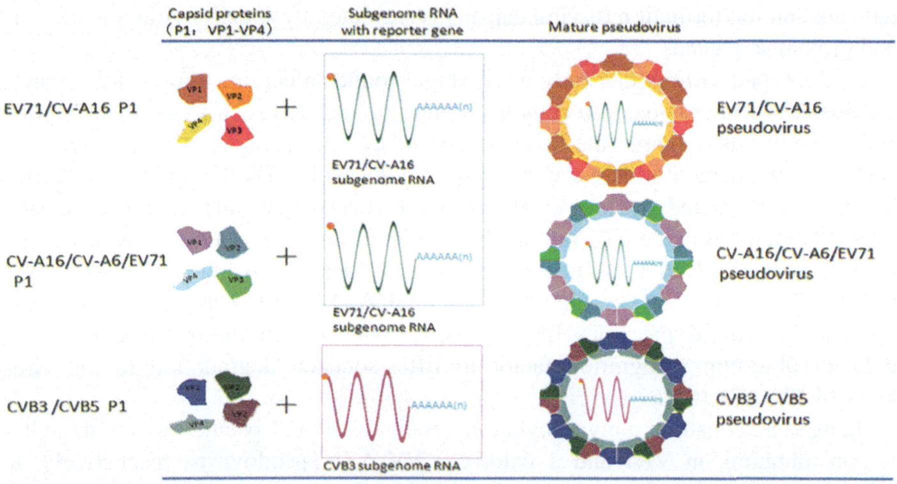

Fig. 11.4 Packaging compatibility for pseudotyped enterovirus. CVA16 capsid could package the EV-A71 replicon RNA into infectious CVA16 pseudovirus, and both EV-A71 and CVA16 replicon RNA could be packaged by CVA6 capsid into infectious CVA6 pseudoviruses. CVB5 capsid could package the CVB3 replicon RNA into infectious CVB5 pseudovirus. However, CVA6 capsid could not package CVB3 and CVB5 replicon RNA into infectious CVA6 pseudovirus

this virus [61]. The result also demonstrates the incompatibility of protease and its substrates among different enterovirus species.

11.4 The Application of Pseudotyped Enterovirus

Since the P1 coding region for structural proteins is substituted with a luciferase reporter gene in the genome, the pseudotyped enterovirus could only elicit a single round infection. With much less biosafety concern, it can be used as a surrogate for the wild-type live virus and serve as a valuable tool in investigating host-virus interactions, screening susceptible cell lines as well as NtAb research, vaccine development, and antiviral drug development.

11.4.1 An Useful Tool of Studying Molecular Virology of Enterovirus

For the first time in 1998, Jia et al. replaced the structural gene P1 from polio RNA with a luciferase reporter gene and built pseudotyped poliovirus [62] and found the encapsidation efficiency of the replicon RNA by heterologous capsid proteins was significantly lower than that of using homologous poliovirus capsids. The results indicate that the formation of viral capsids is regulated by specific viral nucleic acid and protein sequences.

Pseudotyped viruses can only elicit single-round infection, thus could separate the process of viral entry into cells from other procedures such as virus replication and assembly, and enable the investigation of the mechanisms of virus entry into cells and the interaction between the virus and host cells. The EV-A71 pseudovirus has been constructed by Pan et al., who use reverse genetics to introduce site-directed mutations in the VP1 region of EV-A71. Their examination of the molecular interaction between VP1 and its viral receptor SCARB2 (scavenger receptor class B member 2) indicates the key sites on VP1 and SCARB2 [34]. Yan et al. use the same approach to reveal the roles of intracellular valosin-containing protein and its cofactor ubiquitin recognition factor in ER-associated degradation in the virus entry of EV-A71 [63].

Jiang et al. construct a myristoylation-deficient EV-A71 pseudovirus with a Gly-to-Ala mutation on VP4 and a wild-type EV-A71 pseudovirus, respectively, to investigate the replication and assembly of EV-A71 viral genomic RNA [64].

11.4.2 An Useful Tool for Detection of NtAb

11.4.2.1 The Advantage of the NtAb Detection Based on Pseudotyped Enterovirus

NtAb detection plays a critical role in the evaluation of the effectiveness of vaccines against viruses, and it is essential for monitoring an epidemic and evaluating the immunogenicity of a vaccine. Some enteroviruses such as poliovirus, EV-A71, CVB3, and EV-D68 can infect the central nervous system and induce severe neural diseases and complications; therefore, the use of pseudotyped virus is of great significance in ensuring biosafety in scientific research. The third edition of *Global Action Plan (GAPIII) issued by the World Health Organization in* May 2015 *advocate for minimizing poliovirus facility-associated risk after type-specific eradication of wild polioviruses and sequential* cessation of oral polio vaccine use [65]. One of the major challenges in implementing GAP III, however, is to maintain the surveillance of poliovirus NtAb in serum while limiting the use of live poliovirus. Our research shows pseudotyped poliovirus can be used as a surrogate for infectious live poliovirus [66] in quantitative analysis of antiviral effect of feces during infection and NtAb titer against poliovirus in feces and serum [67].

Enterovirus species have high mutation and recombination rates, which can lead to the emergence of new pathogenic strains. Hence, it is necessary to conduct research on the cross-neutralizing activity across different subtypes in the development of vaccines to infer the cross-protective effects of vaccines and provide strong evidence for the selection of vaccine strains. Pseudotyped virus constructed by reverse genetics can overcome the obstacles in the acquisition, transportation, and preservation of live virus, as pseudotyped viruses, compared to live viruses, are more stable, which promotes its application worldwide, especially in resource-constrained regions [68, 69].

11.4.2.2 The Principle of pvNA

The principles of high-throughput pseudotyped virus neutralization assay (pvNAs) are as follows: when incubated with pseudoviruses, NtAb in the serum samples would prevent the pseudotyped virus from entering into host cells, resulting in less amount of expressed luciferases in cells. The concentration of NtAb could be calculated based on the luciferase activity measured in infected cell lysate.

Compared with the conventional CPE method which typically needs 5–7 days, pvNAs shorten assay time to 15–48 hours, although pvNAs may vary slightly across different enteroviruses. In short, serially diluted sera are incubated with diluted pseudotyped viruses with optimized concentration at 37°C for 1–2 hours in 96-well plates. Then cell suspension is added into the wells and incubated at CO_2 incubator. After optimal incubation time, cells are lysed and the expression of reporter gene is measured in relative light units (RLUs). During the experiment, both cells control and pseudotyped virus control should be set in each plate. In order to ensure the sensitivity and robustness of the assay, parameters such as cell density, incubation time, and pseudotyped virus content should be optimized. As the golden standard assay, sera are tested in duplicates in CPE method and the NtAb titer is determined as the reciprocal of the highest dilution of serum at which over 50% of wells shows complete inhibition of CPE. Samples with titers of ≥8 are considered positive and otherwise negative. For pvNAs assay, the cutoff values should be chosen when pvNAs produces the best overall balance for agreement, sensitivity, specificity, as well as the highest Youden's index (Y=Sensitivity + Specificity-1), when compared with CPE method [70]. For assays using known serum as standard reference of NtAb, the standard curve is plotted according to the relationship between the standard-antibody concentration and the RLU of report gene, and the antibody of test sample is extrapolated by fitting algorithm of the standard curve, After correcting for dilution, the concentration of the individual values is then determined.

Currently, the assay methods for detecting NtAb of polio, EV-A71, CVA10, CVA16, CVB3, and CVB5 pseudoviruses have been established (Table 11.5).

Evaluation for pvNAs in pseudotyped enterovirus studies are listed in Table 11.5. The results show that pvNAs established by different laboratories exhibit good agreement and correlation with the "gold standard" CPE-based method and could be used as surrogates for the CPE in NtAb monitoring, cross-neutralizing activity research, and vaccine evaluations. After comparing a series of detection methods from different laboratories, pseudotyped virus is found to be more sensitive to neutralizing than live virus [71, 72] and can adapt to multiple serum screening by choosing different reporter genes [68, 69]. Thus, pseudotyped virus has provided a strong and quick method for comparing the neutralizing activity of different strains.

Table 11.5 Evaluation of NtAb assay system for pseudotyped enterovirus

Virus	Threshold value set for NtAB titers	Qualitative compliance rate (%)	Quantitative correlation (r)	Reference
PV	The highest dilution that inhibited type 1, 2, and 3 pseudovirus infection at ≤20%, 2.5%, and 10% were set for NtAb titers	–	PV1-0.92 PV2-0.83 PV3-0.79	66, 71, 72
PV	The highest dilution that type 1, 2, and 3 pseudovirus infection cells at 85%, 75%, and 88% were set for NtAb titers	–	PV1-0.99 PV2-0.98 PV3-0.99	40, 73
PV	The dilution that inhibited pseudovirus infection at ≤50% were set for NtAb titers	–	PV1-0.90 PV2-0.85 PV3-0.80	32
EV-A71	The concentration was extrapolated according to the curve between the standard-antibody concentration and the luciferase activity	97.2	0.91	70
EV-A71	The dilution that inhibited pseudotyped virus infection at ≤50% were set for NtAb titers	–	B4-0.8772 B5-0.7761 C2-0.6150 C4a-0.8229 C5-0.6725	35
CVA16	The threshold value of 20% CV-A16-luc infectivity was determined to be the NtAb titer	–	0.877	33
CVA10	The titer was determined as the reciprocal of the dilution at which 50% of the complete pseudovirus neutralization (pNT50)	0.957	0.83	41
CVA6		0.876	0.93	37
CVB3		0.891	0.804	38
CVB5		0.604	0.85	39

11.4.2.3 Clinical Application of pvNA for Measurement of NtAb in Human Samples

Minetaro et al. developed a pvNA assay with type 1, 2, and 3 pseudotyped poliovirus, and 131 human sera collected from vaccine-immune population were tested for serosurveillance of poliovirus. The correlation coefficient r value between pvNAs and CPE results was 0.956, 0.912, and 0.890 for poliovirus type 1, 2, and 3, respectively. The results suggest that pvNA with pseudotyped poliovirus (Sabin) could serve as an alternative to CPE with wild-type poliovirus.

Wu et al. developed a pvNA assay for qualitatively measuring NtAb against EV-A71. A total of 144 samples from EV-A71 vaccine clinical trials who received one dose of vaccine were used to validate the clinical applicability of pvNAs. Samples were divided into three groups according to the sources of inactivated EV-A71 vaccines. All samples were detected using pvNAs and CPE assay, and the anti-EV-A71 geometric mean antibody concentrations (GMCs) were measured by the two methods were compared. For the 18 paired clinical samples from individuals immunized with vaccine A, the GMCs increased from 485 U/ml(CPE) and 306 U/ml(pvNAs) to 1687 and 2176 U/ml after immunization, detected by the pvNAs and CPE assay, respectively. No statistical significance was found before and after immunization (before immunization $p = 0.27$, after immunization $p = 0.86$). Similarly, 27 clinically paired serum samples from 9 participants in phase III clinical trials of EV-A71 are also compared. The test results show a similar pattern of antibodies before and after immunization by the two detection methods. No statistical significance in the titer of antibodies at different blood drawing times was found between the two detection methods ($p>0.05$) [70].

Jiang et al. constructed pseudoviruses of different EV-A71 subgenotypes and developed pvNAs to detect NtAb titers of serum samples from patients inoculated with EV-A71 vaccine of C4a strain. The results demonstrate that EV-A71 vaccination with subgenotype C4a could provide broad cross-protection against different EV-A71 subgenotypes. Their study allows pvNAs to be more widely used in evaluations of different types of vaccines.

pvNAs had also been used in detection of IgA to confirming the mucosal immunity induced by OPV or IPV in Wright et al.'s study. They further investigated the appearance and duration of mucosal protective immune responses by OPV vaccines, aiming to provide reference for vaccine immunization strategy [73, 74].

To sum up with the aforementioned pvNA studies, the NtAb results obtained from pvNA and the CPE assay are consistent among findings from previous studies. However, these pvNA assays used different laboratory procedures, cell lines, virus strains, and different threshold values and reference materials among laboratories, showing the lack of harmonization of the pvNA assays. It needs collaborative efforts to standardize these pvNA assays for enterovirus. For the comparison of inter- and intralaboratory serologic data, Jiang et al. suggest to establish an international standard protocol for measuring NtAbs. Similarly, Arya et al. recommend international units rather than an arbitrary dilution figure in describing the NtAb content [35, 75]. Like the case of detection of COVID-19 NtAb, it is necessary to unify the reference materials, titration of pseudotyped virus, detection reagents, and the definition of titer, so as to increase the comparability of test results among different countries and laboratories.

11.4.3 A Safe, Sensitive, and Visualizing Model with Pseudotyped Enterovirus

Zhou et al. applied the pseudotyped enterovirus to construct the pEV-A71(FY)-Luc/R26-hSCARB2 mouse models for the vaccine evaluation [76]. The high sensitivity and efficiency of luciferase-luciferin bioluminescent system was employed in the detection of bioluminescence in mice attacked by pEV-A71(FY)-Luc via in vivo imaging system (IVIS). The results show the pseudovirus carrying the reporter gene which was safe and could be used to observe viral infections in vivo. The results show the pseudotyped virus carrying the reporter gene is safer and more stable and has high sensitivity in substituting live virus to construct the infection model and to be used in investigating the protective effect of active and passive immunity to EV-A71.

Wu et al. develop a IVIS BALB/c model for the evaluation of CVA6 vaccine based on the pseudotyped virus with reporter gene. The results show that antibodies and inactivated vaccines for vaccine evaluation could inhibit pseudotyped CVA6 virus infection in mice and a dose-dependent manner was exhibited in in vivo imaging luminescence in the AID50 study [77]. The infection model constructed in this study could also be used in the investigation of viral infection mechanisms, living drugs, especially used in inhibiting virus from binding to receptors and blocking virus infection pathways, and screening of antibodies and other drugs.

11.4.4 Screening and Evaluation of Anti-enterovirus Drugs

Jiang et al. use pseudotyped EV-A71 and CVA16 viruses to demonstrate the antiviral activity of luteolin, galanin, and quercetin. The study suggests that luteolin target the post-attachment stage of EV-A71 and CA16 infection by inhibiting viral RNA replication and might serve as a lead compound to develop potent anti-EV-A71 and CA16 drugs [78].

11.5 Summary

Viruses in genus *Enterovirus* are non-enveloped viruses and often require a custom-made pseudotyping system as the commonly used pseudotyping systems for enveloped viruses are ineffective. Not only does the pseudotyped enterovirus has superior biosafety, but also their reporter gene like luciferase makes it easy to quantify the viral entry events at cell level. Thus, pseudotyped enteroviruses provide a useful tool for serological surveillance, vaccine evaluation, virology researches, and antiviral drug development.

Although several pseudotyped enteroviruses have been successfully made, the reason that the similar pseudotyped virus construction strategy works for some enteroviruses but not for some others remains unclear, which calls for more understanding of the molecular mechanism of the life cycle of enteroviruses. More studies on the sophisticated architecture of viral genomes as well as the viral-host interaction at molecular level during infections are needed for developing a general applicable pseudotyping strategy for all enteroviruses.

References

1. Nikonov, O.S., Chernykh, E.S., Garber, M.B., Nikonova, E.Y.: Enteroviruses: classification, diseases they cause, and approaches to development of antiviral drugs. *Biochem.* **82**, 1615–1631 (2017). https://doi.org/10.1134/s0006297917130041
2. Royston, L., Tapparel, C.: Rhinoviruses and respiratory enteroviruses: not as simple as ABC. *Viruses.* **8** (2016). https://doi.org/10.3390/v8010016
3. Jubelt, B., Lipton, H.L.: Enterovirus/picornavirus infections. *Handb. Clin. Neurol.* **123**, 379–416 (2014). https://doi.org/10.1016/b978-0-444-53488-0.00018-3
4. Suresh, S., Forgie, S., Robinson, J.: Non-polio enterovirus detection with acute flaccid paralysis: a systematic review. *J. Med. Virol.* **90**, 3–7 (2018). https://doi.org/10.1002/jmv.24933
5. Pons-Salort, M., Grassly, N.C.: Serotype-specific immunity explains the incidence of diseases caused by human enteroviruses. *Science.* **361**, 800–803 (2018). https://doi.org/10.1126/science.aat6777
6. Oberste, M.S.: Progress of polio eradication and containment requirements after eradication. *Transfusion.* **58(Suppl 3)**, 3078–3083 (2018). https://doi.org/10.1111/trf.15018
7. Kramer, R., Lina, B., Shetty, J.: Acute flaccid myelitis caused by enterovirus D68: case definitions for use in clinical practice. *Eur. J. Paediatr. Neurol.* **23**, 235–239 (2019). https://doi.org/10.1016/j.ejpn.2019.01.001
8. Ayers, T., et al.: Acute flaccid myelitis in the United States: 2015–2017. *Pediatrics.* **144** (2019). https://doi.org/10.1542/peds.2019-1619
9. Cui, A., et al.: An outbreak of aseptic meningitis caused by coxsackievirus A9 in Gansu, the People's Republic of China. *Virol. J.* **7**, 72 (2010). https://doi.org/10.1186/1743-422x-7-72
10. Kang, H.J., et al.: A different epidemiology of Enterovirus A and Enterovirus B co-circulating in Korea, 2012–2019. *J. Pediatric Infect. Dis. Soc.* **10**, 398–407 (2021). https://doi.org/10.1093/jpids/piaa111
11. Esposito, S., Bosis, S., Niesters, H., Principi, N.: Enterovirus D68 infection. *Viruses.* **7**, 6043–6050 (2015). https://doi.org/10.3390/v7112925
12. Brotons, P., et al.: The positive rhinovirus/enterovirus detection and SARS-CoV-2 persistence beyond the acute infection phase: an intra-household surveillance study. *Viruses.* **13** (2021). https://doi.org/10.3390/v13081598
13. Novikov Iu, I., Stulova, M.A., Lavrova, I.K.: Myocarditis caused by Coxsackie B viruses in adults. *Ter. Arkh.* **56**, 37–43 (1984)
14. Lee, W.S., et al.: Acute viral myopericarditis presenting as a transient effusive-constrictive pericarditis caused by coinfection with coxsackieviruses A4 and B3. *Korean J. Intern. Med.* **27**, 216–220 (2012). https://doi.org/10.3904/kjim.2012.27.2.216
15. Shanmugam, J., Raveendranath, M., Balakrishnan, K.G.: Isolation of ECHO virus type-22 from a child with acute myopericarditis—a case report. *Indian Heart J.* **38**, 79–80 (1986)
16. Fukuhara, T., et al.: Myopericarditis associated with ECHO virus type 3 infection—a case report. *Jpn. Circ. J.* **47**, 1274–1280 (1983). https://doi.org/10.1253/jcj.47.1274
17. Leendertse, M., et al.: Pleurodynia caused by an echovirus 1 brought back from the tropics. *J. Clin. Virol.* **58**, 490–493 (2013). https://doi.org/10.1016/j.jcv.2013.06.039
18. Rodriguez-Calvo, T.: Enterovirus infection and type 1 diabetes: unraveling the crime scene. *Clin. Exp. Immunol.* **195**, 15–24 (2019). https://doi.org/10.1111/cei.13223
19. Liu, H., et al.: Characterization of Coxsackievirus A6 strains isolated from children with hand, foot, and mouth disease. *Front. Cell. Infect. Microbiol.* **11**, 700191 (2021). https://doi.org/10.3389/fcimb.2021.700191
20. Mirand, A., et al.: A large-scale outbreak of hand, foot and mouth disease, France, as at 28 September 2021. *Euro Surveill.* **26** (2021). https://doi.org/10.2807/1560-7917.Es.2021.26.43.2100978
21. Wright, P.W., Strauss, G.H., Langford, M.P.: Acute hemorrhagic conjunctivitis. *Am. Fam. Physician.* **45**, 173–178 (1992)

22. Langford, M.P., Anders, E.A., Burch, M.A.: Acute hemorrhagic conjunctivitis: anti-coxsackievirus A24 variant secretory immunoglobulin A in acute and convalescent tear. *Clin. Ophthalmol.* **9**, 1665–1673 (2015). https://doi.org/10.2147/opth.S85358
23. ICTV Picornaviridae, Genus Enterovirus. https://talk.ictvonline.org/ictv-reports/ictv_online_report/positive-sense-rna-viruses/w/picornaviridae/681/genus-enterovirus. (2022)
24. Zell, R., et al.: ICTV virus taxonomy profile: picornaviridae. *J. Gen. Virol.* **98**, 2421–2422 (2017). https://doi.org/10.1099/jgv.0.000911
25. Oberste, M.S., Maher, K., Kilpatrick, D.R., Pallansch, M.A.: Molecular evolution of the human enteroviruses: correlation of serotype with VP1 sequence and application to picornavirus classification. *J. Virol.* **73**, 1941–1948 (1999). https://doi.org/10.1128/jvi.73.3.1941-1948.1999
26. Palmenberg, A.C., et al.: Sequencing and analyses of all known human rhinovirus genomes reveal structure and evolution. *Science*. **324**, 55–59 (2009). https://doi.org/10.1126/science.1165557
27. Simmonds, P., et al.: Proposals for the classification of human rhinovirus species C into genotypically assigned types. *J. Gen. Virol.* **91**, 2409–2419 (2010). https://doi.org/10.1099/vir.0.023994-0
28. Belsham, G.J., Jackson, R.J.: Translation initiation on picornavirus RNA. In: Translational control of gene expression, pp. 869–900. Cold Spring Harbor (2000)
29. Toyoda, H., et al.: A second virus-encoded proteinase involved in proteolytic processing of poliovirus polyprotein. *Cell.* **45**, 761–770 (1986). https://doi.org/10.1016/0092-8674(86)90790-7
30. Mousnier, A., et al.: Fragment-derived inhibitors of human N-myristoyltransferase block capsid assembly and replication of the common cold virus. *Nat. Chem.* **10**, 599–606 (2018). https://doi.org/10.1038/s41557-018-0039-2
31. Baggen, J., Thibaut, H.J., Strating, J., van Kuppeveld, F.J.M.: The life cycle of non-polio enteroviruses and how to target it. *Nat. Rev. Microbiol.* **16**, 368–381 (2018). https://doi.org/10.1038/s41579-018-0005-4
32. Jiang, Z., et al.: A simple and safe antibody neutralization assay based on polio pseudoviruses. *Hum. Vaccin. Immunother.* **15**, 349–357 (2019). https://doi.org/10.1080/21645515.2018.1526553
33. Jin, J., et al.: Development of a Coxsackievirus A16 neutralization assay based on pseudoviruses for measurement of neutralizing antibody titer in human serum. *J. Virol. Methods.* **187**, 362–367 (2013). https://doi.org/10.1016/j.jviromet.2012.11.014
34. Chen, P., et al.: Molecular determinants of enterovirus 71 viral entry: cleft around GLN-172 on VP1 protein interacts with variable region on scavenge receptor B 2. *J. Biol. Chem.* **287**, 6406–6420 (2012). https://doi.org/10.1074/jbc.M111.301622
35. Zhang, H., et al.: Analysis of cross-reactive neutralizing antibodies in human HFMD serum with an EV-A71 pseudovirus-based assay. *PLoS One*. **9**, e100545 (2014). https://doi.org/10.1371/journal.pone.0100545
36. Hao, X.T., et al.: Establishment and preliminary application of a pseudovirus luciferase assay for quantitative detection of coxsackievirus A16 neutralizing antibody. Chin. J. Viral Dis. **6**, 6–11 (2016). https://doi.org/10.16505/j.2095-0136.2016.01.002
37. Su, Y., et al.: A surrogate assay for measuring Coxsackievirus A6 neutralizing antibodies. *Hum. Vaccin. Immunother.* **14**, 3034–3040 (2018). https://doi.org/10.1080/21645515.2018.1504540
38. Chen, P., et al.: A rapid and quantitative assay for measuring neutralizing antibodies of Coxsackievirus B3. *J. Virol. Methods*. **232**, 1–7 (2016). https://doi.org/10.1016/j.jviromet.2016.02.010
39. Chen, P., et al.: Development of a pseudovirus based assay for measuring neutralizing antibodies against coxsackievirus B5. *J. Virol. Methods*. **246**, 21–26 (2017). https://doi.org/10.1016/j.jviromet.2017.04.005
40. Liu, S., et al.: Comparison of the neutralizing activities of antibodies in clinical sera against both Sabin and wild-type polio pseudoviruses. *J. Virol. Methods*. **300**, 114376 (2022). https://doi.org/10.1016/j.jviromet.2021.114376

41. Li, K., et al.: Development of a pseudovirus-based assay for measuring neutralizing antibodies against Coxsackievirus A10. *Hum. Vaccin. Immunother.* **16**, 1434–1440 (2020). https://doi.org/10.1080/21645515.2019.1691404
42. Dong, F.Y., et al.: Construction of enterovirus D68 pseudovirus. Prog. Microbiol. Immunol. **47**, 8–14 (2019). https://doi.org/10.13309/j.cnki.pmi.2019.06.002
43. Kräusslich, H.G., Wimmer, E.: Viral proteinases. *Annu. Rev. Biochem.* **57**, 701–754 (1988). https://doi.org/10.1146/annurev.bi.57.070188.003413
44. Pallansch, M.A., et al.: Protein processing map of poliovirus. *J. Virol.* **49**, 873–880 (1984). https://doi.org/10.1128/jvi.49.3.873-880.1984
45. Ypma-Wong, M.F., Filman, D.J., Hogle, J.M., Semler, B.L.: Structural domains of the poliovirus polyprotein are major determinants for proteolytic cleavage at Gln-Gly pairs. *J. Biol. Chem.* **263**, 17846–17856 (1988)
46. Sommergruber, W., et al.: Cleavage specificity on synthetic peptide substrates of human rhinovirus 2 proteinase 2A. *J. Biol. Chem.* **267**, 22639–22644 (1992)
47. Pallai, P.V., et al.: Cleavage of synthetic peptides by purified poliovirus 3C proteinase. *J. Biol. Chem.* **264**, 9738–9741 (1989)
48. van Kuppeveld, F.J., Galama, J.M., Zoll, J., Melchers, W.J.: Genetic analysis of a hydrophobic domain of coxsackie B3 virus protein 2B: a moderate degree of hydrophobicity is required for a cis-acting function in viral RNA synthesis. *J. Virol.* **69**, 7782–7790 (1995). https://doi.org/10.1128/jvi.69.12.7782-7790.1995
49. van Kuppeveld, F.J., Galama, J.M., Zoll, J., van den Hurk, P.J., Melchers, W.J.: Coxsackie B3 virus protein 2B contains cationic amphipathic helix that is required for viral RNA replication. *J. Virol.* **70**, 3876–3886 (1996). https://doi.org/10.1128/jvi.70.6.3876-3886.1996
50. Donohue, R.C., Pfaller, C.K., Cattaneo, R.: Cyclical adaptation of measles virus quasispecies to epithelial and lymphocytic cells: to V, or not to V. *PLoS Pathog.* **15**, e1007605 (2019). https://doi.org/10.1371/journal.ppat.1007605
51. Holland, J.J.: Transitions in understanding of RNA viruses: a historical perspective. *Curr. Top. Microbiol. Immunol.* **299**, 371–401 (2006). https://doi.org/10.1007/3-540-26397-7_14
52. Domingo, E., Sheldon, J., Perales, C.: Viral quasispecies evolution. *Microbiol Mol Biol Rev.* **76**, 159–216 (2012). https://doi.org/10.1128/mmbr.05023-11
53. Sánchez-Campos, S., et al.: Differential Shape of geminivirus mutant spectra across cultivated and wild hosts with invariant viral consensus sequences. *Front. Plant Sci.* **9**, 932 (2018). https://doi.org/10.3389/fpls.2018.00932
54. Domingo, E., Perales, C.: Quasispecies and virus. *EBJ.* **47**, 443–457 (2018). https://doi.org/10.1007/s00249-018-1282-6
55. Xiao, Y., et al.: RNA recombination enhances adaptability and is required for virus spread and virulence. *Cell Host Microbe.* **19**, 493–503 (2016). https://doi.org/10.1016/j.chom.2016.03.009
56. Moreno, E., et al.: Internal disequilibria and phenotypic diversification during replication of Hepatitis C virus in a noncoevolving cellular environment. *J. Virol.* **91** (2017). https://doi.org/10.1128/jvi.02505-16
57. Diamond, D.C., et al.: Antigenic variation and resistance to neutralization in poliovirus type 1. *Science.* **229**, 1090–1093 (1985). https://doi.org/10.1126/science.2412292
58. Blondel, B., et al.: Mutations conferring resistance to neutralization with monoclonal antibodies in type 1 poliovirus can be located outside or inside the antibody-binding site. *J. Virol.* **57**, 81–90 (1986). https://doi.org/10.1128/jvi.57.1.81-90.1986
59. Domingo, E., Perales, C.: Viral quasispecies. *PLoS Genet.* **15**, e1008271 (2019). https://doi.org/10.1371/journal.pgen.1008271
60. Porter, D.C., et al.: Demonstration of the specificity of poliovirus encapsidation using a novel replicon which encodes enzymatically active firefly luciferase. *Virology.* **243**, 1–11 (1998). https://doi.org/10.1006/viro.1998.9046
61. Blom, N., Hansen, J., Blaas, D., Brunak, S.: Cleavage site analysis in picornaviral polyproteins: discovering cellular targets by neural networks. *Protein Sci.* **5**, 2203–2216 (1996). https://doi.org/10.1002/pro.5560051107

62. Jia, X.Y., Van Eden, M., Busch, M.G., Ehrenfeld, E., Summers, D.F.: Trans-encapsidation of a poliovirus replicon by different picornavirus capsid proteins. *J. Virol.* **72**, 7972–7977 (1998). https://doi.org/10.1128/jvi.72.10.7972-7977.1998
63. Yan, J., et al.: Involvement of VCP/UFD1/Nucleolin in the viral entry of Enterovirus A species. *Virus Res.* **283**, 197974 (2020). https://doi.org/10.1016/j.virusres.2020.197974
64. Cao, J., et al.: Myristoylation of EV-A71 VP4 is essential for infectivity and interaction with membrane structure. *Virol. Sin.* **35**, 599–613 (2020). https://doi.org/10.1007/s12250-020-00226-1
65. Global Polio Eradication Initiative. http://www.polioeradication.org/Portals/0/Document/Resources/PostEradication/GAPIII_2014.pdf
66. Arita, M., Iwai-Itamochi, M.: Evaluation of antigenic differences between wild and Sabin vaccine strains of poliovirus using the pseudovirus neutralization test. *Sci. Rep.* **9**, 11970 (2019). https://doi.org/10.1038/s41598-019-48534-1
67. Brickley, E.B., et al.: Intestinal immune responses to type 2 Oral Polio Vaccine (OPV) challenge in infants previously immunized with Bivalent OPV and either high-dose or standard inactivated polio vaccine. *J. Infect. Dis.* **217**, 371–380 (2018). https://doi.org/10.1093/infdis/jix556
68. Wright, E., et al.: Virus neutralising activity of African fruit bat (Eidolon helvum) sera against emerging lyssaviruses. *Virology.* **408**, 183–189 (2010). https://doi.org/10.1016/j.virol.2010.09.014
69. Molesti, E., et al.: Multiplex evaluation of influenza neutralizing antibodies with potential applicability to in-field serological studies. *J Immunol Res.* **2014**, 457932 (**2014**). https://doi.org/10.1155/2014/457932
70. Wu, X., et al.: Development and evaluation of a pseudovirus-luciferase assay for rapid and quantitative detection of neutralizing antibodies against enterovirus 71. *PLoS One.* **8**, e64116 (2013). https://doi.org/10.1371/journal.pone.0064116
71. Heyndrickx, L., et al.: International network for comparison of HIV neutralization assays: the NeutNet report II. *PLoS One.* **7**, e36438 (2012). https://doi.org/10.1371/journal.pone.0036438
72. Louder, M.K., et al.: HIV-1 envelope pseudotyped viral vectors and infectious molecular clones expressing the same envelope glycoprotein have a similar neutralization phenotype, but culture in peripheral blood mononuclear cells is associated with decreased neutralization sensitivity. *Virology.* **339**, 226–238 (2005). https://doi.org/10.1016/j.virol.2005.06.003
73. Wright, P.F., et al.: Intestinal immunity is a determinant of clearance of poliovirus after oral vaccination. *J. Infect. Dis.* **209**, 1628–1634 (2014). https://doi.org/10.1093/infdis/jit671
74. Wright, P.F., et al.: Vaccine-induced mucosal immunity to poliovirus: analysis of cohorts from an open-label, randomised controlled trial in Latin American infants. *Lancet Infect. Dis.* **16**, 1377–1384 (2016). https://doi.org/10.1016/s1473-3099(16)30169-4
75. Arya, S.C., Agarwal, N.: Poliovirus neutralization test with poliovirus pseudovirus to measure neutralizing antibody in humans. *CVI.* **19**, 458; author reply 459 (2012). https://doi.org/10.1128/cvi.05568-11
76. Zhou, S., et al.: A safe and sensitive enterovirus A71 infection model based on human SCARB2 knock-in mice. *Vaccine.* **34**, 2729–2736 (2016). https://doi.org/10.1016/j.vaccine.2016.04.029
77. Wu, X., et al.: Establishment of bioluminescent imaging mouse model for coxsackievirus A6 vaccine using the pseudovirus system. Chin. J. Viral Dis. **8**, 515–520 (2018). https://doi.org/10.16505/j.2095-0136.2018.0108
78. Xu, L., et al.: Identification of luteolin as enterovirus 71 and coxsackievirus A16 inhibitors through reporter viruses and cell viability-based screening. *Viruses.* **6**, 2778–2795 (2014). https://doi.org/10.3390/v6072778

Chapter 12
Pseudotyped Viruses for Orthohantavirus

Tingting Ning, Weijin Huang, Li Min, Yi Yang, Si Liu, Junxuan Xu, Nan Zhang, Si-An Xie, Shengtao Zhu, and Youchun Wang

Abstract Orthohantaviruses, members of the *Orthohantavirus* genus of *Hantaviridae* family of the *Bunyavirales* order, are enveloped, negative-sense, single-stranded, tripartite RNA viruses. They are emerging zoonotic pathogens carried by small mammals including rodents, moles, shrews, and bats and are the etiologic agents of hemorrhagic fever with renal syndrome (HFRS) and hantavirus cardiopulmonary syndrome (HCPS) among humans. With the characteristics of low biological risk but strong operability, a variety of pseudotyped viruses have been constructed as alternatives to authentic orthohantaviruses to help delineate the roles of host factors in viral entry and other virus-host interactions, to assist in deciphering mechanisms of immune response and correlates of protection, to enhance our understanding of viral antigenic property, to characterize viral entry inhibitors, and to be developed as vaccines. In this chapter, we will discuss the general property of orthohantavirus, construction of pseudotyped orthohantaviruses based on different packaging systems, and their current applications.

T. Ning · L. Min · Y. Yang · S. Liu · J. Xu · N. Zhang · S.-A. Xie · S. Zhu (✉)
Department of Gastroenterology, Beijing Friendship Hospital, Capital Medical University, National Clinical Research Center for Digestive Disease, Beijing Digestive Disease Center, Beijing Key Laboratory for Precancerous Lesion of Digestive Disease, Beijing, China
e-mail: zhushengtao@ccmu.edu.cn

W. Huang
Division of HIV/AIDS and Sexually Transmitted Virus Vaccines, National Institutes for Food and Drug Control (NIFDC), Beijing, China

Y. Wang (✉)
Chinese Academy of Medical Sciences & Peking Union Medical College, Beijing, China

Institute of Medical Biology, Chinese Academy of Medicine Sciences & Peking Union Medical College, Kunming, China
e-mail: wangyc@nifdc.org.cn; wangyc@imbcams.com.cn

Y. Wang (ed.), *Pseudotyped Viruses*, Advances in Experimental Medicine and Biology 1407, https://doi.org/10.1007/978-981-99-0113-5_12

Keywords Orthohantavirus · HTNV · ANDV · Pseudotyped virus · Neutralization

Abbreviations

ANDV	Andes virus
BSL	Biosafety level
CHOV	Choclo virus
DOBV	Dobrava-Belgrade virus
GFP	Green fluorescent protein
GPC	Glycoprotein precursor
GRFT	Griffithsin
HCPS	Hantavirus cardiopulmonary syndrome
HFRS	Hemorrhagic fever with renal syndrome
HIV-1	Human immunodeficiency virus type 1
HTNV	Hantaan virus
L	Large segment
Luc	Luciferase
LV	Lentivirus
M	Medium segment
MLV	Murine leukemia virus
N	Nucleocapsid protein
NE	Nephropathia epidemica
NWH	New World hantaviruses
OWH	Old World hantaviruses
PBNA	Pseudovirus-based neutralization assay
PCDH-1	Protocadherin-1
PHV	Prospect Hill virus
PRNT	Plaque reduction neutralization test
PUUV	Puumala virus
RdRp	RNA-dependent RNA polymerase
rVSVs	Recombinant VSVs
S	Small segment
SEOV	Seoul virus
SIV	Simian immunodeficiency virus
SNV	Sin Nombre virus
TULV	Tula virus
vRNA	Viral RNA
VSV	Vesicular stomatitis virus

12.1 Introduction

Orthohantaviruses are enveloped, negative-sense, single-stranded RNA viruses belonging to the genus *Orthohantavirus*, family *Hantaviridae*, and order *Bunyavirales* [1]. The orthohantavirus genome is composed of three RNA segments (small, medium, large), respectively, encoding a nucleocapsid protein (N), two surface glycoproteins, and an RNA-dependent RNA polymerase (RdRp) [2]. Orthohantaviruses are emerging zoonotic etiologies carried by small mammal animals including rodents, shrews, bats, and moles [3]. Currently, 38 species of orthohantavirus are known, of which 26 are rodent-borne, 4 are mole-borne, 7 are shrew-borne, and 1 is bat-borne.

Orthohantavirus spreads to humans when a person inhales aerosols or dust particles of urine, feces, or saliva contaminated by orthohantavirus. Generally humans do not spread orthohantavirus efficiently and are therefore considered dead-end hosts, although limited person-to-person transmission has been found for Andes virus (ANDV) [4, 5]. Notably, only 15 rodent-borne orthohantaviruses are recognized to cause human diseases including hemorrhagic fever with renal syndrome (HFRS) in the Eurasia and hantavirus cardiopulmonary syndrome (HCPS) in the Americas [6, 7]. And as shown in Fig. 12.1, rodent-borne orthohantaviruses are classified into three clades based on their reservoir hosts: (1) New World hantaviruses (NWH) (e.g., ANDV and Sin Nombre virus (SNV)) that are pathogens of HCPS and are harbored in members of the Sigmodontinae *or* Neotominae subfamily, (2) Old World hantaviruses (OWH) (e.g., Seoul virus (SEOV), Hantaan virus (HTNV), and Dobrava-Belgrade virus (DOBV)) that are pathogens of HFRS and are carried by Murinae rodents, and (3) viruses found in the New or Old World that are related to nephropathia epidemica (NE, a mild form of HFRS) (e.g., Puumala virus (PUUV)) or are non-virulent (e.g., Prospect Hill virus (PHV) and Tula virus (TULV)) and are harbored in members of Arvicolinae subfamily [8]. A summary of the host affiliation and geographic distribution of the current 38 orthohantavirus species and the diseases they trigger are given in Table 12.1.

12.2 General Property of Orthohantavirus

12.2.1 Orthohantavirus Particle and Genome

Orthohantaviruses are enveloped viruses and form spherical structure with a diameter of 80–160 nm (Fig. 12.2) [9, 10]. The negative-strand RNA genomes of orthohantavirus contain three segments: a 6.5–6.6 kb large segment (L), a 3.7–3.8 kb medium segment (M), and a 1.8–2.1 kb small segment (S) [11]. Each segment contains highly conserved 3′ and 5′ untranslated regions consisting of complementary nucleotides [10, 12]. And these untranslated regions recognize each other to constitute panhandle structures which entitle the RNA segments a

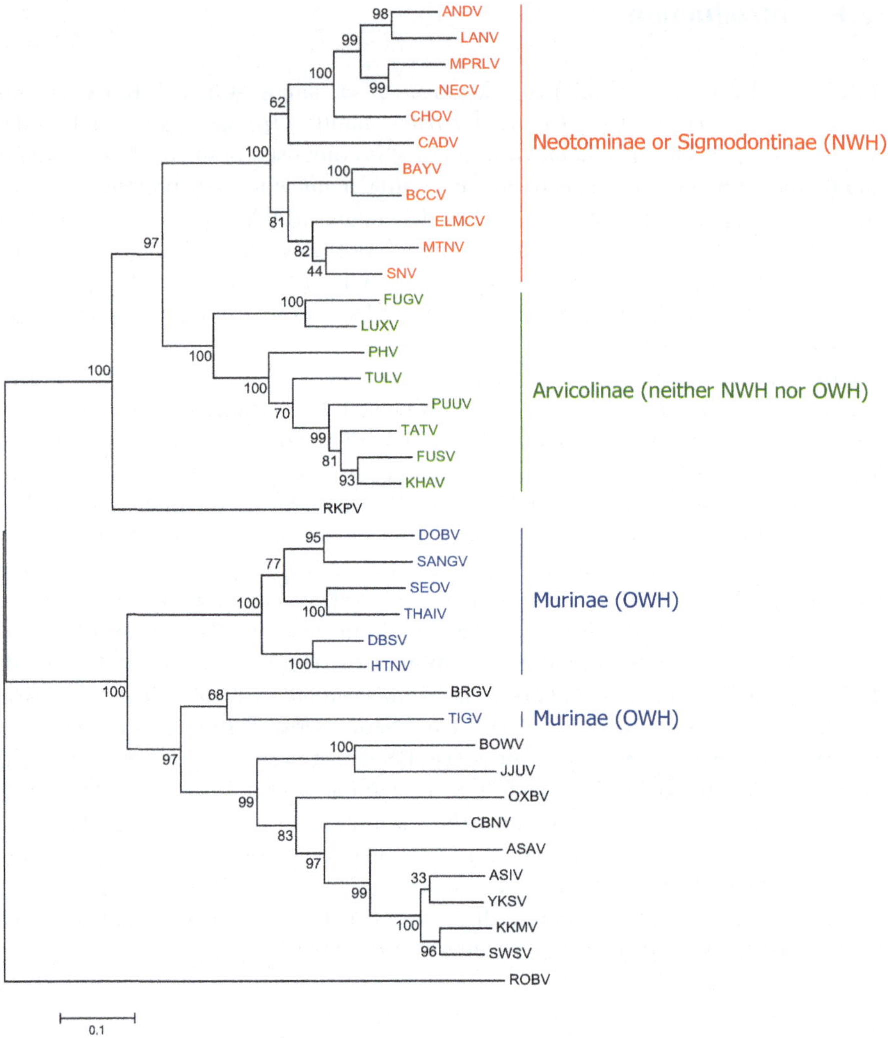

Fig. 12.1 Phylogeny of orthohantaviruses based on the glycoprotein sequences (taken from GenBank) and inferred using the maximum likelihood method available in MEGA6, excluding poorly aligned regions. Colors represent the primary rodent species groups as follows: Neotominae or Sigmodontinae (red); Arvicolinae (green); Murinae (blue)

microscopically circular appearance [6]. These panhandle structures are proved to regulate viral replication and transcription, as proven by other negative-strand RNA viruses, such as influenza A virus and vesicular stomatitis virus (VSV), and are thought to be a highlight of bunyaviruses [13–15]. Viral RNA (vRNA) segments are each relevant to an RdRp and are encased in N [11, 16]. These are encapsulated in a lipid envelope that is derived from host and decorated with two surface glycoproteins, Gc (C-terminal subunit) and Gn (N-terminal subunit).

Table 12.1 The geographic distribution and host affiliation of the current 38 orthohantavirus species and the diseases they cause

Virus	Species	Reservoir host				Geographic distribution	Disease	GenBank of glycoprotein
		Species	Subfamily	Family	Order			
Sangassou virus (SANGV)	*Sangassou orthohantavirus*	*Hylomyscus simus*	Murinae	Muridae	Rodentia	Africa	HFRS	AEZ02947.1
Dobrava virus (DOBV)	*Dobrava-Belgrade orthohantavirus*	*Apodemus flavicollis*	Murinae	Muridae	Rodentia	Europe	HFRS	ADA68891.1
Hantaan virus (HTNV)	*Hantaan orthohantavirus*	*Apodemus agrarius*	Murinae	Muridae	Rodentia	Eastern Asia	HFRS	CAA68456.1
Seoul virus (SEOV)	*Seoul orthohantavirus*	*Rattus*	Murinae	Muridae	Rodentia	Worldwide	HFRS	AAB19434.1
Dabieshan virus (DBSV)	*Dabieshan orthohantavirus*	*Niviventer confucianus*	Murinae	Muridae	Rodentia	Asia	Unknown	AFG26305.1
Thailand virus (THAIV)	*Thailand orthohantavirus*	*Bandicota indica*	Murinae	Muridae	Rodentia	Asia	HFRS	QZA58117.1
Tigray virus (TIGV)	*Tigray orthohantavirus*	*Stenocephalemys albipes*	Murinae	Muridae	Rodentia	Africa	Unknown	AOR07001.1
Khabarovsk virus (KHAV)	*Khabarovsk orthohantavirus*	*Microtus fortis*	Arvicolinae	Cricetidae	Rodentia	Asia	Unknown	AIL25317.1
Puumala virus (PUUV)	*Puumala orthohantavirus*	*Clethrionomys glareolus*	Arvicolinae	Cricetidae	Rodentia	Europe, Asia	HFRS/NE	AJC50718.1
Luxi virus (LUXV)	*Luxi orthohantavirus*	*Eothenomys miletus*	Arvicolinae	Cricetidae	Rodentia	China	HFRS	ADV33303.1
Fugong virus (FUGV)	*Fugong orthohantavirus*	*Eothenomys eleusis*	Arvicolinae	Cricetidae	Rodentia	China	Unknown	AMB43173.1
Prospect Hill virus (PHV)	*Prospect Hill orthohantavirus*	*Microtus pennsylvanicus*	Arvicolinae	Cricetidae	Rodentia	North America	Unknown	CAA38922.1
Tula virus (TULV)	*Tula orthohantavirus*	*Microtus arvalis*	Arvicolinae	Cricetidae	Rodentia	Europe, Russia, Asia	Unknown	NP_942586.1

(continued)

Table 12.1 (continued)

Virus	Species	Reservoir host				Geographic distribution	Disease	GenBank of glycoprotein
		Species	Subfamily	Family	Order			
Fusong virus (FUSV)	*Fusong orthohantavirus*	*Clethrionomys rufocanus*	Arvicolinae	Cricetidae	Rodentia	Asia	Unknown	ABV80308.1
Tatenale virus (TATV)	*Tatenale orthohantavirus*	*Microtus agrestis*	Arvicolinae	Cricetidae	Rodentia	Europe	Unknown	QIA61110.1
Montano virus (MTNV)	*Montano orthohantavirus*	*Peromyscus beatae*	Neotominae	Cricetidae	Rodentia	North America	Unknown	BAK08520.1
El Moro Canyon virus (ELMCV)	*El Moro Canyon orthohantavirus*	*Reithrodontomys megalotis*	Neotominae	Cricetidae	Rodentia	North America	HCPS	AAA87198.1
Sin Nombre virus (SNV)	*Sin Nombre orthohantavirus*	*Peromyscus maniculatus*	Neotominae	Cricetidae	Rodentia	North America	HCPS	NP_941974.1
Necocli virus (NECV)	*Necocli orthohantavirus*	*Zygodontomys cherriei*	Sigmodontinae	Cricetidae	Rodentia	South America	Unknown	AHJ38538.1
Maporal virus (MPRLV)	*Maporal orthohantavirus*	*Oligoryzomys delicatus*	Sigmodontinae	Cricetidae	Rodentia	South America	HCPS	AAR14889.1
Laguna Negra virus (LANV)	*Laguna Negra orthohantavirus*	*Calomys laucha*	Sigmodontinae	Cricetidae	Rodentia	South America	HCPS	AAB87603.1
Cano Delgadito virus (CADV)	*Cano Delgadito orthohantavirus*	*Sigmodon alstoni*	Sigmodontinae	Cricetidae	Rodentia	South America	Unknown	ABB88646.1
Andes virus (ANDV)	*Andes orthohantavirus*	*Oligoryzomys longicaudatus*	Sigmodontinae	Cricetidae	Rodentia	South America	HCPS	NP_604472.1
Choclo virus (CHOV)	*Choclo orthohantavirus*	*Oligoryzomys fulvescens*	Sigmodontinae	Cricetidae	Rodentia	North America	HCPS	ABB90558.1
Bayou virus (BAYV)	*Bayou orthohantavirus*	*Oryzomys palustris*	Sigmodontinae	Cricetidae	Rodentia	North America	HCPS	AAA61690.1
Black Creek Canal virus (BCCV)	*Black Creek Canal orthohantavirus*	*Sigmodon hispidus*	Sigmodontinae	Cricetidae	Rodentia	North America	HCPS	L39950.1

Rockport virus (RKPV)	*Rockport orthohantavirus*	*Scalopus aquaticus*	Scalopinae	Talpidae	Soricomorpha	North America	Unknown	AEA11485.1
Oxbow virus (OXBV)	*Oxbow orthohantavirus*	*Neurotrichus gibbsii*	Talpinae	Talpidae	Soricomorpha	North America	Unknown	ACT68338.1
Bruges virus (BRGV)	*Bruges orthohantavirus*	*Talpa europaea*	Talpinae	Talpidae	Soricomorpha	Europe	Unknown	AOC84247.1
Asama virus (ASAV)	*Asama orthohantavirus*	*Urotrichus talpoides*	Talpinae	Talpidae	Soricomorpha	Asia	Unknown	ACI28508.1
Yakeshi virus (YKSV)	*Yakeshi orthohantavirus*	*Sorex isodon*	Soricinae	Soricidae	Soricomorpha	Asia	Unknown	AGI62349.1
Kenkeme virus (KKMV)	*Kenkeme orthohantavirus*	*Sorex roboratus*	Soricinae	Soricidae	Soricomorpha	Russia	Unknown	AIL25322.1
Asikkala virus (ASIV)	*Asikkala orthohantavirus*	*Sorex minutus*	Soricinae	Soricidae	Soricomorpha	Europe	Unknown	AGK36762.1
Seewis virus (SWSV)	*Seewis orthohantavirus*	*Sorex araneus*	Soricinae	Soricidae	Soricomorpha	Europe, Asia	Unknown	ATX68098.1
Cao Bang virus (CBNV)	*Cao Bang orthohantavirus*	*Crocidura shantungensis*	Crocidurinae	Soricidae	Soricomorpha	Asia	Unknown	AIZ66322.1
Jeju virus (JJUV)	*Jeju orthohantavirus*	*Crocidura shantungensis*	Crocidurinae	Soricidae	Soricomorpha	Asia	Unknown	AEX56232.1
Bowe virus (BOWV)	*Bowe orthohantavirus*	*Crocidura douceti*	Crocidurinae	Soricidae	Soricomorpha	Africa	Unknown	AGW23848.1
Robina virus (ROBV)	*Robina orthohantavirus*	*Pteropus alecto*	Pteropodidae	Pteropodidae	*Chiroptera*	Australia	Unknown	QED22047.1

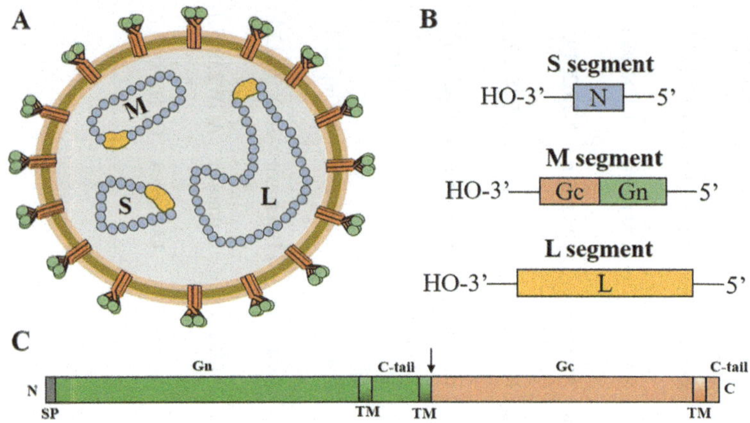

Fig. 12.2 Schematic representation of orthohantavirus particle and genes. (**A**) Orthohantaviruses carry a tri-segmented negative-sense RNA genome encased in a lipid envelop that is studded with surface spikes comprising the Gn and Gc glycoproteins. (**B**) Small (S), medium (M), and large (L) genome segments encode nucleoprotein, Gn/Gc glycoprotein, and RNA-dependent RNA polymerase, respectively. (**C**) Linear representation of orthohantavirus Gn/Gc. Gn is proteolytically cleaved from Gc at the conserved WAASA site by the cellular signal protease complex (represented by arrow). SP, signal peptide; TM, transmembrane region; C-tail, cytoplasmic tail

The S and L segments respectively encode the 50 KD N and the 250 KD RdRp [12, 17, 18]. The N encapsulates the vRNA to shelter it from cellular nucleases and cooperates with RdRp to make sure that the viral genome could replicate efficiently [16, 19]. The RdRp participates in replication and transcription of the orthohantavirus genome based on its replicase, transcriptase, and endonuclease activities [20, 21].

The M segment encodes a glycoprotein precursor (GPC) of approximately 1133–1158 amino acids [22]. The N-terminus of GPC contains a signal peptide which guides the translating ribosomes to the endoplasmic reticulum, where GPC is co-translationally cleaved at a conserved WAASA sequence by the cellular signal peptidase complex to produce Gc and Gn [23, 24]. Notably, Gc and Gn undergo N- and O-glycosylations, move together to the Golgi apparatus, and then incorporate into the viral particles. Gc and Gn are the only viral proteins protruding from the surface of orthohantavirus virions and direct viral entry into several susceptible target cells.

12.2.2 Orthohantavirus Entry Pathway

Although orthohantaviruses could cause human diseases around the world, our understanding of their entry mechanisms is still deprived. This lack of knowledge seriously impede the development of new therapeutic approaches. The first indispensable step to establish infection is viral attachment to the cell surface. And a variety of cellular proteins have been involved in the process of OWH and NWH attachment to cells, including $\alpha_5\beta_1$ and $\alpha_V\beta_3$ integrins [25]. After attachment, orthohantaviruses are internalized through several endocytic strategies like dynamin-independent pathways and clathrin-mediated endocytosis, hinging on certain cell type and experimental conditions. Following internalization, virions are transported to early endosomes and ultimately moved to late endosomal and lysosomal compartments. Viral entry ends with membrane fusion of the viral particles with the endosome, followed by traffic of the viral nucleocapsid into the cytoplasm. The membrane fusion of orthohantavirus is driven by large conformational changes in the Gc protein. After membrane fusion, the viral nucleocapsid is assumed to traffic into the cytoplasm. The detailed molecular mechanisms of viral "uncoating" and the traffic of viral nucleocapsid to intracellular sites of replication and transcription remain unclear, as is the possible participation of host factors in these processes [6].

12.2.3 Structure and Function of Orthohantavirus Glycoproteins

Orthohantavirus Gn/Gc are essential and enough for cellular entry. Gn and Gc form a grid-like structure on the viral surface that is a hallmark of orthohantavirus [26, 27]. Gn constitutes a square shape at the distal end of the spike and is involved in attachment and fusion control [28]. Gc, a class II fusion protein, mediates membrane fusion of virus and host cell through conformational change under low pH in the late endosome [29–31]. Besides its roles in host-cell entry, the Gn/Gc complex is the dominating target of the neutralizing antibody-mediated immune response. Gc may be under less humoral immune pressure because it is sheltered partly by Gn, especially in the region containing highly conserved fusion loop [28, 32]. Based on peptide scanning, B cell epitopes have been mapped to Gn and Gc. However which epitopes are immunodominant or associated to recognition by neutralizing antibodies is still unknown [33–36].

12.3 Construction of Pseudotyped Orthohantaviruses

Studies of live orthohantaviruses are bound by biosafety level (BSL)-3 laboratories, which have brought large obstacles to orthohantavirus study and influenced the development of corresponding therapeutics and vaccines. With the advantages of high detection efficiency, strong operability, low biohazard, and high sensitivity, a variety of pseudotyped viruses have been constructed as alternatives to authentic orthohantaviruses. According to the replication capability, the pseudotyped orthohantaviruses can be divided into two types, one is replication-deficient and another is replication-competent. The replication-deficient pseudotyped orthohantaviruses are normally constructed with VSV, lentivirus (LV), and murine leukemia virus (MLV) as backbone vector while the replication-competent one is constructed with VSV backbone vector. Each has its benefits and drawbacks and will be elaborated in detail in the following chapter. As shown in Table 12.2, a variety of pseudotyped orthohantaviruses have been successfully constructed.

12.3.1 Construction of Replication-Deficient Pseudotyped Orthohantaviruses

12.3.1.1 The VSV-Based Packaging System

At present, the VSV packaging system is the most widely used pseudotyped orthohantavirus packaging system [37–42, 44–46, 64]. Briefly, cells are transfected with orthohantavirus glycoprotein expression plasmid. And 24h post transfection, cells are then infected with VSVΔG*G, with luciferase (Luc) or green fluorescent protein (GFP) as reporting genes. When this recombinant VSVΔG*G system is first used to incorporate orthohantavirus glycoprotein, it efficiently generates VSVΔG*G-HTNV and VSVΔG*G-SEOV with titers of 10^5 to 10^6 IU/ml, which are higher than those of live HTNV and SEOV with titers of 10^4 to 10^5 focus-forming units/ml [41]. Besides the aforementioned two pseudotyped viruses, VSVΔG*G-DOBV, VSVΔG*G-PUUV, VSVΔG*G-ANDV, and VSVΔG*G-SNV have been successfully generated based on the VSV pseudotype system. Significantly, when the VSV packaging system is applied to produce the pseudotyped orthohantaviruses, the generated pseudovirions may be mixed with residual VSVs, which may affect the neutralization assay. Best of all, the residual VSV could be minimized through improving experimental conditions. Nevertheless, if excessive VSV interfere with pseudovirus-based assay, pre-treating virus preparation with a VSV neutralizing antibody can be implemented.

12.3.1.2 The LV-Based Packaging System

Generally, the LV-based vectors comes from human immunodeficiency virus type 1 (HIV-1), which is the best known LV. The HIV packaging system is divided into two-plasmid, three-plasmid, and four-plasmid systems, based on the quantity of plasmids involved in the system [65]. Recently the two-plasmid system including an expression plasmid and a packaging plasmid has been preferably applied to generate pseudotyped HTNV, SEOV, and ANDV. Briefly, cells are transfected with orthohantavirus glycoprotein expression plasmid and packaging plasmid pNL4-3.Luc.RE, which encodes the HIV genome containing the firefly Luc gene [49, 50]. The three-plasmid system has also been used to construct pseudotyped HTNV, including expression plasmid pLVX-M together with packaging plasmids pGag/Pol (coding for HIV-1 Gag-Pol) and pRev (coding for HIV-1 Rev) [51]. To date, HIV-1 pseudotyped with HTNV, SEOV, and ANDV glycoprotein have been

Table 12.2 Construction and application of pseudotyped orthohantaviruses

Packaging system		Application	Virus	Plasmids	Reporting gene	Reference
Replication-deficient	VSV	Viral entry and infection	HTNV, ANDV, or PUUV	VSVΔG*G, pWRG/HTNV-M, pWRG/PUUV-M or pWRG/ANDV-M	Luc	[37, 38]
			HTNV or ANDV	VSVΔG*G, pWRG/HTNV-M or pI18	Luc	[39, 40]
		Quantification of neutralizing antibodies	HTNV or SEOV	VSVΔG*G, pCAGGS/MCS-HTNV-M or pCAGGS/MCS-SEOV-M	GFP	[41]
			HTNV, ANDV, or PUUV	VSVΔG*G, pWRG/HTNV-M, pWRG/PUUV-M or pWRG/ANDV-M	Luc	[37, 38]
			SNV, ANDV, PUUV, DOBV, HTNV, or SEOV	VSVΔG*G, pWRG/SNV-M, pWRG/ANDV-M, pWRG/PUUV-M, pWRG/DOBV-M, pWRG/HTNV-M, pWRG/SEOV-M	GFP	[42]
		Antigenic property study	PUUV	VSVΔG*G, pCAGGS/MCS-PUUV-M	GFP	[43]
			HTNV or SEOV	VSVΔG*G, pcDNA3.1(+)/HTNV-M, pcDNA3.1(+)/SEOV-M	Luc	[44]
		Identification of viral entry inhibitors	HTNV or ANDV	VSVΔG*G, pWRG/HTNV-M or pI18	GFP	[45]
			HTNV	VSVΔG*G, pcDNA3.1/HNTV-M	Luc	[46, 47]
		Vaccine approach	HTNV or SEOV	VSVΔG*G, pCAGGS/MCS-HTNV-M	GFP	[48]
	LV-HIV-1	Quantification of neutralizing antibodies	HTNV or SEOV	pNL4-3.Luc.RE	Luc	[49]
		Identification of viral entry inhibitors	ANDV	pNL4-3.Luc.RE	Luc	[50]
		Vaccine approach	HTNV	pLV-M, pGag/Pol, pRev	ZsGreen1	[51]
	LV-SIV	Viral entry and infection	ANDV	pSIV3+, pGAE1.0, pI.18	GFP	[52–54]
	MLV	Viral entry and infection	HTNV	pcGP, pcnβG, pcHTNVM	β-gal	[55]

(continued)

Table 12.2 (continued)

Packaging system		Application	Virus	Plasmids	Reporting gene	Reference
Replication-competent	rVSV	Viral entry and infection	ANDV, SNV, or HTNV	rVSV-ANDV-M, rVSV-SNV-M, rVSV-HTNV-M, rVSV-CHOV-M, rVSV-PUUV-M	mNeonGreen	[56–61]
			ANDV, SNV, HTNV, SEOV, DOBV, MPRLV, or PHV	rVSV-ANDV-M, rVSV-SNV-M, rVSV-HTNV-M, rVSV-SEOV-M, rVSV-DOBV-M, rVSV-MPRLV-M, rVSV-PHV-M	GFP	[56]
		Vaccine approach	HTNV, SNV, or HTNV	rVSV-ANDV-M, rVSV-SNV-M, rVSV-HTNV-M	None	[53, 54, 62, 63]

successfully generated and have been used for developing vaccines [51], evaluating neutralizing antibodies [49], and discovering drugs [50].

The simian immunodeficiency virus (SIV)-based lentiviral packing system has been used to construct pseudotyped ANDV, and the psedovirions have been applied to the mechanistic study for the viral entry and infection [52, 64, 66]. ANDV Gn/Gc-pseudotyped SIV particles are prepared using the three-plasmid co-transfection method. Briefly, cells are transfected with the following plasmids: pSIV3+ (coding for SIV Gag-Pol), pGAE1.0 (coding for the GFP), and pI.18/GPC (coding for ANDV glycoprotein) constructs. 48h after transfection, supernatants containing the pseudotyped virus particles are collected and purified by ultracentrifugation [52–54].

12.3.1.3 The MLV-Based Packaging System

The MLV packaging system, also known as retroviral system, is widely applied to produce pseudotyped viruses. To produce MLV pseudotyped with HTNV glycoprotein, plasmids pcHTNV-M, pcGP, and pcnβG are transfected into cells. Notably, the first plasmid pcHTNV-M is the expression plasmid encoding glycoproteins of HTNV. The second plasmid pcGP is the packaging plasmid encoding MLV Gag-Pol. The third plasmid pcnβG consists of MLV LTR, resistance gene, and reporter gene [67]. 48h post transfection, the supernatants containing the pseudovirions are harvested and centrifuged [55]. However, the titers of pseudovirions generated by this system are only around 10^3 IU/ml, indicating the low expression level or instability of HTNV glycoproteins and/or ineffective virion assembly [55].

Generally, each of the three aforementioned systems has its own special benefits and drawbacks. Firstly, in the aspect of virus yield, MLV and VSV packaging systems have higher virus yields than LV system. Secondly, in the aspect of operational simplicity, the MLV and LV packaging systems are time-saving and less complicated compared to VSV system [68]. Compared with MLV and HIV pseudotyped virions, another benefit of VSV pseudotyped virus is that the expression of its reporter gene is robustly driven by the potent intracellular replication of the VSV genome so as to be detected just several hours after infection [69]. Finally, in terms of safety, VSV and MLV systems are much safer over the LV system, given authentic VSV and MLV are less virulent than LV.

12.3.2 Construction of Replication-Competent Pseudotyped Orthohantaviruses

Although the aforementioned replication-deficient single-cycle pseudotyped viruses have improved our knowledge of orthohantavirus Gn/Gc complex assembly, viral attachment and entry, and antiviral immune responses, they are laborious to generate in quantity. By contrast, self-replicating, recombinant VSVs (rVSVs), whose genomes have been manipulated to replace its own glycoprotein gene with orthohantavirus M gene (encoding Gn and Gc glycoproteins), are more likely to generate in high yield and be applied as a fast-acting, efficacious vaccine candidate for orthohantavirus disease, including rVSV-HTNV, rVSV-ANDV, and rVSV-SNV [53, 54, 62, 63]. Besides, the rVSV platform has also been used to construct a variety of replicated pseudotyped orthohantaviruses to study the mechanism of viral entry and infection [56–61], as shown in Table 12.2.

Replication-competent rVSVs bearing orthohantavirus glycoprotein are generated using a plasmid-based rescue system [53]. Briefly, cells are transfected with the helper plasmids encoding the VSV ribonucleoprotein constituents and the full-length VSV genomic plasmid containing the orthohantavirus ANDV GPC sequence, all under the control of the T7 promoter. After incubated at 37°C for 6 days, the supernatants were blind passaged onto fresh Vero E6 cells.

12.4 Applications of Pseudotyped Orthohantaviruses

At present, pseudotyped orthohantaviruses have greatly assisted in revealing the effects of host factors on viral attachment/entry and other virus-host interactions [37, 38, 56–61], helped elucidate mechanisms of antiviral immune response and correlates of protection [37, 38, 49, 70, 71], enhanced our knowledge of viral antigenic property [43, 44], and have successfully been used to characterize viral entry inhibitors and developed as vaccines [45, 46, 50, 72].

12.4.1 Mechanistic Study for Viral Entry and Infection

A variety of rodent-borne orthohantaviruses result in zoonotic diseases along with serious disease and death. However, the evaluation of zoonotic risk together with the exploration of therapeutics are trapped by our poor understanding of the specific mechanisms of orthohantavirus infection. Notably, viral attachment and entrance are the foremost steps in zoonotic spread and infection of orthohantavirus. Pseudotyped orthohantaviruses have been widely applied to perform in vitro research on virus-host interaction, such as the identities of cellular receptors or factors and their influence on viral virulence and host range.

12.4.1.1 Identification of Cellular Receptors and Factors

Pseudotyped viruses have been widely used to identify cellular receptors and factors involved in orthohantavirus entry and infection. VSV pseudotyped with New and

Old World ANDV and HTNV glycoprotein have been used to investigate cellular factors associated with entry of the pathogenic ANDV and HTNV into human respiratory epithelial cells [39, 40]. Cellular phosphatidylserine receptors of the T-cell immunoglobulin and mucin protein [40] and macropinocytosis [39] are firstly demonstrated to play a role in ANDV and HTNV infection. Notably, pseudotyped ANDV and HTNV depend on a series of non-overlapping and known cellular conductor of macropinocytosis to entry, suggesting that pathogenic orthohantaviruses can recruit specific regulators, depending on their needs [39]. Moreover, pseudotyped HTNV are reported to make use of $\alpha_v\beta_3$ integrin and glycosaminoglycan heparan sulfate as co-receptors, firstly demonstrating that orthohantaviruses use apoptotic mimicry to infect human respiratory epithelial cells, which may provide an explanation why these viruses could readily across the species barrier [40]. Although it has been reported that pathogenic orthohantaviruses interact with $\alpha_v\beta_3$ integrin [25], both pseudotyped PUUV and HTNV could potently infect cell lines without detectable $\alpha_v\beta_3$ expression. While SupT1 cells could stably express the $\alpha_v\beta_3$ integrin heterodimer but could not rescue virus infection, indicating that extra host factors are necessary to help orthohantavirus entry into certain cell lines [38]. Based on the VSV pseudotypes, to NWH ANDV, pH dependency is the requirement for entry [37]. While to OWH HTNV and PUUV, only mildly acidic pH is required for entry [38]. And when they are incubated at or below pH 6.0 in the absence of host cells, they are rapidly inactivated [38].

SIV vector pseudotyped with the ANDV glycoprotein has also been generated to investigate early steps of ANDV infection. Exhaustion of cholesterol from host cells potently weaken cell infection, demonstrating a possible effect of lipid rafts on ANDV cell entry [66].

To systematically identify host factors involved in orthohantavirus entry, the rVSV bearing the ANDV Gn/Gc glycoproteins (rVSV-ANDV Gn/Gc) has been applied to conduct a loss-of-function genetic screen in HAP1 haploid human cells [57, 59]. Multiple genes associated with the sterol regulatory element binding protein pathway have been identified as determining factors of viral entrance into endothelial cells [57], indicating that membrane cholesterol has important effects on orthohantavirus membrane fusion [59]. Similarly, the rVSV has been used to perform whole gene disruption among human cells [58]. And protocadherin-1 (PCDH1), a protein belonging to the cadherin superfamily, has been demonstrated as an important determinant for cell entry and infection by several NWHs, such as SNV and ANDV [58]. Moreover, based on rVSVs bearing orthohantavirus glycoprotein, Dieterle et al. have confirmed that loss of β_1 integrin, β_3 integrin, and/or decay-accelerating factor makes little or no impact on entrance by a large set of orthohantaviruses. By contrast, loss of PCDH-1, a newly discovered cellular receptor for some orthohantaviruses, significantly reduces NWH entry and infection but not the OWH. All the results have been independently corroborated with authentic orthohantaviruses, indicating the robustness of the replication-competent pseudotyped virus platform [56].

12.4.1.2 Identification of Key Amino Acid in Gn/Gc

Based on pseudotyped viruses, certain amino acids in Gn/Gc have been reported to affect orthohantavirus entry and infection. By using ANDV pseudotyped SIV particles, Barriga et al. have found the entrance of ANDV into host cells can be prevented by the Gc fusion protein fragments that are analogous to inhibitory fragments of class II fusion proteins [64]. Munoz et al. have found SIV particle infectivity can be impaired by amino acid substitutions of Gc fusion protein at positions W115 and N118 [52]. These results indicate the importance of Gc in orthohantavirus-cell membrane fusion and infection, especially the aromatic and polar residues in Gc. Moreover, by using rVSV bearing wild-type and mutant Gn/Gc (I352K/S1094L) glycoproteins of HTNV, Slough et al. have found I352K/S1094L substitutions in HTNV Gn and Gc, respectively, have a synergistic effect on increasing infectivity through transporting HTNV Gn/Gc from the Golgi complex to the cell surface, consequently enhancing Gn/Gc insertion into budding VSV particles [60].

12.4.1.3 Cell Tropism

MLV pseudotypes of HTNV [55], VSV pseudotypes of ANDV [37], and VSV pseudotypes of HTNV and PUUV [38] have been used for analysis of cell tropism. A wide variety of sensitive target cells is expected because orthohantaviruses have been demonstrated by immunohistochemistry to infect so many different sorts of human tissues, such as the brain, intestine, lymph nodes, urinary bladder, adrenal, adipose tissue, pancreas, heart, kidney, spleen, and skeletal muscle [73].

12.4.2 Quantification of Neutralizing Antibodies

Orthohantavirus neutralizing antibody measurement is important for both serotyping and the analysis of the immune response triggered by vaccination, viral infection, or passive application of investigative monoclonal/polyclonal antibodies. The traditional method for the detection of orthohantavirus-specific neutralizing antibody is the plaque reduction neutralization test (PRNT) that is featured with authentic infectious orthohantaviruses and restricted to BSL-3 laboratory. Nevertheless, the drawbacks of PRNT are obvious, such as manipulation of the fatal virions and time-consuming procedures (1–2 weeks) [74, 75]. Thus many pseudovirus-based neutralization assays (PBNA) employing VSV pseudotypes decorated with orthohantavirus glycoprotein were developed because of the reduced biohazard, cost, and time to conduct. Pseudotyped virus system has been used to detect neutralizing antibodies from patient sera or vaccinated animal sera, which appears to have comparable titers but higher sensitivity in comparison with PRNT featured with live infectious orthohantaviruses [37, 38, 49, 70, 71].

Based on the Choclo virus (CHOV) glycoprotein pseudotyped system, neutralizing antibodies against CHOV have been detected in only a few acute phase samples but in most asymptomatic samples. The fact that the asymptomatic population presents high levels of neutralizing antibodies along with the upregulation of IL-4 supports the assumption that CHOV infection have caused strong immune response bypassing the pathogenic inflammatory response and ultimately causing no symptoms. Undoubtedly, such assumption is extremely difficult to be verified [61].

Engdahl et al. have used pseudotyped viruses to broaden the understanding of the natural humoral immune response to NWH infection based on a panel of human monoclonal antibodies derived from ANDV or SNV infected individuals. Several ANDV-reactive antibodies exhibit exclusively reactivity to ANDV, while most SNV-reactive antibodies display wide identification and cross-neutralization to both NWH and OWH. It is worth noting that four monoclonal antibodies against NWH protect hamsters against ANDV challenge at clinically relevant doses. These results demonstrate the human antibody an astrictive and potently neutralizing reactivity to NWHs and indicate therapeutic potential against HCPS of the human monoclonal antibodies [42].

12.4.3 Antigenic Property Study

The orthohantavirus mutants have emerged and accumulated with the passage of time. It is necessary to ascertain whether their antigenic property has changed, consequently making the existing vaccines incompetence. However, standard PRNT requires huge amounts of authentic mutant strains to examine their antigenic property evolution, which significantly decreases the speed of fast antigenic determination of emerging/circulating mutants [44].

Pseudotyped VSV bearing the full-length glycoprotein of HTNV and SEOV has been used to monitor the antigenic property change of evolving HTNV and SEOV strains [44]. Results generated by pseudotyped viruses are highly in consistent with those by live viruses but with obvious benefits in the light of safety, time-saving, and high-throughput ability. Fifty-three HTNV mutant pseudotypes and 46 SEOV mutants have been developed successfully. And antisera vaccinated with HTNV or SEOV vaccine have been found to be able to neutralize their whole corresponding mutant strains except for the K816I. This mutant strain presents a decreased reactivity by fivefold in comparison with the reference virion. Considering that K816I has been found to contribute to the antigenic property of PUUV neutralized by the monoclonal antibody 1C9, the biological significance of this amino acid substitution should be further studied [76]. VSV pseudotyped with the surface glycoproteins of PUUV have been created to perform site-directed mutagenesis to determine neutralizing epitopes recognized by monoclonal antibodies 4G2, 1C9, and 5A2 [43]. Besides D272V mutation that contributes to the change of PUUV phenotype to escape-type [77], another three mutant strains have been generated by substituting single bulky-aromatic amino acid residues with alanines, including F70A, Y274A, and F915A [43].

12.4.4 Identification of Viral Entry Inhibitors

To orthohantaviruses, the only existing vaccine, Hantavax®, is merely approved in some Asian countries but not in the USA or Europe, and the treatment strategy is now restricted to supporting treatment [78, 79]. Considering that the orthohantaviruses are distributed around the world, there is an urgent need to develop new medical measures for prevention and treatment. To zoonotic virus spread and infection, viral attachment and entry are the first and most basic procedures, which therefore provides new targeting points for the therapeutic intervention so as to prevent viruses before they can control the target cell and generate viral progeny. However, it is significantly hard to explore potent inhibitors against orthohantavirus on a cellular level, for orthohantaviruses are usually slow-growing even in the most susceptible target cells and thereby lead to little or no cytopathic effect. Another limitation is that there is no applicable model to investigate the impact of inhibitors on animals [46]. As a close correlation of pseudotyped viruses with the authentic ones, the pseudotyped virus platform has been widely used to uncover entry factors of orthohantaviruses [45, 58].

VSV pseudotype harboring the glycoprotein (Gn and Gc) of the ANDV or HTNV and GFP reporter has been generated, validated, and applied to screen a phytochemical library consisting of 320 natural compounds to find novel antiviral compounds against ANDV and HTNV. Two inhibitors, including emetine dihydrochloride and tetrandrine, have been selected and validated with live infectious HTNV under BSL-3 conditions, confirming the robustness of the pseudotyped virus screening platform [45]. Similarly, by using VSV pseudotype bearing HNTV glycoprotein and encoding Fluc, Wen et al. have screened a library containing 556 small molecular compounds and 1813 approved medicines derived from traditional Chinese medicine. And cepharanthine has been identified to be efficacious against pseudotyped HTNV infection both in vivo and in vitro. The effect of the drug has been primitively validated with authentic HTNV in vitro, indicating the robustness of the pseudotyped virus screening platform [47]. By using the same pseudotyped virus system with HTNV glycoprotein on its surface, Sokolova et al. have studied and confirmed the antiviral activity of Camphene Derivatives 3a, 2a, and 7a [46]. HIV-1 pseudotyped particles have been used to demonstrate the in vitro antiviral activity of Griffithsin (GRFT) and its synthetic trimeric tandemer 3mGRFT against ANDV and SNV [50].

12.4.5 Vaccine Approach

Given the high mortality rates of HFRS or HCPS but lack of therapeutic approaches nevertheless, there is a pressing need to develop novel prophylactic vaccines. Although certain inactivated vaccines against HFRS have been approved in China [80], there are still opportunities to explore more effective and safer vaccines through technology innovation [81].

Pseudotyped viruses have been widely used as vaccine candidates against NWHs ANDV and SNV. Replication-competent rVSV-based vaccine that expresses the glycoproteins of ANDV has been developed [53]. This vaccine is significantly effective in the Syrian hamster model of HCPS when administered a few days before lethal ANDV challenge, or when administered around the period of challenge (peri-exposure), and even is efficacious when given post-exposure, proving fast protective pre- and post-exposure effect against challenge with lethal ANDV in the Syrian hamster model [53]. Moreover, it is demonstrated that even a single dose of this vaccine is effective in protecting Syrian hamster from lethal ANDV challenge 6 months after vaccination. However, such protection is not efficacious after one year of vaccination [54]. Notably, rVSV-based vaccines for ANDV and SNV have been reported to be able to trigger a cross-reactive antibody response [62]. Both vaccines are effective in protecting animal models from both homologous and heterologous challenge with SNV and ANDV and prevent HCPS in a lethal ANDV challenge model, suggesting that the administration of a single vaccine against HCPS-causing orthohantavirus could offer significant protection against several pathogens. Besides, it has been demonstrated that vaccination of deer mice with rVSV-based SNV vaccines, particularly oral vaccination, could be an efficacious way to prevent infection of SNV following direct exposure to infected deer mice, indicating the administration of bait style vaccines to block the transmission of rodent-borne orthohantaviruses [63]. Pseudotyped viruses have also been used as vaccine candidates against OWHs HTNV. A VSV pseudotype including HTNV glycoprotein has been reported to stimulate neutralizing antibodies and can offer protective immunity for orthohantavirus challenge in the mouse model [48]. Similarly, a pseudotyped LV has been successfully constructed and applied as an HTNV vaccine [51]. The pseudotyped LV stimulates higher levels of neutralizing antibody than the inactivated vaccine and offers protective immunity for orthohantavirus challenge in the mouse model [51]. These results demonstrate that pseudotyped viruses can stimulate more efficacious immune responses than the inactivated viruses. One possible explanation is that the pseudotyped viruses represent viral envelope glycoproteins with a more natural conformation.

12.5 Conclusions

Orthohantaviruses can cause HFRS in Eurasia and HCPS in the Americas. Recently, repeated outbreaks of orthohantavirus disease have drawn global attention and have constituted serious public health burden. Understanding the natural features and infection mechanism of the virus is urgently needed. However, orthohantavirus study is inaccessible to most research laboratories worldwide, for studies of authentic orthohantaviruses are constrained by BSL-3 conditions, which severely hinder the exploration of both medicines and vaccines against orthohantaviruses. As a result, based on the VSV, LV, MLV, and rVSV packaging system, a variety of pseudotyped orthohantaviruses have been constructed because of the reduced cost, biosafety requirements, and time to complete. And these pseudotyped viruses have been widely used for the study of orthohantaviruses, including mechanistic study for viral entry and infection, evaluating immune responses, monitoring viral antigenic property, characterizing viral entry inhibitors, and developing as vaccines or gene delivery.

Acknowledgements This work was supported by the Young Scientists Fund of the National Natural Science Foundation of China (grant number 82100592), the Research Foundation of Beijing Friendship Hospital, Capital Medical University (grant number yyzz202030), and Natural Science Foundation of Capital Medical University (grant number PYZ20142).

References

1. ICTV. *Virus taxonomy: 2020 release*, https://talk.ictvonline.org/taxonomy/ (2020)
2. Vaheri, A., et al.: Uncovering the mysteries of hantavirus infections. *Nat. Rev. Microbiol.* **11**, 539–550 (2013). https://doi.org/10.1038/nrmicro3066
3. Witkowski, P.T., et al.: Phylogenetic analysis of a newfound bat-borne hantavirus supports a laurasiatherian host association for ancestral mammalian hantaviruses. *Infect. Genet. Evol.* **41**, 113–119 (2016). https://doi.org/10.1016/j.meegid.2016.03.036
4. Martinez, V.P., et al.: Person-to-person transmission of Andes virus. *Emerg. Infect. Dis.* **11**, 1848–1853 (2005). https://doi.org/10.3201/eid1112.050501
5. Martinez-Valdebenito, C., et al.: Person-to-person household and nosocomial transmission of Andes hantavirus, Southern Chile, 2011. *Emerg. Infect. Dis.* **20**, 1629–1636 (2014). https://doi.org/10.3201/eid2010.140353
6. Mittler, E., et al.: Hantavirus entry: perspectives and recent advances. *Adv. Virus Res.* **104**, 185–224 (2019). https://doi.org/10.1016/bs.aivir.2019.07.002
7. Forbes, K.M., Sironen, T., Plyusnin, A.: Hantavirus maintenance and transmission in reservoir host populations. *Curr. Opin. Virol.* **28**, 1–6 (2018). https://doi.org/10.1016/j.coviro.2017.09.003
8. Meyer, B.J., Schmaljohn, C.S.: Persistent hantavirus infections: characteristics and mechanisms. *Trends Microbiol.* **8**, 61–67 (2000). https://doi.org/10.1016/s0966-842x(99)01658-3
9. Lee, S.H., et al.: Dynamic circulation and genetic exchange of a shrew-borne Hantavirus, Imjin virus, in the Republic of Korea. *Sci. Rep.* **7**, 44369 (2017). https://doi.org/10.1038/srep44369
10. Mir, M.A.: Hantaviruses. *Clin. Lab. Med.* **30**, 67–91 (2010). https://doi.org/10.1016/j.cll.2010.01.004

11. Hepojoki, J., Strandin, T., Lankinen, H., Vaheri, A.: Hantavirus structure—molecular interactions behind the scene. *J. Gen. Virol.* **93**, 1631–1644 (2012). https://doi.org/10.1099/vir.0.042218-0
12. Schmaljohn, C.S., Hasty, S.E., Harrison, S.A., Dalrymple, J.M.: Characterization of Hantaan virions, the prototype virus of hemorrhagic fever with renal syndrome. *J. Infect. Dis.* **148**, 1005–1012 (1983). https://doi.org/10.1093/infdis/148.6.1005
13. Wertz, G.W., Whelan, S., LeGrone, A., Ball, L.A.: Extent of terminal complementarity modulates the balance between transcription and replication of vesicular stomatitis virus RNA. *Proc. Natl. Acad. Sci. U. S. A.* **91**, 8587–8591 (1994). https://doi.org/10.1073/pnas.91.18.8587
14. Lee, Y.S., Seong, B.L.: Mutational analysis of influenza B virus RNA transcription in vitro. *J. Virol.* **70**, 1232–1236 (1996). https://doi.org/10.1128/JVI.70.2.1232-1236.1996
15. Tiley, L.S., Hagen, M., Matthews, J.T., Krystal, M.: Sequence-specific binding of the influenza virus RNA polymerase to sequences located at the 5′ ends of the viral RNAs. *J. Virol.* **68**, 5108–5116 (1994). https://doi.org/10.1128/JVI.68.8.5108-5116.1994
16. Mir, M.A., Brown, B., Hjelle, B., Duran, W.A., Panganiban, A.T.: Hantavirus N protein exhibits genus-specific recognition of the viral RNA panhandle. *J. Virol.* **80**, 11283–11292 (2006). https://doi.org/10.1128/JVI.00820-06
17. Schmaljohn, C.S., Jennings, G.B., Hay, J., Dalrymple, J.M.: Coding strategy of the S genome segment of Hantaan virus. *Virology.* **155**, 633–643 (1986). https://doi.org/10.1016/0042-6822(86)90223-0
18. Schmaljohn, C.S., Dalrymple, J.M.: Analysis of Hantaan virus RNA: evidence for a new genus of bunyaviridae. *Virology.* **131**, 482–491 (1983). https://doi.org/10.1016/0042-6822(83)90514-7
19. Mir, M.A., Panganiban, A.T.: Characterization of the RNA chaperone activity of hantavirus nucleocapsid protein. *J. Virol.* **80**, 6276–6285 (2006). https://doi.org/10.1128/JVI.00147-06
20. Kukkonen, S.K., Vaheri, A., Plyusnin, A.: L protein, the RNA-dependent RNA polymerase of hantaviruses. *Arch. Virol.* **150**, 533–556 (2005). https://doi.org/10.1007/s00705-004-0414-8
21. Muyangwa, M., Martynova, E.V., Khaiboullina, S.F., Morzunov, S.P., Rizvanov, A.A.: Hantaviral proteins: structure, functions, and role in Hantavirus infection. *Front. Microbiol.* **6**, 1326 (2015). https://doi.org/10.3389/fmicb.2015.01326
22. Schmaljohn, C.S., Schmaljohn, A.L., Dalrymple, J.M.: Hantaan virus M RNA: coding strategy, nucleotide sequence, and gene order. *Virology.* **157**, 31–39 (1987). https://doi.org/10.1016/0042-6822(87)90310-2
23. Kamrud, K.I., Schmaljohn, C.S.: Expression strategy of the M genome segment of Hantaan virus. *Virus Res.* **31**, 109–121 (1994). https://doi.org/10.1016/0168-1702(94)90074-4
24. Lober, C., Anheier, B., Lindow, S., Klenk, H.D., Feldmann, H.: The Hantaan virus glycoprotein precursor is cleaved at the conserved pentapeptide WAASA. *Virology.* **289**, 224–229 (2001). https://doi.org/10.1006/viro.2001.1171
25. Gavrilovskaya, I.N., Shepley, M., Shaw, R., Ginsberg, M.H., Mackow, E.R.: beta3 Integrins mediate the cellular entry of hantaviruses that cause respiratory failure. *Proc. Natl. Acad. Sci. U. S. A.* **95**, 7074–7079 (1998). https://doi.org/10.1073/pnas.95.12.7074
26. Huiskonen, J.T., et al.: Electron cryotomography of Tula hantavirus suggests a unique assembly paradigm for enveloped viruses. *J. Virol.* **84**, 4889–4897 (2010). https://doi.org/10.1128/JVI.00057-10
27. Martin, M.L., Lindsey-Regnery, H., Sasso, D.R., McCormick, J.B., Palmer, E.: Distinction between Bunyaviridae genera by surface structure and comparison with Hantaan virus using negative stain electron microscopy. *Arch. Virol.* **86**, 17–28 (1985). https://doi.org/10.1007/BF01314110
28. Serris, A., et al.: The Hantavirus surface glycoprotein lattice and its fusion control mechanism. *Cell.* **183**, 442–456 (2020). https://doi.org/10.1016/j.cell.2020.08.023

29. Allen, E.R., et al.: A protective monoclonal antibody targets a site of vulnerability on the surface of Rift Valley fever virus. *Cell Rep.* **25**, 3750–3758 (2018). https://doi.org/10.1016/j.celrep.2018.12.001
30. Guardado-Calvo, P., et al.: Mechanistic insight into Bunyavirus-induced membrane fusion from structure-function analyses of the Hantavirus envelope glycoprotein Gc. *PLoS Pathog.* **12**, e1005813 (2016). https://doi.org/10.1371/journal.ppat.1005813
31. Willensky, S., et al.: Crystal structure of glycoprotein C from a Hantavirus in the post-fusion conformation. *PLoS Pathog.* **12**, e1005948 (2016). https://doi.org/10.1371/journal.ppat.1005948
32. Li, S., et al.: A molecular-level account of the antigenic hantaviral surface. *Cell Rep.* **16**, 278 (2016). https://doi.org/10.1016/j.celrep.2016.06.039
33. Arikawa, J., Schmaljohn, A.L., Dalrymple, J.M., Schmaljohn, C.S.: Characterization of Hantaan virus envelope glycoprotein antigenic determinants defined by monoclonal antibodies. *J. Gen. Virol.* **70(Pt 3)**, 615–624 (1989). https://doi.org/10.1099/0022-1317-70-3-615
34. Engdahl, T.B., Crowe Jr., J.E.: Humoral immunity to Hantavirus infection. *mSphere.* **5** (2020). https://doi.org/10.1128/mSphere.00482-20
35. Heiskanen, T., et al.: Phage-displayed peptides mimicking the discontinuous neutralization sites of Puumala Hantavirus envelope glycoproteins. *Virology.* **262**, 321–332 (1999). https://doi.org/10.1006/viro.1999.9930
36. Koch, J., Liang, M., Queitsch, I., Kraus, A.A., Bautz, E.K.: Human recombinant neutralizing antibodies against Hantaan virus G2 protein. *Virology.* **308**, 64–73 (2003). https://doi.org/10.1016/s0042-6822(02)00094-6
37. Ray, N., Whidby, J., Stewart, S., Hooper, J.W., Bertolotti-Ciarlet, A.: Study of Andes virus entry and neutralization using a pseudovirion system. *J. Virol. Methods.* **163**, 416–423 (2010). https://doi.org/10.1016/j.jviromet.2009.11.004
38. Higa, M.M., Petersen, J., Hooper, J., Doms, R.W.: Efficient production of Hantaan and Puumala pseudovirions for viral tropism and neutralization studies. *Virology.* **423**, 134–142 (2012). https://doi.org/10.1016/j.virol.2011.08.012
39. Torriani, G., et al.: Macropinocytosis contributes to hantavirus entry into human airway epithelial cells. *Virology.* **531**, 57–68 (2019). https://doi.org/10.1016/j.virol.2019.02.013
40. Mayor, J., Torriani, G., Rothenberger, S., Engler, O.: T-cell immunoglobulin and mucin (TIM) contributes to the infection of human airway epithelial cells by pseudotype viruses containing Hantaan virus glycoproteins. *Virology.* **543**, 54–62 (2020). https://doi.org/10.1016/j.virol.2020.02.002
41. Ogino, M., et al.: Use of vesicular stomatitis virus pseudotypes bearing Hantaan or Seoul virus envelope proteins in a rapid and safe neutralization test. *Clin. Diagn. Lab. Immunol.* **10**, 154–160 (2003). https://doi.org/10.1128/cdli.10.1.154-160.2003
42. Engdahl, T.B., et al.: Broad and potently neutralizing monoclonal antibodies isolated from human survivors of New World hantavirus infection. *Cell Rep.* **35**, 109086 (2021). https://doi.org/10.1016/j.celrep.2021.109086
43. Levanov, L., Iheozor-Ejiofor, R.P., Lundkvist, A., Vapalahti, O., Plyusnin, A.: Defining of MAbs-neutralizing sites on the surface glycoproteins Gn and Gc of a hantavirus using vesicular stomatitis virus pseudotypes and site-directed mutagenesis. *J. Gen. Virol.* **100**, 145–155 (2019). https://doi.org/10.1099/jgv.0.001202
44. Ning, T., et al.: Monitoring neutralization property change of evolving Hantaan and Seoul viruses with a novel pseudovirus-based assay. *Virol. Sin.* **36**, 104–112 (2021). https://doi.org/10.1007/s12250-020-00237-y
45. Mayor, J., Torriani, G., Engler, O., Rothenberger, S.: Identification of novel antiviral compounds targeting entry of Hantaviruses. *Viruses.* **13** (2021). https://doi.org/10.3390/v13040685
46. Sokolova, A.S., et al.: Synthesis and antiviral activity of camphene derivatives against different types of viruses. *Molecules.* **26** (2021). https://doi.org/10.3390/molecules26082235

47. Wen, X., et al.: Screening and identification of HTNVpv entry inhibitors with high-throughput pseudovirus-based chemiluminescence. *Virol. Sin.* (2022). https://doi.org/10.1016/j.virs.2022.04.015
48. Lee, B.H., et al.: A pseudotype vesicular stomatitis virus containing Hantaan virus envelope glycoproteins G1 and G2 as an alternative to hantavirus vaccine in mice. *Vaccine*. **24**, 2928–2934 (2006). https://doi.org/10.1016/j.vaccine.2005.12.040
49. Li, W., et al.: Comparison of serological assays to titrate Hantaan and Seoul hantavirus-specific antibodies. *Virol. J.* **14**, 133 (2017). https://doi.org/10.1186/s12985-017-0799-0
50. Shrivastava-Ranjan, P., et al.: Hantavirus infection is inhibited by Griffithsin in cell culture. *Front. Cell. Infect. Microbiol.* **10**, 561502 (2020). https://doi.org/10.3389/fcimb.2020.561502
51. Yu, L., et al.: A recombinant pseudotyped lentivirus expressing the envelope glycoprotein of Hantaan virus induced protective immunity in mice. *Virol. J.* **10**, 301 (2013). https://doi.org/10.1186/1743-422X-10-301
52. Cifuentes-Munoz, N., Barriga, G.P., Valenzuela, P.D., Tischler, N.D.: Aromatic and polar residues spanning the candidate fusion peptide of the Andes virus Gc protein are essential for membrane fusion and infection. *J. Gen. Virol.* **92**, 552–563 (2011). https://doi.org/10.1099/vir.0.027235-0
53. Brown, K.S., Safronetz, D., Marzi, A., Ebihara, H., Feldmann, H.: Vesicular stomatitis virus-based vaccine protects hamsters against lethal challenge with Andes virus. *J. Virol.* **85**, 12781–12791 (2011). https://doi.org/10.1128/JVI.00794-11
54. Prescott, J., DeBuysscher, B.L., Brown, K.S., Feldmann, H.: Long-term single-dose efficacy of a vesicular stomatitis virus-based Andes virus vaccine in Syrian hamsters. *Viruses*. **6**, 516–523 (2014). https://doi.org/10.3390/v6020516
55. Ma, M., et al.: Murine leukemia virus pseudotypes of La Crosse and Hantaan Bunyaviruses: a system for analysis of cell tropism. *Virus Res.* **64**, 23–32 (1999). https://doi.org/10.1016/s0168-1702(99)00070-2
56. Dieterle, M.E., et al.: Genetic depletion studies inform receptor usage by virulent hantaviruses in human endothelial cells. *elife*. **10** (2021). https://doi.org/10.7554/eLife.69708
57. Petersen, J., et al.: The major cellular sterol regulatory pathway is required for Andes virus infection. *PLoS Pathog.* **10**, e1003911 (2014). https://doi.org/10.1371/journal.ppat.1003911
58. Jangra, R.K., et al.: Protocadherin-1 is essential for cell entry by New World hantaviruses. *Nature*. **563**, 559–563 (2018). https://doi.org/10.1038/s41586-018-0702-1
59. Kleinfelter, L.M., et al.: Haploid genetic screen reveals a profound and direct dependence on cholesterol for Hantavirus membrane fusion. *MBio*. **6**, e00801 (2015). https://doi.org/10.1128/mBio.00801-15
60. Slough, M.M., Chandran, K., Jangra, R.K.: Two point mutations in old world Hantavirus glycoproteins afford the generation of highly infectious recombinant vesicular stomatitis virus vectors. *MBio*. **10** (2019). https://doi.org/10.1128/mBio.02372-18
61. Salinas, T.P., et al.: Cytokine profiles and antibody response associated to choclo orthohantavirus infection. *Front. Immunol.* **12**, 603228 (2021). https://doi.org/10.3389/fimmu.2021.603228
62. Warner, B.M., et al.: Vesicular stomatitis virus-based vaccines provide cross-protection against Andes and Sin Nombre Viruses. *Viruses*. **11** (2019). https://doi.org/10.3390/v11070645
63. Warner, B.M., et al.: Oral vaccination with recombinant vesicular stomatitis virus expressing Sin Nombre Virus glycoprotein prevents Sin Nombre virus transmission in deer mice. *Front. Cell. Infect. Microbiol.* **10**, 333 (2020). https://doi.org/10.3389/fcimb.2020.00333
64. Barriga, G.P., et al.: Inhibition of the Hantavirus fusion process by predicted domain III and stem peptides from glycoprotein Gc. *PLoS Negl. Trop. Dis.* **10**, e0004799 (2016). https://doi.org/10.1371/journal.pntd.0004799
65. Xiang, Q., Li, L., Wu, J., Tian, M., Fu, Y.: Application of pseudovirus system in the development of vaccine, antiviral-drugs, and neutralizing antibodies. *Microbiol. Res.* **258**, 126993 (2022). https://doi.org/10.1016/j.micres.2022.126993

66. Cifuentes-Munoz, N., Darlix, J.L., Tischler, N.D.: Development of a lentiviral vector system to study the role of the Andes virus glycoproteins. *Virus Res.* **153**, 29–35 (2010). https://doi.org/10.1016/j.virusres.2010.07.001
67. Soneoka, Y., et al.: A transient three-plasmid expression system for the production of high titer retroviral vectors. *Nucleic Acids Res.* **23**, 628–633 (1995). https://doi.org/10.1093/nar/23.4.628
68. Li, Q., Liu, Q., Huang, W., Li, X., Wang, Y.: Current status on the development of pseudoviruses for enveloped viruses. *Rev. Med. Virol.* **28** (2018). https://doi.org/10.1002/rmv.1963
69. Schmidt, F., et al.: Measuring SARS-CoV-2 neutralizing antibody activity using pseudotyped and chimeric viruses. *bioRxiv*. (2020). https://doi.org/10.1101/2020.06.08.140871
70. Vial, C., et al.: Comparison of VSV pseudovirus and focus reduction neutralization assays for measurement of anti-Andes orthohantavirus neutralizing antibodies in patient samples. *Front. Cell. Infect. Microbiol.* **10**, 444 (2020). https://doi.org/10.3389/fcimb.2020.00444
71. Spengler, J.R., et al.: Experimental Andes virus infection in deer mice: characteristics of infection and clearance in a heterologous rodent host. *PLoS One*. **8**, e55310 (2013). https://doi.org/10.1371/journal.pone.0055310
72. Qian, Z., et al.: Targeting vascular injury using Hantavirus-pseudotyped lentiviral vectors. *Mol. Ther.* **13**, 694–704 (2006). https://doi.org/10.1016/j.ymthe.2005.11.016
73. Zaki, S.R., et al.: Hantavirus pulmonary syndrome. Pathogenesis of an emerging infectious disease. *Am. J. Pathol.* **146**, 552–579 (1995)
74. Lee, P.W., Gibbs Jr., C.J., Gajdusek, D.C., Yanagihara, R.: Serotypic classification of hantaviruses by indirect immunofluorescent antibody and plaque reduction neutralization tests. *J. Clin. Microbiol.* **22**, 940–944 (1985). https://doi.org/10.1128/jcm.22.6.940-944.1985
75. Chu, Y.K., et al.: Cross-neutralization of hantaviruses with immune sera from experimentally infected animals and from hemorrhagic fever with renal syndrome and hantavirus pulmonary syndrome patients. *J. Infect. Dis.* **172**, 1581–1584 (1995). https://doi.org/10.1093/infdis/172.6.1581
76. de Carvalho Nicacio, C., et al.: A neutralizing recombinant human antibody Fab fragment against Puumala hantavirus. *J. Med. Virol.* **60**, 446–454 (2000). https://doi.org/10.1002/(sici)1096-9071(200004)60:4<446::aid-jmv13>3.0.co;2-v
77. Horling, J., Lundkvist, A.: Single amino acid substitutions in Puumala virus envelope glycoproteins G1 and G2 eliminate important neutralization epitopes. *Virus Res.* **48**, 89–100 (1997). https://doi.org/10.1016/s0168-1702(97)01436-6
78. Brocato, R.L., Hooper, J.W.: Progress on the prevention and treatment of Hantavirus disease. *Viruses*. **11** (2019). https://doi.org/10.3390/v11070610
79. Iannetta, M., et al.: Viral hemorrhagic fevers other than Ebola and Lassa. *Infect. Dis. Clin. N. Am.* **33**, 977–1002 (2019). https://doi.org/10.1016/j.idc.2019.08.003
80. Cho, H.W., Howard, C.R., Lee, H.W.: Review of an inactivated vaccine against hantaviruses. *Intervirology*. **45**, 328–333 (2002). https://doi.org/10.1159/000067925
81. Zhang, F.L., et al.: The expression and genetic immunization of chimeric fragment of Hantaan virus M and S segments. *Biochem. Biophys. Res. Commun.* **354**, 858–863 (2007). https://doi.org/10.1016/j.bbrc.2007.01.020

Chapter 13
Pseudotyped Viruses for *Phlebovirus*

Jiajing Wu, Weijin Huang, and Youchun Wang

Abstract Rift Valley fever virus (RVFV) is a member of the *Phlebovirus* genus, one of the 20 genera in the *Phenuiviridae* family. RVFV causes disease in animals and humans and is transmitted by sandflies or ticks. However, research into RVFV is limited by the requirement for biosafety level 3 (BSL-3) containment. Pseudotyped virus overcomes this limitation as it can be handled in a BSL-2 environment. Pseudotyped RVFV possesses an identical envelope protein structure to that of the authentic virus, simulating the same process of receptor binding and membrane fusion to host cells. Pseudotyped phleboviruses are therefore useful tools to study the infection mechanism of these viruses and for the screening of inhibitory drugs and the development of therapeutic monoclonal antibodies.

Keywords Rift Valley fever virus · Pseudotyped virus · Neutralization · In vitro · In vivo

Abbreviations

RVFV	Rift Valley fever virus
BSL	Biosafety level
vRNA	Viral RNA

J. Wu
Beijing Yunling Biotechnology Co. Ltd, Beijing, China

W. Huang
Division of HIV/AIDS and Sex-transmitted Virus Vaccines, National Institutes for Food and Drug Control (NIFDC) and WHO Collaborating Center for Standardization and Evaluation of Biologicals, Beijing, China

Y. Wang (✉)
Chinese Academy of Medical Sciences & Peking Union Medical College, Beijing, China

Institute of Medical Biology, Chinese Academy of Medicine Sciences & Peking Union Medical College, Kunming, China
e-mail: wangyc@nifdc.org.cn; wangyc@imbcams.com.cn

Y. Wang (ed.), *Pseudotyped Viruses*, Advances in Experimental Medicine and Biology 1407, https://doi.org/10.1007/978-981-99-0113-5_13

PBNA	Pseudotyped virus-based neutralizing antibody assay
RdPp	RNA-dependent RNA polymerase
RNP	Ribonucleoprotein
VSV	Vesicular stomatitis virus
HIV	Human immunodeficiency virus

13.1 Biological Characteristics of *Phlebovirus*

Currently, the *Phenuiviridae* family is divided into 20 genera, including 137 species capable of causing severe disease in animals and humans. Among them, the *Phlebovirus* genus comprises 66 species transmitted by sandflies, ticks, mosquitoes, and grasshoppers, including Rift Valley fever virus (RVFV), and other phleboviruses such as Saint Floris, Candiru, Gordil, and Viola. RVFV has a wide geographical distribution and has attracted particular research interest.

RVFV mainly infects ruminant livestock, causing high mortality rates among newborn animals and an "abortion storm" among pregnant mothers, which has a devastating effect on animal health. Most human infections present with self-limiting flu-like symptoms, but some cases progress to severe disease. With global climate change, economic globalization, and increased international travel, the threat of further global spread of RVFV is amplified. RVFV is listed as a Category A priority pathogen, and experimental manipulations therefore must be performed in biosafety level (BSL)-3 controlled facilities. There are currently no commercially available vaccines or therapeutics for RVFV. At present, RVFV serological testing mainly uses the authentic virus, so there is an urgent need to establish a safer and more effective alternative assessment method.

13.1.1 Structure of Rift Valley Fever Virus (RVFV)

RVFV is a single-stranded negative-sense RNA virus, which has a virus envelope that is spherical in shape and three nucleic acid segments (~11.9 kb in total) with diameters of 90 ~ 100 nm that are located in the nucleocapsid and possess the S, M, and L genes, respectively. Each RNA segment has identical 3′ and 5′ terminal complementary sequences (3′-UGUGUUUC, 5′-ACACAAG). The three segments form a pot-handle-like structure through complementary 3′ and 5′ ends, forming non-covalent closed circular RNAs [1]. Genomic viral RNA (vRNA) combines with nucleoprotein (N) to form independent L, M, and S ribonucleoprotein complexes (ribonucleoproteins, RNPs), and many RNA-dependent RNA polymerases (RdPp) are attached to RNPs. These proteins play important roles in the viral life cycle (including adsorption, replication, transcription, assembly, and budding), pathogenicity, and host immune responses.

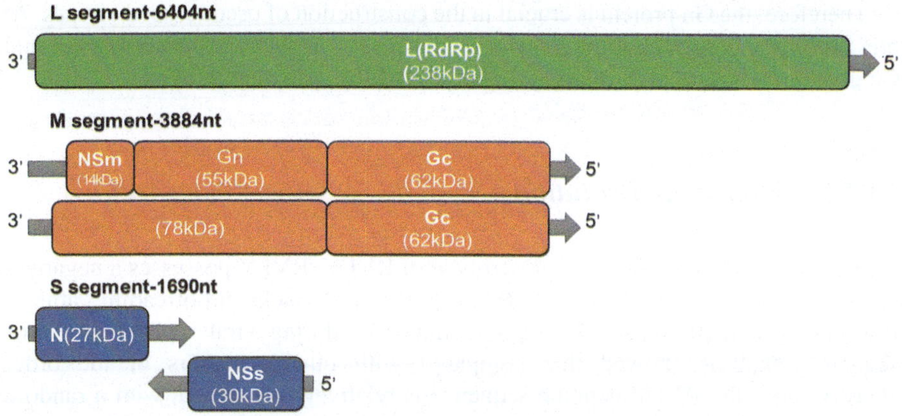

Fig 13.1 RVFV genome structure and the encoded proteins

The L segment is approximately 6404 bp in length and encodes the largest viral protein (Fig. 13.1), the L protein, which mainly functions as an RNA polymerase and plays an important role in viral replication and transcription [2–4].

The M segment is approximately 3885 bp in length with only one open reading frame, but multiple initiation codons (AUG) at the 5′ end of the mRNA from which the virus can selectively initiate translation. The polypeptide precursor produced by translation is cleaved at different sites to form four proteins (Gn, Gc, NSm1, and NSm2), of which Gn (encoded by the N-terminal region of the polypeptide precursor) and Gc (encoded by the C-terminal region of the polypeptide precursor) are glycoproteins, and NSm1 and NSm2 are nonstructural proteins [1, 5–7].

During the life cycle of RVFV, glycoproteins Gn and Gc play an important role in the processes of virus adsorption, entry, assembly, and budding. In addition, glycoproteins can induce the body to produce neutralizing protective antibodies, which are the main mechanism by which the body exerts humoral immunity [8].

Gn and Gc glycoproteins mainly recognize cell surface receptors to promote virus entry into cells. The maturation site of RVFV in cells is the Golgi apparatus [9]. Gn and Gc proteins play an important role in localization to the Golgi apparatus. Studies have shown that the Golgi localization signal of RVFV is a 48-amino acid (aa) region located at the C-terminus of the Gn protein, and the localization of RVFV depends on the interaction between Gn protein and Gc protein. In 2021, American scholars confirmed that Gn protein can directly interact with vRNA in infected cells [10]. The four epitopes contained in Gn glycoprotein were located using monoclonal antibodies. The amino acid sequence positions of the four epitopes were I (105–138 aa), II (229–239 aa), IV (127–146 aa), and III (362–375 aa), respectively [11]. Among them, epitope III is a non-neutralizing epitope, while epitopes I, II, and IV are neutralizing epitopes. Epitope IV is greatly affected by the conformation of the glycoprotein, and the antigenic determinants denatured by SDS cannot react with the corresponding monoclonal antibody IV.

Therefore, the Gn protein is crucial in the construction of pseudotyped viruses. At present, in the construction of pseudotyped viruses, the M genes encoding Gn and Gc glycoproteins are indispensable regardless of the system employed.

13.1.2 Molecular Evolution

Currently, there is only one known serotype of RVFV. RVFV possesses a negative-stranded RNA genome, and the RNA polymerase lacks proofreading ability, resulting in a high frequency of point mutations during viral replication. Virus sequence analysis showed that compared with other members of the order *Bunyavirales*, the RVFV genome sequence is relatively conserved, with a random distribution of point mutations and no obvious nucleotide hypervariable regions. Bird et al. conducted a molecular evolution study on the L, M, and S segments of the whole genomes of 33 viruses isolated from 1944 to 2000 [12]. The results revealed that the nucleotide and amino acid differences among strains were small, and the similarity levels for nucleotide and amino acid sequences between strains were 95% and 98%, respectively. All 33 RVFV M segments were found to be highly conserved. Phylogenetic analysis of the 33 strains identified a total of seven lineages, A~G, associated with endemic regions and epidemic time periods. These viruses may have originated from a common ancestral virus that circulated between 1880 and 1890.

RVFV is a segmented virus, and gene reassortment is another important evolutionary mechanism of this virus, which can be observed both in nature and in the laboratory [13]. At present, more than ten strains of gene reassortment viruses that occur under natural conditions have been identified, involving the L, M, and S segments. Gene reassortment increases the genetic diversity of RVFV, but its impact on virus replication, transmission, pathogenicity, and host adaptation is unclear.

13.2 Construction of Pseudotyped RVFV

Because RVFV is considered a high-risk pathogen, the use of authentic virus for research is restricted for safety reasons. Pseudotyped RVFV offers an alternative research tool that is safer and allows for consistent and comparable studies to be performed under BSL-2 conditions. At present, there are three commonly used packaging backbone vectors for pseudotyped RVFVs, including human immunodeficiency virus (HIV-1)-based lentiviral vectors, vesicular stomatitis virus (VSV)-based vectors, and self-assembly virus-like particles.

13.2.1 Construction of Pseudotyped RVFV Using Lentiviral Vectors

The HIV-1-based lentiviral system is the most commonly used packaging system for the construction of pseudotyped RVFVs. Li et al. constructed pseudotyped RVFVs based on the HIV system in 2018 [14]. In this study, RVFV pseudotyped virus carrying the GFP reporter gene was obtained by co-transfection of 293T cells with M fragment expression plasmid pcDNA3.1-M-rvfv (RVFV ZH-548 strain, GenBank accession no. NC014396), lentiviral expression vector pLV-EGFP-C, and PH1 plasmid.

The packaging process of pseudotyped RVFV based on lentiviral vectors is generally conducted as follows:

1. Seed 293T cells one day before transfection.
2. Incubate the cells with DMEM (high glucose) containing 10% FBS for 12–18 h. The cells should be 80% confluent before transfection.
3. Co-transfect M fragment expression plasmid and lentivirus-based packaging vector plasmid/plasmids using Lipofectamine 2000 or other transfection reagents. Dilute the plasmid with Opti-MEM if necessary.
4. Incubate at 37°C for 4–6 h.
5. Remove the medium, add fresh DMEM with 10% FBS, and incubate at 37°C overnight for another 42–44 h.
6. Harvest the culture supernatant. Filter using a 0.45 μM pore-size filter and store at −70°C until use.

All work involving pseudotyped RVFV should be performed in a BSL-2 facility.

13.2.2 Construction of Pseudotyped RVFV Using VSV-Based Vectors

Since the RVFV envelope glycoprotein (encoded by the M segment) can recognize and bind to cell surface receptors that mediate virus entry into cells, the strategy for constructing an envelope protein expression plasmid, which is necessary for the preparation of RVFV pseudotyped viruses, is generally to clone the M gene into eukaryotic expression vectors. In 2019, our laboratory constructed a pseudotyped RVFV ZH-548 strain (GenBank accession no. NC014396) using the VSV vector system [15]. In general, it only takes three days to obtain the RVFV pseudotyped virus after the envelope protein expression plasmid has been synthesized and the VSVΔG pseudotyped backbone virus has been constructed.

The packaging process of RVFV pseudotyped virus based on VSV vectors is generally conducted as follows:

Day 1: HEK 293T Cell Preparation

1. Seed 293T cells, at 95% confluence, into T75 tissue culture flasks at a ratio of 1: 2.5 to ensure 80%–90% confluence the following day.
2. Incubate the cells with DMEM (high glucose) containing 10% FBS overnight.

Day 2: Transfection of Envelope Proteins

1. Observe the cells under an inverted microscope to ensure the desired confluency for transfection.
2. Transfect 30 μg of RVFV envelope protein expression plasmid using Lipofectamine 3000 or other transfection reagents. Dilute the plasmid with Opti-MEM if necessary. Please follow the instructions provided with the selected transfection reagent for specific operations.
3. Remove the medium, add fresh DMEM with 10% FBS, and incubate at 37°C overnight (~18 h).

Day 3: Infection of Transfected Cells with VSVΔG/GFP*-G, VSVΔG/Luci*-G, or VSVΔG/SEAP*-G

1. In a BSL-2 facility, remove the medium and wash with phosphate-buffered saline (PBS) or serum-free DMEM three times. Add serum-free DMEM.
2. Inoculate with a multiplicity of infection = 4 of VSVΔG/GFP*-G, VSVΔG/Luci*-G, or VSVΔG/SEAP*-G. Gently swirl the flask to ensure that the VSV backbone virus is evenly distributed on the cell monolayer.
3. Adsorb the viruses at 37°C for 1 h.
4. Remove the medium and wash with PBS or serum-free DMEM 3–5 times.
5. Add 15 mL of the medium and incubate at 37°C overnight (17–18 h).
6. Harvest the culture supernatant. Filter using a 0.45 μM pore-size filter and store at −70°C until use.

One of the most common problems that can arise during pseudotyped virus generation is the presence of residual VSVΔG/GFP*-G, VSVΔG/Luci*-G, or VSVΔG/SEAP*-G pseudotyped viruses in the collected supernatant. Washing cells twice with 1×PBS and DMEM containing anti*-G antibody completely removes excess VSVΔG/GFP*-G, VSVΔG/Luci*-G, or VSVΔG/SEAP*-G that has not infected cells [16].

13.2.3 Construction of Pseudotyped RVFV Using the Self-Assembly System

Kortekaas et al. constructed RVFV non-replicating virus-like particles (strain 35/74, GenBank accession no. JF784387.1) using a self-assembly system in 2011 [17]. This study was performed using a three-plasmid system to construct pseudotyped

RVFVs. Briefly, RVFV pseudotyped virus particles can be harvested in the supernatant by co-transfecting BHK-T7/5 cells with plasmids expressing L, S-eGFP, and M fragments, respectively (with mammalian codon optimization). Since it is difficult to obtain high titers of pseudotyped virus using the three-plasmid co-transfection method, BHK-Rep cells stably expressing L and S-eGFP fragments were used as packaging cells. Subsequently, higher titers of RVFV pseudotyped virus could be obtained by transfecting only the M fragment expression plasmid or adding the pseudotyped virion RRRs and the M fragment expression plasmid to BHK-Rep cells at the same time.

In conclusion, we found that the high-level expression of the envelope glycoproteins Gn and Gc is the key factor for the successful preparation of pseudotyped RVFV. This may be because RVFV buds into the Golgi apparatus, while HIV/VSV particles that served as the backbone of the pseudotyped viruses assemble at the plasma membrane [18]. Therefore, only when RVFV glycoprotein levels reach the normal maturation pathway can a sufficient amount of glycoprotein reach the assembly site to form higher titer pseudotyped RVFV.

13.3 Application of Pseudotyped RVFV

13.3.1 Neutralizing Assay Based on Pseudotyped RVFV

The main application of pseudotyped RVFV is the establishment of a pseudotyped virus-based RVFV neutralizing antibody assay (PBNA). After the sample to be tested interacts with the pseudotyped RVFV, the titer of the neutralizing antibody can be detected by determining the expression of the reporter gene carried by the pseudotyped virus.

Using a pseudotyped RVFV carrying a firefly luciferase reporter gene based on the VSV vector system constructed in our laboratory, we established a PBNA that can be performed in a BSL-2 facility [15]. In the process of establishing this method, we selected a number of commonly used cell lines to identify the cell tropism of pseudotyped RVFV. Finally, Huh-7 cells were determined as a model cell line for the in vitro detection of RVFV neutralizing antibodies, which may make the detection results more stable and accurate. In general, PBNA-based sample detection can be completed within two days.

Compared with traditional neutralizing antibody detection methods, the PBNA established in our laboratory has the advantages of safety, rapidity, high-throughput capability, and objective interpretation of the results. It should be noted that in neutralization experiments, the amount of virus inoculum is critical. Moreover, the pseudotyped virus only has the ability to infect during a single round, and the amount of inoculum has a more significant impact on the test results. If the selection is not appropriate, it will cause deviation in the evaluation of the neutralization inhibition ability of the antibody. Therefore, during the establishment of PBNA, it is important to optimize the amount of pseudotyped virus added.

13.3.2 Visual In Vivo Neutralizing Antibody Evaluation Model

To date, rodent [19, 20] and non-human primate [21, 22] models have been established to study RVFV infection of the authentic virus. The most commonly used rodent model requires BSL-3 facilities, which limits its application. The non-human primate model also requires specialist facilities and is complex and expensive to operate. A simple in vivo experimental model for RVFV was therefore needed.

Our laboratory established a visual pseudotyped RVFV mouse infection model [15], only requiring BSL-2 facilities that had the advantages of high sensitivity, simple operation, and convenient real-time monitoring of virus infection. After infecting mice with pseudotyped RVFV, the fluorescence distribution and luminescence detected by injecting substrate reagents were reflected by the expression of firefly luciferase. The infection site and the degree of infection in the mice were measured by dynamic observations. Anatomical observations showed that the main infection sites were the liver, spleen, and lungs, which were consistent with the infection sites reported for the authentic virus [23]. This result indicated that the model is a viable alternative to using authentic virus in some studies, especially for evaluation of the in vivo efficacy of vaccines and antibodies. Image detection of ions could be performed within six hours of infection, which greatly shortens the experimental time.

After establishing a visual pseudotyped RVFV infection model in mice, the recombinant RVFV-M DNA vaccine was evaluated in vivo. The results showed that when the ID_{50} value of neutralizing antibody in vitro exceeded 1:2500, no firefly luciferase was detected in mice. This result directly shows that when the antibodies in the mice reach a certain level, they completely protect against virus infection. Combined with in vivo and in vitro neutralization assays, the protective effect of vaccines and antibodies can be assessed more accurately and clearly using this model.

13.3.3 The Mechanism of Viral Infection

DC-SIGN, a C-type lectin that is primarily restricted to interstitial dendritic cells and certain tissue macrophages, has been identified as a receptor for RVFV. However, the broad cell tropism of RVFV suggests that other receptors or polysaccharides on the cell surface are important for viral entry into cells lacking DC-SIGN expression. In 2012, de Boer et al. reported that the presence of heparan sulfate on the cell surface significantly promoted the entry of pseudotyped RVFV into cells [24]. They confirmed this by pre-incubation of pseudotyped RVFV carrying the eGFP reporter gene with highly sulfated heparin, artificial enzymatic removal of heparan sulfate from cells, or the infection of pseudotyped RVFV virus into cells genetically lacking heparan sulfate synthesis.

In the same year, the team also used this pseudotyped RVFV carrying the eGFP reporter gene to discover that H778, H857, and H1087 histidines in Gc may jointly contribute to pH sensing, initiation, and propagation of conformational rearrangements toward the postfusion structure [25].

13.3.4 Pseudotyped RVFV as a Candidate Vaccine

At present, there is no approved RVFV vaccine. Vaccines developed using different technologies, including inactivated vaccines [26], live attenuated vaccines [27–29], viral vector vaccines [30, 31], recombinant subunit vaccines [32, 33], and DNA vaccines [34], are undergoing further clinical evaluations to confirm their safety and efficacy. Because it has the same envelope protein as the authentic virus, pseudotyped RVFV can also be considered as a candidate vaccine.

In 2011, Moormann et al. used a self-assembly system to construct an RVFV pseudotyped virus (named RRPs). After confirming its ability to infect cells and neutralize antibodies, it was used as a vaccine to immunize mice. Groups of 10 mice were immunized once or twice with 50 μl of inoculum containing 10^6 $TCID_{50}$ RRPs via the subcutaneous or intramuscular route at 3-week intervals [25]. Nine unvaccinated mice served as a control group. Mice were challenged on day 42 with a known lethal dose of RVFV strain 35/74. All unvaccinated mice exhibited significant clinical signs and weight loss, and all died within 12 days of challenge. The survival rate of mice inoculated via the subcutaneous route was 60%, compared with 100% for the intramuscularly vaccinated group. These mice did not show any clinical signs or weight loss throughout the experiment.

13.3.5 Neutralization Sensitivity Analysis of Natural and Artificial RVFV Variants

As an RNA virus, RVFV is prone to mutate to escape neutralization by antibodies. However, the isolation of authentic viruses is limited by biosafety risks, and the immunological evaluation of mutations occurring in nature often cannot be performed in a timely manner. These problems can be addressed by constructing pseudotyped mutant viruses.

Our laboratory has investigated the broad spectrum of neutralizing antibodies induced by DNA vaccine candidates by constructing mutant strains of different pseudotyped RVFVs. First, the amino acid sequences encoded by the M gene of RVFV were obtained from databases such as GenBank, and these sequences were compared and the naturally occurring amino acid substitution sites were identified.

Then, the effect of amino acid substitution on protein function was analyzed by SIFT [35–37] and PROVEAN [38–40] software, and meaningful mutations were further screened. Finally, 39 natural amino acid mutation sites (32 Gn sites, 7 Gc sites) that may affect protein function were identified, and 42 pseudotyped RVFV mutants were further constructed. Analysis and comparison of these pseudotyped RVFV mutants by PBNA found no significant changes in neutralization susceptibility. This may be because a single amino acid change has a limited effect on the protein or that the serum induced by the recombinant DNA vaccine based on the M gene of strain RVFV ZH-548 has a broad spectrum of activity.

13.4 Conclusion

To date, a variety of RVFV vaccines have been developed, and the detection of neutralizing antibodies in vitro and in vivo using pseudotyped viruses may be a safe and effective method for evaluating vaccine efficacy. Regarding RVFV infectivity, there are still many issues to be resolved, such as the discovery of receptors other than DC-SIGN and the mechanism of action of non-structural proteins. Pseudotyped viruses provide new tools for such studies in the future.

Pseudotyped viruses can be constructed for most enveloped viruses, including RVFV, but not all pseudotyped viruses can be produced at high titer. Further exploration of the virus packaging system and the creation of a novel pseudotyped virus packaging system will be the focus of our future research. An ultimate aim is to systematically replace authentic virus in the evaluation of candidate vaccines and antiviral drugs in all experiments. In conclusion, pseudotyped virus constitutes an important tool for future biological research on this pathogen.

Acknowledgment We thank Liwen Bianji (Edanz) (https://www.liwenbianji.cn) for editing the language of a draft of this book chapter.

References

1. Bracci, N., et al.: Rift Valley fever virus Gn V5-epitope tagged virus enables identification of UBR4 as a Gn interacting protein that facilitates Rift Valley fever virus production. *Virology*. **567**, 65–76 (2022). https://doi.org/10.1016/j.virol.2021.12.010
2. Elliott, R.M., Dunn, E., Simons, J.F., Pettersson, R.F.: Nucleotide sequence and coding strategy of the Uukuniemi virus L RNA segment. *J. Gen. Virol.* **73(Pt 7)**, 1745–1752 (1992). https://doi.org/10.1099/0022-1317-73-7-1745
3. Accardi, L., Gro, M.C., Di Bonito, P., Giorgi, C.: Toscana virus genomic L segment: molecular cloning, coding strategy and amino acid sequence in comparison with other negative strand RNA viruses. *Virus Res.* **27**, 119–131 (1993). https://doi.org/10.1016/0168-1702(93)90076-y
4. Wang, X., et al.: Structure of Rift Valley fever virus RNA-dependent RNA polymerase. *J. Virol.* **96**, e0171321 (2022). https://doi.org/10.1128/JVI.01713-21

5. Collett, M.S., et al.: Complete nucleotide sequence of the M RNA segment of Rift Valley fever virus. *Virology*. **144**, 228–245 (1985). https://doi.org/10.1016/0042-6822(85)90320-4
6. Ronnholm, R., Pettersson, R.F.: Complete nucleotide sequence of the M RNA segment of Uukuniemi virus encoding the membrane glycoproteins G1 and G2. *Virology*. **160**, 191–202 (1987). https://doi.org/10.1016/0042-6822(87)90060-2
7. Said, A., Elmanzalawy, M., Ma, G., Damiani, A.M., Osterrieder, N.: An equine herpesvirus type 1 (EHV-1) vector expressing Rift Valley fever virus (RVFV) Gn and Gc induces neutralizing antibodies in sheep. *Virol. J.* **14**, 154 (2017). https://doi.org/10.1186/s12985-017-0811-8
8. Spiegel, M., Plegge, T., Pohlmann, S.: The role of Phlebovirus glycoproteins in viral entry, assembly and release. *Viruses*. **8** (2016). https://doi.org/10.3390/v8070202
9. Gerrard, S.R., Nichol, S.T.: Characterization of the Golgi retention motif of Rift Valley fever virus G(N) glycoprotein. *J. Virol.* **76**, 12200–12210 (2002). https://doi.org/10.1128/jvi.76.23.12200-12210.2002
10. Tercero, B., Narayanan, K., Terasaki, K., Makino, S.: Characterization of the molecular interactions that govern the packaging of viral RNA segments into Rift Valley fever Phlebovirus particles. *J. Virol.* **95**, e0042921 (2021). https://doi.org/10.1128/JVI.00429-21
11. Keegan, K., Collett, M.S.: Use of bacterial expression cloning to define the amino acid sequences of antigenic determinants on the G2 glycoprotein of Rift Valley fever virus. *J. Virol.* **58**, 263–270 (1986). https://doi.org/10.1128/JVI.58.2.263-270.1986
12. Bird, B.H., Khristova, M.L., Rollin, P.E., Ksiazek, T.G., Nichol, S.T.: Complete genome analysis of 33 ecologically and biologically diverse Rift Valley fever virus strains reveals widespread virus movement and low genetic diversity due to recent common ancestry. *J. Virol.* **81**, 2805–2816 (2007). https://doi.org/10.1128/JVI.02095-06
13. Elliott, R.M.: Orthobunyaviruses: recent genetic and structural insights. *Nat. Rev. Microbiol.* **12**, 673–685 (2014). https://doi.org/10.1038/nrmicro3332
14. Li, Y., et al.: Packaging of Rift Valley fever virus pseudoviruses and establishment of a neutralization assay method. *J. Vet. Sci.* **19**, 200–206 (2018). https://doi.org/10.4142/jvs.2018.19.2.200
15. Ma, J., et al.: In vitro and in vivo efficacy of a Rift Valley fever virus vaccine based on pseudovirus. *Hum. Vaccin. Immunother.* **15**, 2286–2294 (2019). https://doi.org/10.1080/21645515.2019.1627820
16. Almahboub, S.A., Algaissi, A., Alfaleh, M.A., ElAssouli, M.Z., Hashem, A.M.: Evaluation of neutralizing antibodies against highly pathogenic Coronaviruses: a detailed protocol for a rapid evaluation of neutralizing antibodies using vesicular stomatitis virus Pseudovirus-based assay. *Front. Microbiol.* **11**, 2020 (2020). https://doi.org/10.3389/fmicb.2020.02020
17. Kortekaas, J., et al.: Creation of a nonspreading Rift Valley fever virus. *J. Virol.* **85**, 12622–12630 (2011). https://doi.org/10.1128/JVI.00841-11
18. King, B., Daly, J.: Pseudotypes: your flexible friends. *Future Microbiol.* **9**, 135–137 (2014). https://doi.org/10.2217/fmb.13.156
19. Lang, Y., et al.: Mouse model for the Rift Valley fever virus MP12 strain infection. *Vet. Microbiol.* **195**, 70–77 (2016). https://doi.org/10.1016/j.vetmic.2016.09.009
20. Lorenzo, G., Martin-Folgar, R., Hevia, E., Boshra, H., Brun, A.: Protection against lethal Rift Valley fever virus (RVFV) infection in transgenic IFNAR(-/-) mice induced by different DNA vaccination regimens. *Vaccine*. **28**, 2937–2944 (2010). https://doi.org/10.1016/j.vaccine.2010.02.018
21. Wonderlich, E.R., et al.: Peripheral blood biomarkers of disease outcome in a monkey model of Rift Valley fever encephalitis. *J. Virol.* **92** (2018). https://doi.org/10.1128/JVI.01662-17
22. Smith, D.R., et al.: Attenuation and efficacy of live-attenuated Rift Valley fever virus vaccine candidates in non-human primates. *PLoS Negl. Trop. Dis.* **12**, e0006474 (2018). https://doi.org/10.1371/journal.pntd.0006474
23. Smith, D.R., et al.: The pathogenesis of Rift Valley fever virus in the mouse model. *Virology*. **407**, 256–267 (2010). https://doi.org/10.1016/j.virol.2010.08.016

24. de Boer, S.M., et al.: Heparan sulfate facilitates Rift Valley fever virus entry into the cell. *J. Virol.* **86**, 13767–13771 (2012). https://doi.org/10.1128/JVI.01364-12
25. de Boer, S.M., et al.: Acid-activated structural reorganization of the Rift Valley fever virus Gc fusion protein. *J. Virol.* **86**, 13642–13652 (2012). https://doi.org/10.1128/JVI.01973-12
26. El-Sissi, A.F., Mohamed, F.H., Danial, N.M., Gaballah, A.Q., Ali, K.A.: Chitosan and chitosan nanoparticles as adjuvant in local Rift Valley fever inactivated vaccine. *3 Biotech.* **10**, 88 (2020). https://doi.org/10.1007/s13205-020-2076-y
27. Moutailler, S., Krida, G., Madec, Y., Bouloy, M., Failloux, A.B.: Replication of Clone 13, a naturally attenuated avirulent isolate of Rift Valley fever virus, in Aedes and Culex mosquitoes. *Vector Borne Zoonotic Dis.* **10**, 681–688 (2010). https://doi.org/10.1089/vbz.2009.0246
28. Dungu, B., et al.: Evaluation of the efficacy and safety of the Rift Valley fever Clone 13 vaccine in sheep. *Vaccine.* **28**, 4581–4587 (2010). https://doi.org/10.1016/j.vaccine.2010.04.085
29. Ikegami, T.: Rift Valley fever vaccines: an overview of the safety and efficacy of the live-attenuated MP-12 vaccine candidate. *Expert Rev. Vaccines.* **16**, 601–611 (2017). https://doi.org/10.1080/14760584.2017.1321482
30. Lopez-Gil, E., et al.: A single immunization with MVA expressing GnGc glycoproteins promotes epitope-specific CD8+-T cell activation and protects immune-competent mice against a lethal RVFV infection. *PLoS Negl. Trop. Dis.* **7**, e2309 (2013). https://doi.org/10.1371/journal.pntd.0002309
31. Warimwe, G.M., et al.: Immunogenicity and efficacy of a chimpanzee adenovirus-vectored Rift Valley fever vaccine in mice. *Virol. J.* **10**, 349 (2013). https://doi.org/10.1186/1743-422X-10-349
32. Ikegami, T.: Candidate vaccines for human Rift Valley fever. *Expert. Opin. Biol. Ther.* **19**, 1333–1342 (2019). https://doi.org/10.1080/14712598.2019.1662784
33. Ikegami, T., Makino, S.: Rift valley fever vaccines. *Vaccine.* **27(Suppl 4)**, D69–D72 (2009). https://doi.org/10.1016/j.vaccine.2009.07.046
34. Spik, K., et al.: Immunogenicity of combination DNA vaccines for Rift Valley fever virus, tick-borne encephalitis virus, Hantaan virus, and Crimean Congo hemorrhagic fever virus. *Vaccine.* **24**, 4657–4666 (2006). https://doi.org/10.1016/j.vaccine.2005.08.034
35. Sim, N.L., et al.: SIFT web server: predicting effects of amino acid substitutions on proteins. *Nucleic Acids Res.* **40**, W452–W457 (2012). https://doi.org/10.1093/nar/gks539
36. Ng, P.C., Henikoff, S.: SIFT: predicting amino acid changes that affect protein function. *Nucleic Acids Res.* **31**, 3812–3814 (2003). https://doi.org/10.1093/nar/gkg509
37. Kumar, P., Henikoff, S., Ng, P.C.: Predicting the effects of coding non-synonymous variants on protein function using the SIFT algorithm. *Nat. Protoc.* **4**, 1073–1081 (2009). https://doi.org/10.1038/nprot.2009.86
38. Choi, Y., Chan, A.P.: PROVEAN web server: a tool to predict the functional effect of amino acid substitutions and indels. *Bioinformatics.* **31**, 2745–2747 (2015). https://doi.org/10.1093/bioinformatics/btv195
39. Gao, T., et al.: Identification and functional analysis of the SARS-COV-2 nucleocapsid protein. *BMC Microbiol.* **21**, 58 (2021). https://doi.org/10.1186/s12866-021-02107-3
40. Sandell, L., Sharp, N.P.: Fitness effects of mutations: an assessment of PROVEAN predictions using mutation accumulation data. *Genome Biol. Evol.* **14** (2022). https://doi.org/10.1093/gbe/evac004

Chapter 14
Pseudotyped Virus for Bandavirus

Ruifeng Chen, Weijing Huang, and Youchun Wang

Abstract The genus *Bandavirus*, belonging to family *Phenuiviridae*, order *Bunyavirales*, consists of eight tick-borne bunyaviruses. The Dabie bandavirus, formerly known as severe fever with thrombocytopenia virus (SFTSV), belongs to the genus *Bandavirus*. This emerging pathogen was first identified in central China in 2009. In recent years, the disease has been reported to cause several outbreaks in eastern Asia areas, including China, Japan, Korea, and Vietnam. Tick-to-human transmission is the main route of infection in humans, and transmission via the contact of body fluids from person-to-person was also reported. Despite its high fatality rate, there is currently no vaccine or antiviral therapy available. The therapeutic efficacies of several antiviral agents against Dabie bandavirus are still being evaluated. However, the virus is a potent pathogen with high biosafety experimental conditions. Therefore, replication-incompetent pseudotyped viruses play an important role. In this chapter, we succinctly summarize the basic features concerning Dabie bandavirus, including virion structure, genome characteristics, especially the characteristics of glycoprotein, and probable pathogenic mechanism. And, we put an important part in expounding the construction of pseudoviruses and its application.

Keywords Bandavirus · Severe fever with thrombocytopenia syndrome virus · Dabie bandavirus, Pseudovirus · Pseudotyped virus

---

R. Chen
Immunotech Applied Science Limited, Beijing, China
e-mail: m18790678092@163.com

W. Huang
Division of HIV/AIDS and Sex-transmitted Virus Vaccines, National Institutes for Food and Drug Control (NIFDC) and WHO Collaborating Center for Standardization and Evaluation of Biologicals, Beijing, China

Y. Wang (✉)
Chinese Academy of Medical Science & Peking Union Medical College, Beijing, China

Institute of Medical Biology, Chinese Academy of Medical Science & Peking Union Medical College, Kunming, China
e-mail: wangyc@nifdc.org.cn; wangyc@imbcams.com.cn

Y. Wang (ed.), *Pseudotyped Viruses*, Advances in Experimental Medicine and Biology 1407, https://doi.org/10.1007/978-981-99-0113-5_14

Abbreviations

BSL	Biosafety level
DC-SIGN	Dendritic cell-specific intercellular adhesion molecule-3-grabbing non-integrin
EM	Electron microscopy
ER	Endoplasmic reticulum
Fluc	Firefly luciferase
GFP	Green fluorescent protein
HS	Heparan sulfate
ICTV	The International Committee on Taxonomy of Viruses
NMMHC-IIA	Non-muscle myosin heavy chain IIA
Np	Nucleoprotein
NSs	Non-translated protein
PBS	Phosphate-buffered saline
RdRp	RNA-dependent RNA polymerase
RNP	RNA ribonucleoprotein
SFTS	Severe fever with thrombocytopenia syndrome
SFTSV	Severe fever with thrombocytopenia syndrome virus
VSV	Vesicular stomatitis virus

14.1 Introduction

According to the International Committee on Taxonomy of Viruses (ICTV), the genus *Bandavirus*, classified under the order *Bunyavirales*, family *Phenuiviridae*, comprises eight species, including Dabie bandavirus (Dabie bandavirus), Bhanja virus (BHAV), Guertu virus (GTV), Heartland virus (HRTV), Hunter Island virus (HUIV), Kismaayo virus (KISV), Lone star virus (LSV), and Razdan virus (RAZV) [1]. The general information is given in Table 14.1. And Dabie bandavirus, formerly known as SFTSV, could be potential threats to public health.

In 2009, an unknown infectious disease suddenly made an outbreak in rural areas of Hubei Province, China, killing 5 of 17 patients. After analysis and identification, the China CDC named it the severe fever with thrombocytopenia (SFTS) in 2010. SFTS is an emerging viral hemorrhagic fever caused by SFTSV. Synonymously this virus is known as Dabie bandavirus or Huaiyangshan Banyangvirus [2]. And according to the new nomenclature of ICTV, the term Dabie bandavirus has been the current scientific name.

Dabie bandavirus has been reported mainly in some Asian countries, including China, South Korea, Japan, and most recently in Vietnam, Thailand, and Pakistan [3–7]. As is reported, it is transmitted through the bite of some species of ticks, like *Haemaphysalis longicornis* and *Dermacentor silvarum* [8]. Of which, *Haemaphysalis longicornis* has also been found in Australia, the Pacific regions,

Table 14.1 The general information of the current eight bandavirus species

Species	Organism name	RefSeq genome	RefSeq strain	GenBank of glycoprotein	Geographic distribution	Isolation host	Isolation source	Collection date	Release date
Bhanja bandavirus	Bhanja virus (BHAV)	GCF_001019855.1	ibAr2709	JX961617.1	Nigeria	*Rhipicephalus decoloratus*	N/A	1968	2015-06-04
Dabie bandavirus	Severe fever with thrombocytopenia virus (Dabie bandavirus)	GCF_003087855.1	HNXH	KT254590.1	China	*Homo sapiens*	Blood	2014-10-11	2019-06-28
Guertu bandavirus	Guertu virus (GTV)	GCF_004789915.1	DXM	KT328592.1	China	*Dermacentor nuttalli*	N/A	2013-05	2019-06-28
Heartland bandavirus	Heartland virus (HRTV)	GCF_000922255.1	Patient1	JX005844.1	USA: Missouri	*Homo sapiens*	Blood	2009-06-19	2015-02-22
Hunter Island bandavirus	Hunter Island virus (HUIV)	GCF_001271075.2	CSIRO1568	KM198926.1	Australia	*Ixodes eudyptidis*	N/A	1983	2019-06-21
Kismaayo bandavirus	Kismaayo virus (KISV)	N/A	N/A	N/A	Jubbada Hoose, Somalia	*Rhipicephalus pulchellus*	N/A	N/A	N/A
Lone star bandavirus	Lone star virus (LSV)	GCF_000908395.1	TMA 1381	KC589006.1	USA: Kentucky	*Amblyomma americanum*	N/A	1967-06-20	2015-02-22
Razdan bandavirus	Razdan virus (RAZV)	GCF_000913675.1	LEIV-Arm2741	KC335497.1	Armenia	N/A	N/A	N/A	2015-02-22

Collection date: the date of the sample generated by the sequence
Release date: the date when the sequence was originally released in GenBank

and recently the United States, indicating the possibility of transmission of Dabie bandavirus outside of the Asian region. Dabie bandavirus has a seasonal epidemic character, mostly appearing as sporadic cases in spring and summer. Besides, Dabie bandavirus is also transmitted from person to person and can be contracted through direct contact with the blood or body fluids of infected patients [9]. What's more, Dabie bandavirus may has a number of other reservoirs, such as goats, dogs, cattle, and other certain domestic and wild animals. And some hosts may play a major role in the spread of ticks, thereby increasing the epidemic threat of Dabie bandavirus due to the close interaction of animals and humans [10].

The primary symptoms of this newly emerging zoonotic infectious disease include fever, thrombocytopenia, and leukocytopenia. Some patients may develop other clinical symptoms, like high/persistent fever, anorexia, thrombocytopenia, fatigue, nausea, malaise, or gastrointestinal symptoms [7]. In 2018, the World Health Organization (WHO) declared it as one of the potential pathogens for priority research.

14.2 General Property of Dabie Bandavirus

14.2.1 Virion Structure and Genome Characteristics

Dabie bandavirus are enveloped viruses, spherical in shape with diameter of approximately 80 ~ 100 nm. It has a unit membrane envelope, from which protrude 5 ~ 10 nm long polypeptide spikes. Due to the lack of a matrix protein layer, bandaviruses are very flexible and can be easily distorted in negative-staining electron microscopy (EM) images [11]. As a member of bandaviruses, Dabie bandavirus has a three-segmented RNA genome, consisting of large (L), medium (M), and small (S) segments, of a total length of approximately 11,500 bases [12]. And it is a relatively typical RNA virus of single-stranded negative-sense. The terminal regions of all three segments in Dabie bandavirus, as other members of *Bandavirus* genus, contain short and highly conserved 5′ and 3′ untranslated regions (UTR). Being extensively complementary, they form a loop-like structure of RNA segments [13].

The L segment of Dabie bandavirus consists of 6368 nucleotides, with one open reading frame (ORF) encoding the RNA-dependent RNA polymerase (RdRp). The RdRp functions as the viral transcriptase or replicase. The influenza-like endonuclease domain within its N-terminal is essential for viral cap-dependent transcription.

The M segment is 3378 base pair long, with only one ORF encoding a 1073-amino-acid for the precursor of the glycoprotein. The polypeptide membrane protein precursor produced by translation is cleaved at different sites to mature into two glycoproteins, Gn (encoded by the N-terminal of the polypeptide precursor) and Gc (encoded by the C-terminal of the polypeptide precursor). Gn/ Gc, as the envelope of virus, are translated in the endoplasmic reticulum (ER), associated with the ribonucleoprotein (RNP) complex in the ER-Golgi intermediate compartment (ERGIC) or

Golgi apparatus. Mature Gn and Gc assemble into non-covalent dimers on the Golgi apparatus and are embedded in the viral envelope and exposed on the surface of the virus. Finally, Golgi vesicles containing virus particles are transported to the cell surface, and the infectious virus is released outside the cell via extracellular secretion.

The S segment is relatively conserved between Dabie bandavirus and other bandaviruses, which contains 1744 base pair long. The S segment encodes two proteins in opposite orientations through a double-sense translation mechanism. The antisense RNA encodes nucleoprotein (Np) and the sense RNA encodes non-translated protein (NSs), separated by a 62-bp intergenic region. The Np function in viral RNA is to encapsidate genomic RNA into ribonucleoprotein complexes to protect it from degradation by exogenous nucleases or immune systems in the host cell. And it has an active role in RNA transcription, replication, and virion assembly [14]. The NSs interfere with host interferon production. Nuclear factor-κB (NF-κB) and interferon β are expressed in almost all nucleated cells and involved in almost all cellular life activities. Both the Np and the NSs of Dabie bandavirus inhibit the interferon β and NF-κB signaling to suppress antiviral immune response [15].

14.2.2 Infection Mechanisms of Dabie Bandavirus

As other bunyaviruses, Dabie bandavirus needs to rely on the recognition and binding of envelope proteins to enter the target cells. And once entry, they can assemble progeny particles in the infected host cells. In the process of infection, Dabie bandavirus attaches to the host cell receptor firstly. Cellular attachment of Dabie bandavirus is driven by glycoprotein interactions with host cell factors such as dendritic cell-specific intercellular adhesion molecule-3-grabbing non-integrin (DC-SIGN), heparan sulfate (HS), or non-muscle myosin heavy chain IIA (NMMHC-IIA). Then the binding to the host cell receptor induces uptake. The virus and the cell membrane fuse, and the virus invades into the host cell. Membrane fusion in endosomes is pH-dependent. The fusion of viral and endosomal membranes allows release of the viral ribonucleoprotein complexes into the cytoplasm, which is the site of viral transcription and replication. In the cytoplasm, viral mRNA is released after the fusion of endosomes and lysosomes and then undergoes initial transcription catalyzed by RNA-dependent RNA polymerase. Transcription products are translated into precursor proteins at the rough endoplasmic reticulum (ER) and then cleaved by signal peptidases into Gn and Gc. The viral nucleoprotein and the viral polymerase are synthesized in the cytoplasm,, where they bind to the RNPs via the cytoplasmic tails of Gn during the budding process. At the same time, the replication process starts when sufficient nuclear protein wraps around the newly synthesized antigenome and genome. Using the complementary RNA as a template, viral RNA synthesis is performed and the newly synthesized viral RNA and structural proteins are assembled on the Golgi apparatus to form viral particles. After budding of new virus particles into the Golgi is complete, virus-containing vesicles are transported to the plasma membrane, and then the virions bud out and are released by exocytosis [16].

14.2.3 Genetic Diversity

The genetic diversity of Dabie bandavirus has not been completely elucidated, but point mutations during viral replication, natural reassortment of three segments, and homologous recombination reported may be the probable cause [17]. The evidences of genetic reassortment and homologous recombination of Dabie bandavirus have been found, which facilitate rapid evolution and generation of increased genetic diversity in this virus [18, 19]. Otherwise, because its RNA-dependent RNA polymerase lacks proofreading function, resulting in a high frequency of point mutations, about 10^{-4} mutations/ site/ year, during its replication, which is the basis of its genetic diversity [19]. Genetic diversity caused by mutations might be the potential forces driving Dabie bandavirus evolution. It seems to be a lack of consensus in classification of Dabie bandavirus into genotypes. Some studies clustered Dabie bandavirus into three lineages, and some divided it into five or six genotypes. According to the six well-supported clades, phylogenies of L, M, and S genomic segments of Dabie bandavirus are defined as genotypes A–F. And genotype F was the dominant epidemic genotype of Japan, South Korea, and Zhejiang Province of China.

14.2.4 Characteristics of the Glycoprotein

Viral envelope proteins are known to fuse with plasma or endosomal membranes to endocytose into the host cell. Bunyavirus GPs usually fuse with the endosomal membrane under acidic conditions, causing low pH-dependent fusion with infected cells [20]. Therefore, the glycoprotein is the key receptor-binding protein generating to the infection. It was reported that Dabie bandavirus glycoproteins could bind to the NMMHC-IIA of cell surface protein or DC-SIGN as a receptor for entry into human and animal cell lines [21, 22]. In addition, Gn and Gc are important antigenic components on the surface of the virus and targets recognized by specific neutralizing antibodies. As major targets for vaccine research, they contain antigenic epitopes that can stimulate the body to produce specific neutralizing antibodies [23]. Wu, Y et al. have indicated that domain III of Gn provide epitopes for specific neutralizing antibody, and domain II of Gn is probably an ideal region recognized by a broadly neutralizing antibody [24]. And it proved that Gn is a promising antigen for vaccine development. Overall, glycoproteins of Dabie bandavirus have key roles in viral infection of target cells, viral assembly, viral particle formation, and the development of antiviral strategies..

14.3 Construction of Pseudotyped Dabie Bandavirus

Emerging bunyaviruses endanger animal and human health. Understanding the life cycles of these viruses may inform the development of important antiviral strategies. However, the viral and cellular components essential for Dabie bandavirus entry are not known completely at present, and the high level of biosafety containment required for analysis of authentic Dabie bandavirus played a huge role in hindering for progress. With the characteristics of convenient operability, stability, low biological risk, and high sensitivity, the system of pseudotyped virus makes it possible to integrate the glycoprotein of highly pathogenic viruses like Dabie bandavirus but loses the self-replication ability and can only infect one cell cycle, which has strong biological safety. Because of the difference of systems constructed, the pseudotyped Dabie bandavirus can be divided into three types (Table 14.2).

14.3.1 VSV-Based System

As for the construction of pseudotyped virus based on replication-defective VSV, briefly, 293T cells are transfected with glycoprotein expression plasmid of Dabie bandavirus. After 30 h/48 h transfection, the cells are infected with a replication-defective vesicular stomatitis virus vector (VSVΔG), in which the envelope glycoprotein (encoded by the M segment) gene is replaced with the luciferase gene or green fluorescent protein (GFP) gene and then incubates at 37 °C for another 24 hours [22]. Since then, remove the medium, and wash with phosphate-buffered saline (PBS) or 10% FBS-DMEM to remove residual VSVΔG virus. The following

Table 14.2 Construction of pseudotyped Dabie bandavirus

Packaging system	Packaging vector	Glycoprotein expression plasmid	Application	Reference
VSV	Replication-defective VSVΔG-Luc/GFP	pKS-SFTSV-GP; pCAGGS-SFTSV-Gn/Gc V5; pCAG-SFTSV (YG-1) GP, pCAG-SFTSV (HB29) GP	Functional analysis of the Dabie bandavirus Gn/Gc proteins; study the cell tropism	[22, 25, 26]
	Recombinant VSV (rVSV)	rVSV-SFTSV/AH12-GP; rVSV-eGFP-Dabie bandavirus/AH12-GP; rVSV-eGFP-Dabie bandavirus/YG1-GP; rVSV-Dabie bandavirus/HB29-GP	Vaccine studies	[27, 28]
Lentivirus	p96ZM651gag-opt/pNL4-3E-R-/pCAGGS-	SFTSV-Gn/Gc	Determine the susceptibility of target cells to Dabie bandavirus Gn/Gc-driven infection	[22]

day (or after 24 h), pseudotyped viruses are released into the culture supernatants and collected by centrifugation followed by removal of cellular debris using a 0.45-μm filters and stored at −70 °C or below.

As for replication-competent recombinant VSVs (rVSVs), the genome encoding Dabie bandavirus glycoproteins are in place of the VSV envelope protein [27, 28]. Briefly, a recombinant VSV with a reporter gene lacking the G protein (rVSVΔG* eGFP or Fluc) is constructed by inserting a reporter gene and a single restriction site into the pro-G protein coding region of VSV. The VSV sequence (rVSVΔG* eGFP or Fluc) was then constructed into an expression vector with a promoter, usually the T7 promoter. The sequences expressing Dabie bandavirus membrane protein are inserted into the rVSVΔG* eGFP/Fluc, which was VSVΔG pseudotyped backbone, via a single restriction site to form rVSV- Dabie bandavirus* eGFP or rVSV Dabie bandavirus* Fluc expression vector. The rVSVs can be obtained by co-transfection of the above constructed expression vector plasmids with structural protein expression vectors of VSV, such as plasmids of N, P, M, G, and L, into 293T cells by using transfection promoter, such as polyethylenimine (PEI), Lipofectamine, or calcium phosphate [29]. Collect the pseudotyped viruses at 48–72 h post-transfection when the cytopathogenic effect is recognizable.

14.3.2 Lentiviral-Based System

The lentiviral-based system has been widely used lately in pseudotyped virus studies, and the best developed and characterized lentiviral vector system is based on the human immunodeficiency virus type 1 (HIV-1). As for the production of pseudotyped bandavirus based on lentiviral, in brief, expression, plasmids encoding Dabie bandavirus glycoprotein are co-transfected with structural protein packaging plasmids into 293T cells with the aid of pro-transfection reagents such as calcium phosphate, PEI, or Lipofectamine [30, 31]. At 8 h post-transfection, discard the old culture medium replaced by 10% FBS-DMEM medium, and after 2 days (or about 48 h), pseudotyped viruses are released into the culture supernatants and collected by centrifugation followed by removal of cellular debris using a 0.45-μm filters and stored at −70 °C or below.

14.4 Application of the Pseudotyped Dabie Bandavirus

14.4.1 Pseudotyped Dabie Bandavirus as Vaccine

Dabie bandavirus is an emerging tick-borne bunyavirus that causes fatal zoonotic disease; however, there are no effective antiviral vaccines or therapeutic agents. One of the distinguishing features of VSV is that virus particles are not specifically selected for membrane proteins, so that it can assemble with exogenous viral envelope glycoproteins to form pseudoviral particles, which make it as a very promising attenuated viral vaccine vector. VSV recombinant viruses have shown powerful efficacy in eliciting neutralizing antibodies and protecting living organisms against viral infections, such as Ebola virus, Marburg virus, Lassa virus, Hendra virus, Nipah virus, and Andes virus [32–35]. May 2017, the Democratic Republic of the Congo approved the use of rVSV-ZEBOV vaccine and is currently planning to apply for FDA approval [36]. Some researchers have developed a replication-competent rVSV vector expressing Dabie bandavirus Gn/Gc as a live attenuated virus vaccine. It was proved that a single dose of rVSV pseudotyped expressing the Dabie bandavirus glycoprotein of AH12 virus strain has been shown to elicit high titers of broadly neutralizing antibodies against multiple strains of Dabie bandavirus in both immunocompetent and immunocompromised mice [27].

14.4.2 Analysis of Neutralizing Antibody

The sequences of Dabie bandavirus M segment encode the viral surface glycoproteins, which is a target for recognition and binding by specific neutralizing antibodies. Pseudotyped viruses are infectious virus particles formed by encapsulating non-self virus nucleic acids with viral envelope proteins encoded by M segment. The pseudotyped Dabie bandavirus is one of more representation of the constructed pseudotyped VSVs. The VSV-G protein gene of the parental VSV was knocked out and replaced with the gene encoding the luciferase gene or the green fluorescent protein gene and Dabie bandavirus GPs and transfected with structural proteins in virus-producing cells; then we get Dabie bandavirus-GP pseudotype with reporter gene. Therefore, it is possible to measure whether neutralizing antibodies work by the changes in the expression intensity of the reporter gene. Pseudotyped Dabie bandavirus was neutralized in a dose-dependent fashion by sera from blood samples in the convalescent phase infected by Dabie bandavirus, but not from control patients, demonstrating that the pseudotyped virus system is suitable to detect neutralizing antibodies. Heike Hofmann et al. had quantitatively detected the presence of neutralizing antibodies in human sera after preincubation of lentiviral pseudotyped virus with serum samples from patients in recovery [22]. Therefore, the pseudotyped virus system can be a useful tool of neutralizing antibody to evaluate the antibodies or therapeutic drug.

14.4.3 Analysis of Viral Tropism and Entry

Dabie bandavirus belongs to the enveloped RNA viruses, which infect cells usually by the interaction of the viral membrane with the cell membrane. Therefore, the study of pseudotyped viruses expressing viral envelope proteins based on pseudoviral systems may play an important role in elucidating the tropism and pathogenesis of viruses to possible target cells. It also provides a key basis for the design of vaccines or inhibitors for infections. Although several results have suggested that the glycoprotein of Dabie bandavirus involves in virus entry, its molecular basis for entry into cells is still unclear [7, 15]. The availability of virus strains and the demanding biosafety environment make such research difficult. The pseudotyped virus system solves the key challenge of high-risk level laboratory requirements, allowing for safer and easier in-depth research into Dabie bandavirus infection mechanisms.

Although it is currently not well understood which species or organ-derived cells are susceptible for Dabie bandavirus, the detection of Dabie bandavirus in ticks indicates that the virus can be transmitted from animal reservoir to humans via arthropod vector. Scientific researcher employed pseudotyped Dabie bandavirus based on VSV to demonstrate that Dabie bandavirus receptor might be conserved between species or the different species have equally functional receptors by analyzing Dabie bandavirus Gn/Gc-dependent infection of cell lines from different species. And they showed that Dabie bandavirus Gn/Gc mediates entry into a broad panel of animal and human cells in a pH-dependent fashion. They also plan to determine the susceptibility of target cells to Dabie bandavirus Gn/Gc-driven infection based on lentiviral system, but the infection was relatively inefficient in this system [22]. We speculate that this is most probably because the GPs of Dabie bandavirus are not expressed at the plasma membrane but mainly localized in the ER or Golgi apparatus. Some researchers further have observed that not only DC-SIGN but also DC-SIGN-related (DC-SIGNR) and lymph node sinusoidal endothelial cell C-type lectin (LSECtin) are involved in the entry of Dabie bandavirus with VSV-based pseudotyped packaging system [25].

14.4.4 Infectivity and Neutralization Analysis of Pseudotyped Dabie Bandavirus Mutants

Neutralizing antibody recognition and binding targets are concentrated in the complex structure of the membrane protein of Dabie bandavirus. Because of natural selection, there are many amino acid mutations of the glycoprotein. And in order to develop a broad-spectrum vaccine, it is necessary to study whether these genotypes or mutants affect the neutralization properties of Gn-Gc glycoprotein, or whether vaccines developed with a representative strain in the future can effectively protect different mutant strains. However, it is not easy to obtain representative strains of all genotypes and variants. The pseudovirus packaging system is fully utilized to construct a series of pseudotyped viruses of mutant strains. The mutant GP-cloned plasmids were constructed by site-directed mutagenesis using the expression vector of HB29 glycoprotein as a template. Each mutant GP sequence of the plasmids was packed into pseudotyped viruses of mutants and to examine the infectivity and antigenicity of Dabie bandavirus mutant-GPs. Hideki Tani et al. indicated that serine at residue 962 in the Dabie bandavirus-GP is critical for inducing membrane fusion and viral infection based on this mode [37].

14.5 General Conclusions

Since the first identification of Dabie bandavirus, significant progress has been made over the past decade on various aspects of the virus, including its properties, infection mechanisms, and clinical symptoms. Based on the predictive model of Miao et al., it can be inferred that Dabie bandavirus has already reached its full geographic range suitable for its survival in China [38]. However, no existing standard therapeutic scheme has been established to combat Dabie bandavirus infection, and no commercial vaccine is available..

The structural glycoprotein of Dabie bandavirus serves as the sole target of neutralizing antibody recognition, the main immunogen that induces neutralizing antibody production in the body and the receptor binding protein to viral entry should fully demonstrate their role with pseudoviral packaging system. The pseudotyped virus systems have been used to establish neutralizing antibody assays, analyze the amino acid mutation sites encoded by M fragment, study infection with viruses, and investigate the differences in antigenicity of different genotypic virus strains, etc. of high-level pathogenicity for humans [39–41]. In addition, the systems allow us to focus on the investigation of entry mechanisms mediated by particular viral envelope proteins. In summary, the pseudotyped virus packaging system of Dabie bandavirus is of great significance to analyze the function of the glycoproteins, study the cell tropism of viruses, evaluate the effectiveness of existing research vaccines/antibodies or therapeutic drug, and develop new broad-spectrum vaccines.

References

1. Kuhn, J.H., et al.: 2020 taxonomic update for phylum Negarnaviricota (Riboviria: Orthornavirae), including the large orders Bunyavirales and Mononegavirales. Arch. Virol. **165**, 3023–3072 (2020). https://doi.org/10.1007/s00705-020-04731-2
2. Li, D.X.: Fever with thrombocytopenia associated with a novel bunyavirus in China. Zhonghua shi yan he lin Chuang Bing du xue za zhi, Chin. J. Exp. Clin. Virol. **25**(2), 81–84 (2011)
3. Zohaib, A., et al.: Serologic Evidence of Severe Fever with Thrombocytopenia Syndrome Virus and Related Viruses in Pakistan. Emerg. Infect. Dis. **26**, 1513–1516 (2020). https://doi.org/10.3201/eid2607.190611
4. Tran, X.C., et al.: Endemic Severe Fever with Thrombocytopenia Syndrome, Vietnam. Emerg. Infect. Dis. **25**, 1029–1031 (2019). https://doi.org/10.3201/eid2505.181463
5. Takahashi, T., et al.: The first identification and retrospective study of Severe Fever with Thrombocytopenia Syndrome in Japan. J. Infect. Dis. **209**, 816–827 (2014). https://doi.org/10.1093/infdis/jit603
6. Kim, K.H., et al.: Severe fever with thrombocytopenia syndrome, South Korea, 2012. Emerg. Infect. Dis. **19**, 1892–1894 (2013). https://doi.org/10.3201/eid1911.130792

7. Yu, X.J., et al.: Fever with thrombocytopenia associated with a novel bunyavirus in China. N. Engl. J. Med. **364**, 1523–1532 (2011). https://doi.org/10.1056/NEJMoa1010095
8. Luo, L.M., et al.: Haemaphysalis longicornis Ticks as Reservoir and Vector of Severe Fever with Thrombocytopenia Syndrome Virus in China. Emerg. Infect. Dis. **21**, 1770–1776 (2015). https://doi.org/10.3201/eid2110.150126
9. Liu, Y., et al.: Person-to-person transmission of severe fever with thrombocytopenia syndrome virus. Vector Borne Zoonotic Dis. **12**, 156–160 (2012). https://doi.org/10.1089/vbz.2011.0758
10. Chen, C., et al.: Animals as amplification hosts in the spread of severe fever with thrombocytopenia syndrome virus: A systematic review and meta-analysis. Int. J. Infect. Dis. **79**, 77–84 (2019). https://doi.org/10.1016/j.ijid.2018.11.017
11. Guu, T.S., Zheng, W., Tao, Y.J.: Bunyavirus: structure and replication. Adv. Exp. Med. Biol. **726**, 245–266 (2012). https://doi.org/10.1007/978-1-4614-0980-9_11
12. Zhou, C.M., Yu, X.J.: Unraveling the Underlying Interaction Mechanism Between Dabie bandavirus and Innate Immune Response. Front. Immunol. **12**, 676861 (2021). https://doi.org/10.3389/fimmu.2021.676861
13. Lei, X.Y., Liu, M.M., Yu, X.J.: Severe fever with thrombocytopenia syndrome and its pathogen SFTSV. Microbes Infect. **17**, 149–154 (2015). https://doi.org/10.1016/j.micinf.2014.12.002
14. Zhou, H., et al.: The nucleoprotein of severe fever with thrombocytopenia syndrome virus processes a stable hexameric ring to facilitate RNA encapsidation. Protein Cell. **4**, 445–455 (2013). https://doi.org/10.1007/s13238-013-3901-4
15. Qu, B., et al.: Suppression of the interferon and NF-kappaB responses by severe fever with thrombocytopenia syndrome virus. J. Virol. **86**, 8388–8401 (2012). https://doi.org/10.1128/JVI.00612-12
16. Spiegel, M., Plegge, T., Pohlmann, S.: The Role of Phlebovirus Glycoproteins in Viral Entry, Assembly and Release. Viruses. **8** (2016). https://doi.org/10.3390/v8070202
17. Varsani, A., Lefeuvre, P., Roumagnac, P., Martin, D.: Notes on recombination and reassortment in multipartite/segmented viruses. Curr. Opin. Virol. **33**, 156–166 (2018). https://doi.org/10.1016/j.coviro.2018.08.013
18. He, C.Q., Ding, N.Z.: Discovery of severe fever with thrombocytopenia syndrome bunyavirus strains originating from intragenic recombination. J. Virol. **86**, 12426–12430 (2012). https://doi.org/10.1128/JVI.01317-12
19. Lam, T.T., et al.: Evolutionary and molecular analysis of the emergent severe fever with thrombocytopenia syndrome virus. Epidemics. **5**, 1–10 (2013). https://doi.org/10.1016/j.epidem.2012.09.002
20. Lozach, P.Y., et al.: Entry of bunyaviruses into mammalian cells. Cell Host Microbe. **7**, 488–499 (2010). https://doi.org/10.1016/j.chom.2010.05.007
21. Plegge, T., Hofmann-Winkler, H., Spiegel, M., Pohlmann, S.: Evidence that Processing of the Severe Fever with Thrombocytopenia Syndrome Virus Gn/Gc Polyprotein Is Critical for Viral Infectivity and Requires an Internal Gc Signal Peptide. PLoS One. **11**, e0166013 (2016). https://doi.org/10.1371/journal.pone.0166013
22. Hofmann, H., et al.: Severe fever with thrombocytopenia virus glycoproteins are targeted by neutralizing antibodies and can use DC-SIGN as a receptor for pH-dependent entry into human and animal cell lines. J. Virol. **87**, 4384–4394 (2013). https://doi.org/10.1128/JVI.02628-12
23. Guo, X., et al.: Human antibody neutralizes severe Fever with thrombocytopenia syndrome virus, an emerging hemorrhagic Fever virus. Clin. Vaccine Immunol. **20**, 1426–1432 (2013). https://doi.org/10.1128/CVI.00222-13
24. Wu, Y., et al.: Structures of phlebovirus glycoprotein Gn and identification of a neutralizing antibody epitope. Proc. Natl. Acad. Sci. U. S. A. **114**, E7564–E7573 (2017). https://doi.org/10.1073/pnas.1705176114
25. Tani, H., et al.: Characterization of Glycoprotein-Mediated Entry of Severe Fever with Thrombocytopenia Syndrome Virus. J. Virol. **90**, 5292–5301 (2016). https://doi.org/10.1128/JVI.00110-16

26. Kimura, M., et al.: Characterization of pseudotyped vesicular stomatitis virus bearing the heartland virus envelope glycoprotein. Virology. **556**, 124–132 (2021). https://doi.org/10.1016/j.virol.2020.10.006
27. Dong, F., et al.: Single dose of a rVSV-based vaccine elicits complete protection against severe fever with thrombocytopenia syndrome virus. NPJ Vaccines. **4**, 5 (2019). https://doi.org/10.1038/s41541-018-0096-y
28. Philip Hicks, J.B.W., Tomaz, B.M., Roper, B., Rock, G.L., Boardman, K.M., Blotbter, D.J., Gowen, B.B., Bates, P.: Safety, Immunogenicity and Efficacy of a Recombinant Vesicular Stomatitis Virus Vectored Vaccine Against Severe Fever with Thrombocytopenia Syndrome Virus and Heartland Bandaviruses. bioRxiv. (2021)
29. Carette, J.E., et al.: Ebola virus entry requires the cholesterol transporter Niemann-Pick C1. Nature. **477**, 340–343 (2011). https://doi.org/10.1038/nature10348
30. Connor, R.I., Chen, B.K., Choe, S., Landau, N.R.: Vpr is required for efficient replication of human immunodeficiency virus type-1 in mononuclear phagocytes. Virology. **206**, 935–944 (1995). https://doi.org/10.1006/viro.1995.1016
31. Gao, F., et al.: Codon usage optimization of HIV type 1 subtype C gag, pol, env, and nef genes: in vitro expression and immune responses in DNA-vaccinated mice. AIDS Res. Hum. Retrovir. **19**, 817–823 (2003). https://doi.org/10.1089/088922203769232610
32. Marzi, A., Feldmann, F., Geisbert, T.W., Feldmann, H., Safronetz, D.: Vesicular stomatitis virus-based vaccines against Lassa and Ebola viruses. Emerg. Infect. Dis. **21**, 305–307 (2015). https://doi.org/10.3201/eid2102.141649
33. Kurup, D., Wirblich, C., Feldmann, H., Marzi, A., Schnell, M.J.: Rhabdovirus-based vaccine platforms against henipaviruses. J. Virol. **89**, 144–154 (2015). https://doi.org/10.1128/JVI.02308-14
34. DeBuysscher, B.L., Scott, D., Marzi, A., Prescott, J., Feldmann, H.: Single-dose live-attenuated Nipah virus vaccines confer complete protection by eliciting antibodies directed against surface glycoproteins. Vaccine. **32**, 2637–2644 (2014). https://doi.org/10.1016/j.vaccine.2014.02.087
35. Brown, K.S., Safronetz, D., Marzi, A., Ebihara, H., Feldmann, H.: Vesicular stomatitis virus-based vaccine protects hamsters against lethal challenge with Andes virus. J. Virol. **85**, 12781–12791 (2011). https://doi.org/10.1128/JVI.00794-11
36. Walldorf, J.A., Cloessner, E.A., Hyde, T.B., MacNeil, A., Taskforce, C.D.C.E.E.V.: Considerations for use of Ebola vaccine during an emergency response. Vaccine. **37**, 7190–7200 (2019). https://doi.org/10.1016/j.vaccine.2017.08.058
37. Tani, H., et al.: Identification of the amino acid residue important for fusion of severe fever with thrombocytopenia syndrome virus glycoprotein. Virology. **535**, 102–110 (2019). https://doi.org/10.1016/j.virol.2019.06.014
38. Miao, D., et al.: Epidemiology and Ecology of Severe Fever With Thrombocytopenia Syndrome in China, 20102018. Clin. Infect. Dis. **73**, e3851–e3858 (2021). https://doi.org/10.1093/cid/ciaa1561
39. Tani, H., et al.: Replication-competent recombinant vesicular stomatitis virus encoding hepatitis C virus envelope proteins. J. Virol. **81**, 8601–8612 (2007). https://doi.org/10.1128/JVI.00608-07
40. Takada, A., et al.: A system for functional analysis of Ebola virus glycoprotein. Proc. Natl. Acad. Sci. U. S. A. **94**, 14764–14769 (1997). https://doi.org/10.1073/pnas.94.26.14764
41. Perez, M., et al.: Generation and characterization of a recombinant vesicular stomatitis virus expressing the glycoprotein of Borna disease virus. J. Virol. **81**, 5527–5536 (2007). https://doi.org/10.1128/JVI.02586-06

Chapter 15
Pseudotyped Viruses for Mammarenavirus

Qianqian Li, Weijing Huang, and Youchun Wang

Abstract Mammarenaviruses are classified into New World arenaviruses (NW) and Old World arenaviruses (OW). The OW arenaviruses include the first discovered mammarenavirus-lymphocytic choriomeningitis virus (LCMV) and the highly lethal Lassa virus (LASV). Mammarenaviruses are transmitted to human by rodents, resulting in severe acute infections and hemorrhagic fever. Pseudotyped viruses have been widely used as a tool in the study of mammarenaviruses. HIV-1, SIV, FIV-based lentiviral vectors, VSV-based vectors, MLV-based vectors, and reverse genetic approaches have been applied in the construction of pseudotyped mammarenaviruses. Pseudotyped mammarenaviruses are commonly used in receptor research, neutralizing antibody detection, inhibitor screening, viral virulence studies, functional analysis of N-linked glycans, and studies of viral infection, endocytosis, and fusion mechanisms.

Keywords Mammarenavirus · Arenaviruses · Lassa virus · Lymphocytic choriomeningitis virus · Pseudotyped viruses · Lentiviral vectors · VSV-based vectors · MLV-based vectors

Q. Li
Jiangsu Recbio Technology Co., Ltd., Taizhou, China

W. Huang
Division of HIV/AIDS and Sex-transmitted Virus Vaccines, National Institutes for Food and Drug Control (NIFDC) and WHO Collaborating Center for Standardization and Evaluation of Biologicals, Beijing, China

Y. Wang (✉)
Chinese Academy of Medical Sciences & Peking Union Medical College, Beijing, China

Institute of Medical Biology, Chinese Academy of Medicine Sciences & Peking Union Medical College, Kunming, China
e-mail: wangyc@nifdc.org.cn; wangyc@imbcams.com.cn

Y. Wang (ed.), *Pseudotyped Viruses*, Advances in Experimental Medicine and Biology 1407, https://doi.org/10.1007/978-981-99-0113-5_15

Abbreviations

AMAV	Amapari virus
BSL	Biosafety level
CHPV	Chapare virus
EGFP	Enhanced green fluorescent protein
FIV	Feline immunodeficiency virus
Fluc	Firefly luciferase
FRNT	Focus reduction neutralization
G	Glycoprotein
GFP	Green fluorescent protein
GPC	glycoprotein precursor
GTOV	Guanarito virus
HIV	Human immunodeficiency virus
HS	Heparan sulfate
HTS	High-throughput screening
ICTV	International Committee on Taxonomy of Viruses
IGR	Intergenic region
JUNV	Junin virus
L	Large polymerase protein
LAMP1	Lysosomal-associated membrane protein-1
LASV	Lassa virus
LCM	Lymphocytic choriomeningitis
LCMV	Lymphocytic choriomeningitis virus
LF	Lassa fever
LUJV	Lujo virus
M	Matrix protein
MACV	Machupo virus
MLV	Murine leukemia virus
MOPV	Mopeia virus
N	Nucleoprotein
NPC1	Niemann-Pick C1
P	Phosphoprotein
PICV	Pichinde virus
RBD	Receptor-binding domain
RdRp	RNA-dependent RNA polymerase
RNP	Ribonucleic acid protein
S1P	Cellular Site 1 Protease
SABV	Sabia virus
SIV	Simian immunodeficiency virus
SP	Signal peptide
SSP	Stable-signal peptide
TAMV	Tamiami virus
TCRV	Tacaribe virus

TfR1	Transferrin receptor 1
VSV	Vesicular stomatitis virus
WWAV	Whitewater Arroyo virus
Z	Zinc-binding protein
α-DG	α-Dystroglycan

15.1 Introduction

The genus *Mammarenavirus* contains 40 species that can infect humans and other mammals according to the new 2020 International Committee on Taxonomy of Viruses (ICTV) [1]. The original classification of mammarenaviruses was based mainly on viral antigenic properties and was divided into two groups: the New World (NW) and Old World (OW). The OW group consists of viruses such as Lassa virus (Lassa mammarenavirus, LASV), lymphocytic choriomeningitis virus (lymphocytic choriomeningitis mammarenavirus, LCMV), and Lujo virus (Lujo mammarenavirus, LUJV) [2]. The NW group is composed of viruses such as Junin virus (JUNV), Machupo virus (Machupo mammarenavirus, MACV), Chapare virus (Chapare mammarenavirus, CHPV), Guanarito virus (Guanarito mammarenavirus, GTOV), and Tacaribe virus (Tacaribe mammarenavirus, TCRV) [3]. The results of molecular genetic studies are consistent with the results of comparative serology [4]. To date, there is no licensed vaccine or specific antiviral therapy available for the treatment of mammarenaviruses infection in humans. Currently, the treatment of several mammarenaviruses infection is limited to the off-label use of ribavirin [5], which is only partially effective and is associated with side effects.

15.2 Biological Characteristics of Mammarenavirus

15.2.1 Morphology and Genome Structure

Mammarenavirus belongs to *Arenaviridae* family, *Bunyavirales* order. Mammarenavirus is an enveloped virus containing two ambisense RNAs, and its life cycle is restricted to the cytoplasm [6]. Low-temperature electron microscopy observed that mammarenavirus virions were polymorphic, but basically spherical, with diameters ranging from 40 nm to 200 nm and an average diameter of 110 nm. The surface of mammarenavirus is embedded with evenly spaced glycoproteins (GPs) and the core of viruses contains a large number of ribosomes. GPs on the viral surface are formed by a tetrameric complex consisting of GP1 and GP2. The GP2 protein is anchored within the lipid bilayer and adheres to GP1 by ionic interaction, together forming a distinct bulge.

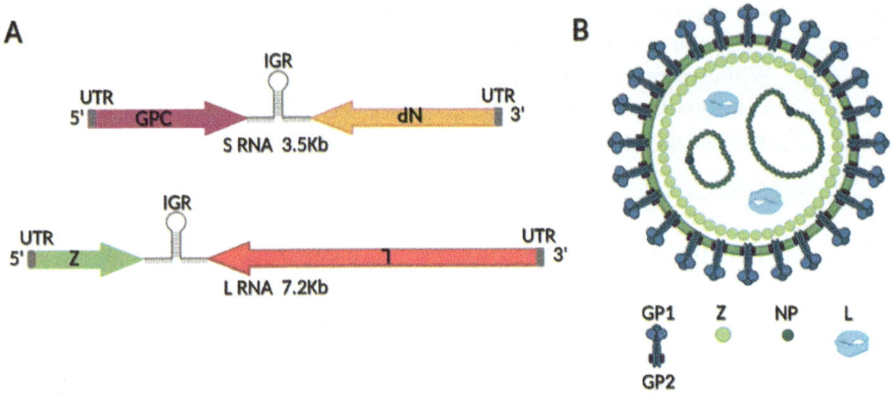

Fig. 15.1 Schematic diagram of mammarenavirus genome and virion structure. (**a**) Schematic diagram of arenavirus genome. The S RNA segment encodes the viral GPC (*purple*) and NP (*orange*) protein. The L RNA segment encodes the viral L (*green*) and Z (*pink*) protein; (**b**) Schematic diagram of arenavirus structure. Arenaviruses are surrounded by a lipid bilayer that contains viral glycoproteins involved in receptor binding (GP1) and viral cell entry (GP2). Beneath the lipid bilayer is a protein layer composed of the Z protein, which plays an important role in viral assembly and budding. The core of the virus consists of the viral ribonucleoprotein complex, which consists of viral genome fragments encapsulated by the viral NP. The L polymerase protein, together with NP, is the minimal component of viral genome replication and gene transcription. (This figure created in BioRender.com)

The genome (Fig. 15.1) contains two double-sense RNAs, which are wrapped by spiral nucleocapsid and form a conical shape. The 3.4 kb small (S) fragment encodes the viral glycoprotein precursor (GPC, 75 kDa) and the viral nucleoprotein (NP, 63 kDa). The 7.2 kb large (L) fragment encodes the RNA-dependent RNA polymerase (RdRp, or L polymerase, 200 kDa) and a zinc-binding protein (Z, 11 kDa) with matrix protein function. The two protein genes are separated by a noncoding intergenic region (IGR) that folds into a stable hairpin structure. GPC protein is cleaved by signal peptidase and cellular Site 1 Protease (S1P) to yield signal peptide (SP) (aa2–58), GP1 (aa59–259), and GP2 (aa260–491). SP proteins play a role in glycoprotein synthesis and membrane localization, low pH-dependent glycoprotein fusion, cleavage of GP1 and GP2, and formation of infectious viral particles. The GP1 protein mediates viral binding to the receptor, thereby inducing the endocytosis of virus. The GP2 protein belongs to class I viral fusion proteins that deliver the capsid into the cytoplasm after viral fusion [7]. The receptor-binding domain (RBD) is located in aa75–237, and the recognized receptors are α-dystroglycan (α-DG) and lysosomal-associated membrane protein-1 (LAMP-1) for LCMV, LASV, and several other mammarenaviruses [8].

15.2.2 Replication of the Viral Genome

After binding to α-DG receptor into the cell, the virus binds to intracellular receptor LAMP1 and then releases the nucleocapsid into the cytoplasm, where transcription and genome replication occur. The mammarenavirus genomic fragments both use an ambisense coding strategy to guide the synthesis of two proteins in opposite directions. Once the viral ribonucleic acid protein (RNP) enters the cytoplasm of infected cells, the polymerase gene associated with the viral RNP located at the 3′ end of the genome initiates transcription and synthesizes messenger RNA (mRNA) for the NP and L protein at the S and L segments, respectively. Subsequently, the viral polymerase can move across intergenic regions using a replicase mode to produce full-length antigenomic RNA copies. This antigenomic RNA will serve as a template for the synthesis of GP (S) and Z protein (L) mRNAs. Each gene can be independently regulated. Viruses assemble at the host cell membrane, and the core of viruses often contains host ribosome.

15.2.3 Pathogenicity

The genus of *Mammarenavirus* contains 40 species. Some of these mammarenaviruses cause hemorrhagic fevers, while others are apparently not pathogenic in humans or rarely infect humans. Diseases caused by mammarenavirus infection include Lassa fever (LF), lymphocytic choriomeningitis (LCM), Argentine hemorrhagic fever, Bolivian and Venezuelan hemorrhagic fevers, etc. Among them, LF and LCM have the most profound impact on human beings.

LF is an acute viral hemorrhagic fever disease caused by LASV infection and transmitted mainly by the multimammate rodent *Mastomys natalensis* [9–11]. There are 100,000 to 300,000 cases of Lassa fever infection in the world every year, causing about 5000 deaths every year, with a 1% mortality rate, and deaths are particularly common in children [12]. Approximately 80% of LASV infections are asymptomatic or mildly symptomatic, and about 25% of survivors develop transient or permanent neurological deafness [13]. Because infection with LASV can cause a high mortality rate, viral culture and animal infection experiments with LASV should be performed in BSL-4 level laboratory.

LCM is an acute viral hemorrhagic fever disease caused by LCMV infection and is usually considered to be aseptic meningitis, with more severe patients occasionally developing central nervous system disease. Approximately one-third of LCMV-infected are asymptomatic, half of patients have the onset of nonspecific febrile illness not involving the nervous system, and the remaining patients present with symptoms of meningitis typically associated with LCMV infection [14]. The mortality rate of LCMV infection was less than 1%, and the hosts are mainly the house mouse. In addition, there is increasing evidence that the globally distributed LCMV is a neglected human pathogen of clinical significance, especially in the case of congenital infection [15]. Besides, the fatal cases of transplant-associated infections caused by LCMV suggest that LCMV poses a great threat to immunocompromised individuals [16]. Infectious experiments such as culture of LCMV virus need to be performed in a BSL-2 level laboratory.

15.2.4 Diversity of LASV and LCMV

LASV originated in Nigeria about 1000 years ago and migrated westward to reach Ivory Coast, Sierra Leone, Guinea, and Liberia, accompanied by high levels of nucleotide diversity in the process, with strain variation up to 32% for the L segments and 25% for the S segments [17]. LASV is a very diverse group with up to seven lineages being currently categorized that tend to cluster to certain geographical areas. LASV lineages I, II, and III are endemic to different regions of Nigeria, whereas lineage IV circulates in Guinea, Liberia, and Sierra Leone [18]. The lineages V and VI were identified in Mali, Ivory Coast, and Nigeria. Two new sequences isolated from Togo were classified as lineage VII [19]. LASV strains from Sierra Leone and from Nigeria have significant differences in genome abundance, codon adaptation index, case fatality rates, and translational efficiency [17]. Besides, the limited degree of amino acid conservation within predicted T-cell epitopes and B-cell epitopes of the LASV GP protein makes it extremely challenging to the vaccine effective against all LASV lineages [20]. And, the high nucleotide diversity affects the accurate diagnosis of LASV [21].

LCMV is widely distributed, mainly in Europe and the United States. The virus also exists in Australia, Asia, and other countries and regions. LCMV is highly genetic diversity and is divided into four distinct lineages (I–IV) [22]. However, these lineages have little correlation between virus genetic lineage and specific geographical location. It may reflect the long and complex phylogeographic history of house mouse host [23]. These lineages can correspond to different host subspecies, lineage I linked to *M. musculus domesticus*, lineage II linked to *M. musculus musculus*, and lineage IV linked to *Apodemus sylvaticus* [22]. Virus lineages I, II, and III have all been associated with severe human disease [23]. The representative strains of lineage I are WE strain and Armstrong strain.

15.3 Construction of Pseudotyped Mammarenaviruses

There are three commonly used packaging backbone vectors of pseudotyped mammarenaviruses, including the lentiviral vector packaging system, the vesicular stomatitis virus (VSV)-based vectors, and the murine leukemia virus (MLV)-based vectors. The lentiviral vector packaging systems include the human immunodeficiency virus-1 (HIV-1)-based vectors, the simian immunodeficiency virus (SIV)-based vectors, and the feline immunodeficiency virus (FIV)-based vectors. In addition, when the pseudotyped virus cannot be constructed by the above packaging backbone vectors, the recombinant pseudotyped virus can be constructed by reverse genetics technology [24, 25].

15.3.1 The Lentiviral Vectors

Lentiviral vectors are often used as the preferred packaging vectors for enveloped virus due to their high packaging efficiency and rapid commercialization process, among which HIV-1-based vector is most used. Lentiviral vectors contain the entire gene sequences for viral transcription, packaging, and integration and provide all proteins except membrane proteins for pseudotyped viruses. It was found that the experimental results of live viruses were highly correlated with those of pseudotyped virus-based experiments [26].

The replication-incompetent HIV pseudotyped with LASV GPC was generated as previously described [27]. Briefly, the LASV-GPC plasmids and HIV-1 backbone plasmids were co-transfected into 293T cells using transfection reagents. Pseudotyped virus supernatant was obtained after 48 hours. After that, the culture supernatant containing the LASV pseudotyped virus was harvested, filtered, aliquoted, and frozen at −70 °C for further use. Besides, the two-plasmid transfection system, three-plasmid transfection system, and four-plasmid transfection system used similar packaging steps, except that the number of plasmids added is different during the plasmid transfection process of 293T.

The packaging structure of SIV-based lentiviral vectors was derived from the BK28 molecular clone of SIVmac251 [28, 29]. To construct SIV-based pseudotyped mammarenaviruses, a three-plasmid system is usually used. Retinal pigment epithelium cells were transduced with SIV-LCMV GP pseudotyped virus constructed from pSIV3 (encoding for Gag-Pol, Tat, Rev, Vif, Vpx, and Vpr) or pSIV4 (encoding for Gag-Pol, Tat, and Rev), the pRMES4SA plasmids (harboring the recombinant SIV-GFP vector genome), and the LCMV-GP expression plasmids, and gene expression was found to be strongest after 3 weeks of selective transduction [30]. SIV-based pseudotyped viruses are constructed in a similar way to the HIV-1-based vectors.

The development of FIV-based lentiviral vector is an attractive alternative to primate-based vectors [31, 32]. FIV-LCMV GP pseudotyped virus was generated by triple transfection of 293T cells with the gag-pol-rev packaging plasmid, the env plasmid, and the vector plasmid. Different variants of LCMV and even a single site mutation can affect the strength of their binding to the α-DG receptor, so that the construction of different variants of FIV-LCMV-GP pseudotyped viruses can achieve specific infection [33]. In addition, FIV-LCMV GP (WE54) can infect mouse neural stem/progenitor cells in vivo [34]. The construction of FIV-based pseudotyped viruses is similar to that of the HIV-1-based vectors.

15.3.2 VSV-Based Vector

For VSV pseudotyped virus system, the backbone was provided by VSV pseudotyped virus (G*ΔG-VSV) that packages expression cassettes for Fluc instead of VSV-G in the VSV genome. Briefly, 293T cells were transfected with mammarenaviruses GPC plasmids using Lipofectamine 3000. Twenty-four hours later, the transfected cells were infected with G*ΔG-VSV. Two hours after infection, cells were washed with PBS three times, and then new complete culture medium was added. Twenty-four hours post infection, pseudotyped mammarenavirus containing culture supernatants were harvested, filtered, and stored at −70 °C in aliquots until use [35]. Compared to other retroviral-based vectors, the VSV-based vectors generally obtained high pseudotyped virus titers and shorten the detection time from 48–72 h to 16–24 h due to the high expression level of reporter protein [36].

15.3.3 MLV-Based Vectors

Both MLV and HIV belong to retroviruses and are the commonly used pseudoviral packaging virus. The efficient MLV packaging system consists of a gag-pol expression plasmid, a plasmid expressing MLV-LTR, packaging sequences and reporter genes, and a heterologous viral GP protein expression plasmid [37]. Pseudotyped mammarenavirus were generated by co-transfection of 293T cells with plasmids encoding LCMV GP, MLV gag-pol, and a retroviral vector encoding the enhanced green fluorescent protein (EGFP) [38]. Supernatants of the packaging cells were harvested, and pseudotyped virus titers were measured by end point dilution on HeLa cells or TE671 cells [38]. In 1999, Miletic et al. successfully constructed the first pseudotyped mammarenaviruses, MLV-LCMV GP pseudotyped virus, which was used to study the cellular infectivity of LCMV [39].

15.3.4 Recombinant Mammarenaviruses

The development of reverse genetic approaches for mammarenaviruses has provided a novel and powerful approach to study the biological properties of mammarenaviruses. Recombinant mammarenaviruses can be classified into two types, single-cycle infectious recombinant mammarenaviruses and replicative recombinant mammarenaviruses.

Single-cycle infectious recombinant mammarenavirus uses green fluorescent protein (GFP) instead of arenaviruses GPC and replicates only in cell lines exogenously transfected plasmids expressing GPC or stably expressing GPC. To generate infectious GP-pseudotyped rLCMVΔGP/GFP virion particles, BHK-21 cells were co-transfected with pPOLI-L plasmid and pPOLI-S ΔGP/GFP plasmid, together with plasmids expressing the minimal transacting factors L (pCAGGs-L), NP (pCAGGs-NP), and GP (pCAGGs-GP) [25]. At 72 h post-transfection, tissue culture supernatants were passage into LCMV GP-expressing BHK-21 cells to amplify rLCMVΔGP/GFP pseudotyped virus [25]. LASV GP-pseudotyped rLCMVΔGP/GFP was generated in the same manner but using LASV GP-expressing BHK-21 cells [25].

Replicative recombinant mammarenaviruses include mammarenaviruses genome-based recombinant viruses and VSV genome-based recombinant viruses [40]. The arenavirus genome-based recombinant viruses include rLASV-LASV GP and rLCMV-LASV GP [41, 42]. The difference between these two types of recombinant viruses is that the GPC protein of the virus uses a heterologous virus GPC. They are constructed in a similar manner, but the rLCMV-LASV GP is much safer [42]. The construction of arenavirus genome-based recombinant viruses can be divided into two systems: recombinant tri-segmented (r3) arenaviruses and recombinant bicistronic arenaviruses (rLCMV/GFP-P2A-NP) [24, 41]. Successful rescue of r3 arenaviruses relies on two S segments and one L segment. The GPC or NP of the S segment is replaced by the reporter gene, thus physically separating the GPC and NP proteins into two distinct S segments (S1 and S2) [24]. Cells are transiently co-transfected with the above plasmids. At 72 h post-transfection, cells are trypsinized and scaled-up. After an additional 72 h incubation period, presence of virus is determined by reporter gene expression [24]. The VSV genome-based recombinant viruses, rVSV-LASV GP, replace the VSV GP with the LASV GP and insert a reporter gene into the genome, thus enabling real-time monitoring of the process of viral entry [40]. Viruses were recovered by co-transfecting BHK-T7 cells with a plasmid carrying a cDNA copy of the recombinant VSV genome and plasmids encoding VSV polymerase, nucleocapsid, phosphoprotein, and glycoprotein proteins [43]. Transfected BHK-T7 cells were overlaid on Vero cells after 48 h, and viruses were passaged onto fresh Vero cells using a low MOI (0.001) to generate subsequent viruses [43].

15.3.5 The Genus of Successful Constructed Pseudotyped Mammarenaviruses

Mammarenaviruses need a high biosafety level. Therefore, pseudotyped mammarenaviruses are often constructed to replace the authentic virus as research tool. Currently, mammarenaviruses that can be successfully constructed include LASV, LCMV, JUNV, LUJV, GTOV, MACV, TCRV, amapari virus (AMAV), etc. [44–47]. Due to the frequent occurrence of LASV outbreaks in African countries in recent years, LASV is the most widely used pseudotyped virus in mammarenaviruses research. In addition, LCMV pseudotyped viruses have been much studied because LCMV has unique neuro-cytophilic properties and is therefore often used as oncolytic virus. Currently, successful LCMV isolates have been constructed including HPI, Arm 53b, Arm Cl13, Arm-F260L, WE, WE54, and WE54-L260F [48, 49].

15.4 Application of the Pseudotyped Mammarenaviruses

Currently, pseudotyped mammarenaviruses have a wide range of applications, mainly in the following six areas, including the study of cell tropism and virus-receptor interactions; the study of viral infection, endocytosis, and fusion mechanisms; the establishment of pseudotyped virus-based neutralizing antibody assays to evaluate immune protection of vaccines; the high-throughput screening of viral inhibitors; the analysis of the relationship between mutations and viral virulence; and the functional analysis of N-linked glycans of GPC.

15.4.1 The Analysis of Virus-Receptor Interactions and Host Range

NW and OW mammarenaviruses have different hosts and cell tropism, so it is hypothesized that mammarenaviruses use different receptors to enter cells. LCMV pseudotyped viruses constructed based on HIV vector and MLV vector can infect cell lines from multiple species, which proves that LCMV pseudotyped viruses has a wide range of cell tropism. Cell lines that can be effectively infected by LCMV pseudotyped viruses include the human-derived epithelial cell lines 293 and HeLa, the fibroblast cell line Te671, the myeloid progenitor cell lines TF-1 and K562, the hepatoma cell line HUH-7, the glioma cell lines nce-G112 and neuroblastoma SH-SY, and the hamster-derived epithelial cell line CHO, the canine-derived thymic stromal cell line Cf2Th, the mouse-derived fibroblast cell line Sc-1a, etc. [38, 39]. α-DG is a receptor for OW mammarenaviruses (including LCMV and LASV) and NW class C mammarenavirus. In addition, the study of LCMV pseudotyped viruses constructed based on HIV vector revealed that the expression of four cell surface molecules, Axl, Tyro3 (from the TAM family), DC-SIGN, and LSECtin (from the C-type lectin family), enhanced HIV-LASV-GP infection of cells [50]. The O-mannosylation of α-DG is important for its receptor function. The study of LCMV pseudotyped viruses constructed based on HIV vector showed that different strains of LCMV have different affinities for O-mannosylated α-DG, among which cl13 and WE54 had high affinity, while ARM53b has low affinity and WE2.2 has no affinity [50]. The study of LCMV pseudotyped viruses constructed based on FIV vector found that the high affinity of WE54 for α-DG receptor was dependent on L260, and its affinity was diminished when the L260F mutation occurred [33]. Furthermore, the study of LCMV pseudotyped viruses constructed based on VSV vector revealed that heparan sulfate (HS) is an alternative receptor for the low-affinity WE HPI strain and that a single point mutation within GP1 (S153F or Y155H) of WE HPI is sufficient to convert viral receptor binding from α-DG to HS [51]. The study of LASV pseudotyped viruses constructed based on VSV vector have shown that LASV can infect α-DG-expressing cell lines, while LASV and LCMV have the same receptor binding domains [52, 53]. In addition, phosphatidylserine-binding receptors Axl and Tyro3, the C-type lectin, and TIM1 receptors can act as cofactors to mediate α-DG-independent LASV entry [42, 54, 55]. The study of mammarenavirus pseudotyped viruses constructed based on MLV vector revealed that transferrin receptor 1 (TfR1) significantly enhanced the infection of MACV, GTOV, and JUNV, but not with those of LASV or LCMV [56], which suggest that TfR1 is the cellular receptor of NW hemorrhagic fever arenaviruses [56]. The study of mammarenavirus pseudotyped viruses constructed based on VSV vector showed that the infectivity of VSV-LUJV-GP was independent of TfR1 and α-DG, indicating that LUJV utilizes an unidentified receptor [57]. Furthermore, cellular NPC1 expression is necessary for efficient VSV-LUJV-GP infection [57].

15.4.2 The Mechanism of Viral Infection, Endocytosis, and Fusion

In addition to being applied to the study of virus-receptor recognition, pseudotyped mammarenaviruses have also been applied in the study of the cleavage mechanism of viral GPC [58], the viral endocytic mechanism [59], and the mechanism of viral fusion [60–62]. The mammarenavirus GPC is translated as a precursor protein that is cleaved into three subunits (SP, G1, G2) by signal peptidase and SKI-1/S1P protease. In a study of the cleavage mechanism of the mammarenavirus GPC, it was found that this endoproteolytic processing is not required for transport to the cell surface but is necessary for infectivity of LCMV GP pseudotyped virus. Further, systematic mutational analysis of the LCMV GP cleavage site using pseudotyped viruses revealed that the consensus motif R-(R/K/H)-L-(A/L/S/T/F)265 is essential for the GPC endoproteolytic processing [58]. In studies of the viral endocytic mechanism, it was found that depletion of cellular cholesterol by treatment with methyl betacyclodextrin or nystatin/progesterone and inhibition of clathrin by an Eps15-dominant negative mutant both resulted in the inhibition of LASV endocytosis. Thus, mammarenaviruses were demonstrated to enter cells through a cholesterol-dependent, non-caveolar, clathrin-mediated endocytic mechanism [59]. Mammarenaviruses enter host cells by GPC mediated low pH-dependent fusion with late endosomes, and there are currently many studies on Lamp1-mediated viral fusion. Using pseudotyped virus-cell surface fusion assays, researchers found that LASV GPC-mediated fusion occurred at a significantly higher pH when Lamp1 was present [60] and that exposure to endosomal pH at physiological temperatures was sufficient to trigger GP-mediated lipid mixing [61]. Lamp1 enhances the efficiency of LASV infection by promoting fusion in less acidic endosomal compartments [60]. However, Lamp1 is not absolutely necessary for LASV cell fusion or entry [62]. In another study, GP-mediated fusion and LASV pseudotyped virus entry were found to be specifically enhanced by an anionic lipid, bis(monoacylglycerol) phosphate (BMP) [62]. BMP is highly enriched in late endosomes and promotes the expansion of the fusion pore and is a cofactor for LASV fusion [62].

15.4.3 Detection of Neutralizing Antibodies and Evaluation of Candidate Vaccines

The presence of neutralizing antibodies is a reliable indicator of protective immunity to arenaviruses and a key monitoring indicator for serological investigation of the epidemic scope of arenaviruses, as well as for evaluation of vaccine immunization efficacy [63]. Pseudotyped mammarenaviruses have been shown to mimic the respective arenaviruses infection, while the neutralizing antibody assay established based on the pseudotyped virus correlates well with the conventional method [25]. The neutralizing antibody assay was performed as follows. After incubating the gradient diluted samples with the pseudotyped virus for one hour, a certain number of cells were added. The expression of the firefly luciferase (Fluc) reporter gene was detected after 2 days of incubation in the incubator. Fluc expression was reduced when the sample to be tested contained neutralizing antibodies to LASV. The neutralizing antibody titer of the tested sample was calculated by using the Reed-Muench formula [27]. The rLASV-GFP-based antibody neutralization assays are currently being used as a standard tool for IRF-Frederick (an international resource) and will accelerate the discovery of anti-LASV medical countermeasures [41].

There are many studies on vaccines against highly pathogenic arenaviruses such as LASV, but no vaccine has been approved for commercial use. The arenavirus vaccines currently under development include live attenuated vaccines, DNA vaccines [64], and replicable vaccines including recombinant vaccinia virus vaccine, MOPV/LASV(ML29) vaccine [65], and recombinant VSV/LASV vaccine [66, 67]. The two most promising vaccine candidates are ML29 and recombinant VSV/LASV vaccines. Pseudotyped virus-based neutralizing antibody assays are often used for neutralizing antibody detection in vaccine immunization sera [27, 68–70].

Epidemiological surveys of serum neutralizing antibodies in the population are necessary to rapidly assess population infection and to minimize the risk and scale of LASV outbreaks. In addition, LASV has seven genetically distinct lineages, so it is equally important to assess cross-reactive and cross-protective humoral immune responses to LASV of distinct lineages following natural infection or immunization. Heinrich et al. examined neutralizing antibodies in the plasma of Lassa fever survivors from Nigeria and Sierra Leone and found that plasma from Nigeria neutralized LASV pseudotyped viruses expressing lineage II GPC better than lineages III and IV GPC expressing pseudotyped viruses, whereas survivors from Sierra Leone did not exhibit lineage bias. This study also demonstrated that neutralization titers based on LASV pseudotyped virus assays significantly correlated with titers determined by plaque reduction with infectious LASV [71].

15.4.4 High-Throughput Screening of Viral Inhibitors

Since the cell attachment and entry of mammarenaviruses are mediated exclusively by the GPC, pseudotyped mammarenaviruses constructed based on the GPC protein can replace live viruses for drug screening in BSL-2 laboratories [72]. Currently, high-throughput screening (HTS) of large compound libraries, small molecule libraries, and approved drugs has rapidly detected multiple viral entry inhibitors, further enabling drug re-positioning [73]. Researchers screened tens of thousands of small molecule compounds and identified multiple inhibitors of viral entry, including compounds 8C1 [44], 17C8 [44], benzimidazole derivative ST-193 [74], LHF-535 (an optimized analog of ST-193) [75], favipiravir [41], ribavirin [41], lacidipine [76], phenothrin [76], isavuconazole [77], clotrimazole derivative TRAM-34 [45], capsaicin [46], arbidol [47], and CP100356 [78]. Mechanistic studies have shown that most inhibitors block pH-dependent membrane fusion, which is the final step of viral entry [76]. Unlike other enveloped viruses, the arenavirus stable-signal peptide (SSP) is unusually long and plays an essential role in glycoprotein maturation and GPC-mediated membrane fusion [79]. The 58-amino-acid SSP contains two hydrophobic structural domains linked by an 8-amino-acid ectodomain loop that interacts with the proximal and transmembrane regions of GP2 to determine the sensitivity of the fusion inhibitor [80]. Further studies showed that the key residues A25, S27, V36, and T40 of SSP and the key residues F427, V431, F434, V435, and V436 of the transmembrane region of GP2 affect the antiviral activity and are the key sites for small molecule compounds to exert their effects [46, 74–77]. In addition, a high-throughput screening method based on a 384-well plate constructed by the HIV system was used to screen the Prestwick Chemical Library, which contains 1200 drugs approved by tje FDA [81] and four hundred and twenty-seven extracts with different polarity, which were obtained from different herb plants [82]. Compounds and traditional Chinese medicine with anti-LASV viral activity were found. Besides, a rLASV-GFP-based high-throughput drug discovery screen platforms is used as a standard tool for studying medical countermeasures to LASV at the IRF-Frederick [41]. However, the pseudotyped virus-based HTS method is limited to screening the entry and fusion inhibitors of arenaviruses and cannot be used to screen the inhibitors of virus replication and proliferation.

15.4.5 Analysis on the Virulence Mechanisms of Viral Mutants

The pathogenicity of arenaviruses in the host varies, and only individual arenaviruses can cause human disease. In a model of non-pathogenic Pichinde virus (PICV) infection in guinea pigs, researchers found that the PICV P2 strain caused a mild febrile reaction in guinea pigs, whereas with passage of the viral strain in guinea pigs, the PICV P18 strain caused severe disease with clinical and pathological features similar to those of human Lassa hemorrhagic fever [83]. The nucleotide sequence comparison revealed that the GPC of P2 and P18 viruses differ at positions 119, 140, and 164. Pseudotyped viruses carrying the P18 GPC could enter cells more efficiently than those carrying the P2 GPC, with the E140 residue playing a critical role in this process [83]. Subsequently, infection of guinea pigs using recombinant viruses containing the E140K mutation demonstrated that E140 is an essential virulence determinant for P18 infection that achieves alterations in viral pathogenicity by enhancing viral entry into target cells [83].

15.4.6 Functional Analysis of N-linked Glycans of GPC

The LASV GPC contains eleven N-linked glycans that play essential roles in GPC cleavage, transport, receptor recognition, epitope shielding, and immune response functions. Zhu et al. used three mutagenesis strategies (asparagine to glutamine, asparagine to alanine, and serine/tyrosine to alanine) to disrupt individual glycan chains on GPC and found that all three mutagenesis strategies resulted in inefficient cleavage of the N89, N119, and N365 glycosylation motif. They then constructed pseudotyped viruses with N-glycosylation deletion (N to Q mutagenesis) of GPC for evaluating the effect of glycosylation deletion on viral infection. It was found that the glycosylation deletion of N89Q and N365Q resulted in the absence of infectivity of the pseudotyped virus, and the glycosylation deletion of N109Q and N119Q resulted in the reduced infectivity of the pseudotyped virus. Furthermore, mice immunized with DNA vaccines based on wild-type and glycosylation-deficient GPCs had fewer neutralizing antibodies against LASV pseudotyped viruses. It is speculated that the glycan residues on GPC provide an immune barrier to LASV, thus hindering vaccine design and development [84].

15.5 Conclusion

In general, with the continuous improvement and promotion of pseudotyped virus construction system, all kinds of mammarenaviruses can be constructed into pseudotyped virus particles. Pseudotyped viruses have the many advantages in virus research and evaluating anti-mammarenaviruses products. Thus, pseudotyped mammarenaviruses play an essential role as one of the useful tools for the biological study of this pathogen.

References

1. Walker, P.J., et al.: Changes to virus taxonomy and to the International Code of Virus Classification and Nomenclature ratified by the International Committee on Taxonomy of Viruses (2021). Arch. Virol. **166**, 2633–2648 (2021). https://doi.org/10.1007/s00705-021-05156-1
2. Briese, T., et al.: Genetic detection and characterization of Lujo virus, a new hemorrhagic fever-associated arenavirus from southern Africa. PLoS Pathog. **5**, e1000455 (2009). https://doi.org/10.1371/journal.ppat.1000455
3. Wiebenga, N.H.: Immunologic studies of Tacaribe, Junin and Machupo viruses. Am. J. Trop. Med. Hyg. **14**, 802–808 (1965). https://doi.org/10.4269/ajtmh.1965.14.802
4. Radoshitzky, S.R., et al.: Past, present, and future of arenavirus taxonomy. Arch. Virol. **160**, 1851–1874 (2015). https://doi.org/10.1007/s00705-015-2418-y
5. McCormick, J.B., et al.: Lassa fever. Effective therapy with ribavirin. N. Engl. J. Med. **314**, 20–26 (1986). https://doi.org/10.1056/NEJM198601023140104
6. Maes, P., et al.: Taxonomy of the order Bunyavirales: second update 2018. Arch. Virol. **164**, 927–941 (2019). https://doi.org/10.1007/s00705-018-04127-3
7. Knipe, D.M., Howley, P.M.: Fields Virology, 6th edn. Lippincott Williams & Wilkins (2013)
8. Jae, L.T., et al.: Virus entry. Lassa virus entry requires a trigger-induced receptor switch. Science. **344**, 1506–1510 (2014). https://doi.org/10.1126/science.1252480
9. McCormick, J.B.: Clinical, epidemiologic, and therapeutic aspects of Lassa fever. Med. Microbiol. Immunol. **175**, 153–155 (1986)
10. Lo Iacono, G., et al.: Using modelling to disentangle the relative contributions of zoonotic and anthroponotic transmission: the case of Lassa fever. PLoS Negl. Trop. Dis. **9**, e3398 (2015). https://doi.org/10.1371/journal.pntd.0003398
11. Coyle, A.L.: Lassa fever. Nursing. **46**, 69–70 (2016). https://doi.org/10.1097/01.NURSE.0000482873.70955.7b
12. Merson, L., et al.: Clinical characterization of Lassa fever: A systematic review of clinical reports and research to inform clinical trial design. PLoS Negl. Trop. Dis. **15**, e0009788 (2021). https://doi.org/10.1371/journal.pntd.0009788
13. Sogoba, N., Feldmann, H., Safronetz, D.: Lassa fever in West Africa: evidence for an expanded region of endemicity. Zoonoses Public Health. **59**(Suppl 2), 43–47 (2012). https://doi.org/10.1111/j.1863-2378.2012.01469.x
14. Laposova, K., Pastorekova, S., Tomaskova, J.: Lymphocytic choriomeningitis virus: invisible but not innocent. Acta Virol. **57**, 160–170 (2013)
15. de la Torre, J.C.: Molecular and cell biology of the prototypic arenavirus LCMV: implications for understanding and combating hemorrhagic fever arenaviruses. Ann. N. Y. Acad. Sci. **1171**(Suppl 1), E57–E64 (2009). https://doi.org/10.1111/j.1749-6632.2009.05048.x
16. Fischer, S.A., et al.: Transmission of lymphocytic choriomeningitis virus by organ transplantation. N. Engl. J. Med. **354**, 2235–2249 (2006). https://doi.org/10.1056/NEJMoa053240
17. Andersen, K.G., et al.: Clinical Sequencing Uncovers Origins and Evolution of Lassa Virus. Cell. **162**, 738–750 (2015). https://doi.org/10.1016/j.cell.2015.07.020
18. Garnett, L.E., Strong, J.E.: Lassa fever: With 50 years of study, hundreds of thousands of patients and an extremely high disease burden, what have we learned? Curr. Opin. Virol. **37**, 123–131 (2019). https://doi.org/10.1016/j.coviro.2019.07.009

19. Whitmer, S.L.M., et al.: New Lineage of Lassa Virus, Togo, 2016. Emerg. Infect. Dis. **24**, 599–602 (2018). https://doi.org/10.3201/eid2403.171905
20. Lukashevich, I.S., Paessler, S., de la Torre, J.C.: Lassa virus diversity and feasibility for universal prophylactic vaccine. F1000Res. **8** (2019). https://doi.org/10.12688/f1000research.16989.1
21. Raabe, V., Koehler, J.: Laboratory Diagnosis of Lassa Fever. J. Clin. Microbiol. **55**, 1629–1637 (2017). https://doi.org/10.1128/JCM.00170-17
22. Fornuskova, A., Hiadlovska, Z., Macholan, M., Pialek, J., de Bellocq, J.G.: New Perspective on the Geographic Distribution and Evolution of Lymphocytic Choriomeningitis Virus, Central Europe. Emerg. Infect. Dis. **27**, 2638–2647 (2021). https://doi.org/10.3201/eid2710.210224
23. Albarino, C.G., et al.: High diversity and ancient common ancestry of lymphocytic choriomeningitis virus. Emerg. Infect. Dis. **16**, 1093–1100 (2010). https://doi.org/10.3201/eid1607.091902
24. Martinez-Sobrido, L., de la Torre, J.C.: Reporter-Expressing, Replicating-Competent Recombinant Arenaviruses. Viruses. **8** (2016). https://doi.org/10.3390/v8070197
25. Rodrigo, W.W., de la Torre, J.C., Martinez-Sobrido, L.: Use of single-cycle infectious lymphocytic choriomeningitis virus to study hemorrhagic fever arenaviruses. J. Virol. **85**, 1684–1695 (2011). https://doi.org/10.1128/JVI.02229-10
26. Wright, E., et al.: Investigating antibody neutralization of lyssaviruses using lentiviral pseudotypes: a cross-species comparison. J. Gen. Virol. **89**, 2204–2213 (2008). https://doi.org/10.1099/vir.0.2008/000349-0
27. Li, Q., et al.: An LASV GPC pseudotyped virus based reporter system enables evaluation of vaccines in mice under non-BSL-4 conditions. Vaccine. **35**, 5172–5178 (2017). https://doi.org/10.1016/j.vaccine.2017.07.101
28. Negre, D., et al.: Characterization of novel safe lentiviral vectors derived from simian immunodeficiency virus (SIVmac251) that efficiently transduce mature human dendritic cells. Gene Ther. **7**, 1613–1623 (2000). https://doi.org/10.1038/sj.gt.3301292
29. Negre, D., Cosset, F.L.: Vectors derived from simian immunodeficiency virus (SIV). Biochimie. **84**, 1161–1171 (2002). https://doi.org/10.1016/s0300-9084(02)00036-6
30. Duisit, G., et al.: Five recombinant simian immunodeficiency virus pseudotypes lead to exclusive transduction of retinal pigmented epithelium in rat. Mol. Ther. **6**, 446–454 (2002). https://doi.org/10.1006/mthe.2002.0690
31. Johnston, J.C., et al.: Minimum requirements for efficient transduction of dividing and nondividing cells by feline immunodeficiency virus vectors. J. Virol. **73**, 4991–5000 (1999). https://doi.org/10.1128/JVI.73.6.4991-5000.1999
32. Wang, G., et al.: Feline immunodeficiency virus vectors persistently transduce nondividing airway epithelia and correct the cystic fibrosis defect. J. Clin. Invest. **104**, R55–R62 (1999). https://doi.org/10.1172/JCI8390
33. Dylla, D.E., Xie, L., Michele, D.E., Kunz, S., McCray Jr., P.B.: Altering alpha-dystroglycan receptor affinity of LCMV pseudotyped lentivirus yields unique cell and tissue tropism. Genet. Vaccines. Ther. **9**, 8 (2011). https://doi.org/10.1186/1479-0556-9-8
34. Stein, C.S., Martins, I., Davidson, B.L.: The lymphocytic choriomeningitis virus envelope glycoprotein targets lentiviral gene transfer vector to neural progenitors in the murine brain. Mol. Ther. **11**, 382–389 (2005). https://doi.org/10.1016/j.ymthe.2004.11.008
35. Nie, J., et al.: Establishment and validation of a pseudovirus neutralization assay for SARS-CoV-2. Emerg. Microbes. Infect. **9**, 680–686 (2020). https://doi.org/10.1080/22221751.2020.1743767
36. Fukushi, S., Tani, H., Yoshikawa, T., Saijo, M., Morikawa, S.: Serological assays based on recombinant viral proteins for the diagnosis of arenavirus hemorrhagic fevers. Viruses. **4**, 2097–2114 (2012). https://doi.org/10.3390/v4102097
37. Soneoka, Y., et al.: A transient three-plasmid expression system for the production of high titer retroviral vectors. Nucleic Acids Res. **23**, 628–633 (1995)

38. Beyer, W.R., Westphal, M., Ostertag, W., von Laer, D.: Oncoretrovirus and lentivirus vectors pseudotyped with lymphocytic choriomeningitis virus glycoprotein: generation, concentration, and broad host range. J. Virol. **76**, 1488–1495 (2002). https://doi.org/10.1128/jvi.76.3.1488-1495.2002
39. Miletic, H., et al.: Retroviral vectors pseudotyped with lymphocytic choriomeningitis virus. J. Virol. **73**, 6114–6116 (1999). https://doi.org/10.1128/JVI.73.7.6114-6116.1999
40. Lay Mendoza, M.F., Acciani, M.D., Levit, C.N., Santa Maria, C., Brindley, M.A.: Monitoring Viral Entry in Real-Time Using a Luciferase Recombinant Vesicular Stomatitis Virus Producing SARS-CoV-2, EBOV, LASV, CHIKV, and VSV Glycoproteins. Viruses. **12** (2020). https://doi.org/10.3390/v12121457
41. Cai, Y., et al.: Recombinant Lassa Virus Expressing Green Fluorescent Protein as a Tool for High-Throughput Drug Screens and Neutralizing Antibody Assays. Viruses. **10** (2018). https://doi.org/10.3390/v10110655
42. Fedeli, C., et al.: Axl Can Serve as Entry Factor for Lassa Virus Depending on the Functional Glycosylation of Dystroglycan. J. Virol. **92** (2018). https://doi.org/10.1128/JVI.01613-17
43. Acciani, M.D., et al.: Ebola Virus Requires Phosphatidylserine Scrambling Activity for Efficient Budding and Optimal Infectivity. J. Virol. **95**, e0116521 (2021). https://doi.org/10.1128/JVI.01165-21
44. Lee, A.M., et al.: Unique small molecule entry inhibitors of hemorrhagic fever arenaviruses. J. Biol. Chem. **283**, 18734–18742 (2008). https://doi.org/10.1074/jbc.M802089200
45. Torriani, G., et al.: Identification of Clotrimazole Derivatives as Specific Inhibitors of Arenavirus Fusion. J. Virol. **93** (2019). https://doi.org/10.1128/JVI.01744-18
46. Tang, K., Zhang, X., Guo, Y.: Identification of the dietary supplement capsaicin as an inhibitor of Lassa virus entry. Acta Pharm. Sin. B. **10**, 789–798 (2020). https://doi.org/10.1016/j.apsb.2020.02.014
47. Herring, S., et al.: Inhibition of Arenaviruses by Combinations of Orally Available Approved Drugs. Antimicrob. Agents Chemother. **65** (2021). https://doi.org/10.1128/AAC.01146-20
48. Steffens, S., et al.: Transduction of human glial and neuronal tumor cells with different lentivirus vector pseudotypes. J. Neuro-Oncol. **70**, 281–288 (2004). https://doi.org/10.1007/s11060-004-6046-8
49. Zhang, C., Hu, B., Xiao, L., Liu, Y., Wang, P.: Pseudotyping lentiviral vectors with lymphocytic choriomeningitis virus glycoproteins for transduction of dendritic cells and in vivo immunization. Hum. Gene. Ther. Methods. **25**, 328–338 (2014). https://doi.org/10.1089/hgtb.2014.105
50. Shimojima, M., Kawaoka, Y.: Cell surface molecules involved in infection mediated by lymphocytic choriomeningitis virus glycoprotein. J. Vet. Med. Sci. **74**, 1363–1366 (2012). https://doi.org/10.1292/jvms.12-0176
51. Volland, A., et al.: Heparan sulfate proteoglycans serve as alternative receptors for low affinity LCMV variants. PLoS Pathog. **17**, e1009996 (2021). https://doi.org/10.1371/journal.ppat.1009996
52. Kunz, S., Rojek, J.M., Perez, M., Spiropoulou, C.F., Oldstone, M.B.: Characterization of the interaction of Lassa fever virus with its cellular receptor alpha-dystroglycan. J. Virol. **79**, 5979–5987 (2005). https://doi.org/10.1128/JVI.79.10.5979-5987.2005
53. Reignier, T., et al.: Receptor use by pathogenic arenaviruses. Virology. **353**, 111–120 (2006). https://doi.org/10.1016/j.virol.2006.05.018
54. Jemielity, S., et al.: TIM-family proteins promote infection of multiple enveloped viruses through virion-associated phosphatidylserine. PLoS Pathog. **9**, e1003232 (2013). https://doi.org/10.1371/journal.ppat.1003232
55. Brouillette, R.B., et al.: TIM-1 Mediates Dystroglycan-Independent Entry of Lassa Virus. J. Virol. **92** (2018). https://doi.org/10.1128/JVI.00093-18

56. Radoshitzky, S.R., et al.: Transferrin receptor 1 is a cellular receptor for New World haemorrhagic fever arenaviruses. Nature. **446**, 92–96 (2007). https://doi.org/10.1038/nature05539
57. Tani, H., et al.: Analysis of Lujo virus cell entry using pseudotype vesicular stomatitis virus. J. Virol. **88**, 7317–7330 (2014). https://doi.org/10.1128/JVI.00512-14
58. Beyer, W.R., Popplau, D., Garten, W., von Laer, D., Lenz, O.: Endoproteolytic processing of the lymphocytic choriomeningitis virus glycoprotein by the subtilase SKI-1/S1P. J. Virol. **77**, 2866–2872 (2003). https://doi.org/10.1128/jvi.77.5.2866-2872.2003
59. Vela, E.M., Zhang, L., Colpitts, T.M., Davey, R.A., Aronson, J.F.: Arenavirus entry occurs through a cholesterol-dependent, non-caveolar, clathrin-mediated endocytic mechanism. Virology. **369**, 1–11 (2007). https://doi.org/10.1016/j.virol.2007.07.014
60. Hulseberg, C.E., Feneant, L., Szymanska, K.M., White, J.M.: Lamp1 Increases the Efficiency of Lassa Virus Infection by Promoting Fusion in Less Acidic Endosomal Compartments. MBio. **9** (2018). https://doi.org/10.1128/mBio.01818-17
61. Bulow, U., Govindan, R., Munro, J.B.: Acidic pH Triggers Lipid Mixing Mediated by Lassa Virus GP. Viruses. **12** (2020). https://doi.org/10.3390/v12070716
62. Markosyan, R.M., Marin, M., Zhang, Y., Cohen, F.S., Melikyan, G.B.: The late endosome-resident lipid bis(monoacylglycero)phosphate is a cofactor for Lassa virus fusion. PLoS Pathog. **17**, e1009488 (2021). https://doi.org/10.1371/journal.ppat.1009488
63. Jahrling, P.B.: Protection of Lassa virus-infected guinea pigs with Lassa-immune plasma of guinea pig, primate, and human origin. J. Med. Virol. **12**, 93–102 (1983). https://doi.org/10.1002/jmv.1890120203
64. Cashman, K.A., et al.: Enhanced Efficacy of a Codon-Optimized DNA Vaccine Encoding the Glycoprotein Precursor Gene of Lassa Virus in a Guinea Pig Disease Model When Delivered by Dermal Electroporation. Vaccine. **1**, 262–277 (2013). https://doi.org/10.3390/vaccines1030262
65. Zapata, J.C., et al.: Genetic variation in vitro and in vivo of an attenuated Lassa vaccine candidate. J. Virol. **88**, 3058–3066 (2014). https://doi.org/10.1128/JVI.03035-13
66. Geisbert, T.W., et al.: Development of a new vaccine for the prevention of Lassa fever. PLoS Med. **2**, e183 (2005). https://doi.org/10.1371/journal.pmed.0020183
67. Lukashevich, I.S., Pushko, P.: Vaccine platforms to control Lassa fever. Expert Rev. Vaccines. **15**, 1135–1150 (2016). https://doi.org/10.1080/14760584.2016.1184575
68. Jiang, J., et al.: Immunogenicity of a protective intradermal DNA vaccine against Lassa virus in cynomolgus macaques. Hum. Vaccin. Immunother. **15**, 2066–2074 (2019). https://doi.org/10.1080/21645515.2019.1616499
69. Wang, M., et al.: Construction and Immunological Evaluation of an Adenoviral Vector-Based Vaccine Candidate for Lassa Fever. Viruses. **13** (2021). https://doi.org/10.3390/v13030484
70. Jiang, J. *et al.* Multivalent DNA Vaccines as A Strategy to Combat Multiple Concurrent Epidemics: Mosquito-Borne and Hemorrhagic Fever Viruses. *Viruses* 13, https://doi.org/10.3390/v13030382 (2021)
71. Heinrich, M.L., et al.: Antibodies from Sierra Leonean and Nigerian Lassa fever survivors cross-react with recombinant proteins representing Lassa viruses of divergent lineages. Sci. Rep. **10**, 16030 (2020). https://doi.org/10.1038/s41598-020-72539-w
72. Basu, A., Mills, D. M. & Bowlin, T. L. High-throughput screening of viral entry inhibitors using pseudotyped virus. Curr. Protoc. Pharmacol. Chapter 13, Unit 13B 13, https://doi.org/10.1002/0471141755.ph13b03s51 (2010)
73. Lee, A.M., Pasquato, A., Kunz, S.: Novel approaches in anti-arenaviral drug development. Virology. **411**, 163–169 (2011). https://doi.org/10.1016/j.virol.2010.11.022
74. Larson, R.A., et al.: Identification of a broad-spectrum arenavirus entry inhibitor. J. Virol. **82**, 10768–10775 (2008). https://doi.org/10.1128/JVI.00941-08
75. Madu, I.G., et al.: A potent Lassa virus antiviral targets an arenavirus virulence determinant. PLoS Pathog. **14**, e1007439 (2018). https://doi.org/10.1371/journal.ppat.1007439
76. Wang, P., et al.: Screening and Identification of Lassa Virus Entry Inhibitors from an FDA-Approved Drug Library. J. Virol. **92** (2018). https://doi.org/10.1128/JVI.00954-18

77. Zhang, X., Tang, K., Guo, Y.: The antifungal isavuconazole inhibits the entry of Lassa virus by targeting the stable signal peptide-GP2 subunit interface of Lassa virus glycoprotein. Antivir. Res. **174**, 104701 (2020). https://doi.org/10.1016/j.antiviral.2019.104701
78. Takenaga, T., et al.: CP100356 Hydrochloride, a P-Glycoprotein Inhibitor, Inhibits Lassa Virus Entry: Implication of a Candidate Pan-Mammarenavirus Entry Inhibitor. Viruses. **13** (2021). https://doi.org/10.3390/v13091763
79. Bederka, L.H., Bonhomme, C.J., Ling, E.L., Buchmeier, M.J.: Arenavirus stable signal peptide is the keystone subunit for glycoprotein complex organization. MBio. **5**, e02063 (2014). https://doi.org/10.1128/mBio.02063-14
80. Shankar, S., et al.: Small-Molecule Fusion Inhibitors Bind the pH-Sensing Stable Signal Peptide-GP2 Subunit Interface of the Lassa Virus Envelope Glycoprotein. J. Virol. **90**, 6799–6807 (2016). https://doi.org/10.1128/JVI.00597-16
81. Wang, J., et al.: A comparative high-throughput screening protocol to identify entry inhibitors of enveloped viruses. J. Biomol. Screen. **19**, 100–107 (2014). https://doi.org/10.1177/1087057113494405
82. Yang, Y., et al.: A cell-based high-throughput protocol to screen entry inhibitors of highly pathogenic viruses with Traditional Chinese Medicines. J. Med. Virol. **89**, 908–916 (2017). https://doi.org/10.1002/jmv.24705
83. Kumar, N., et al.: Characterization of virulence-associated determinants in the envelope glycoprotein of Pichinde virus. Virology. **433**, 97–103 (2012). https://doi.org/10.1016/j.virol.2012.07.009
84. Zhu, X., et al.: Effects of N-Linked Glycan on Lassa Virus Envelope Glycoprotein Cleavage, Infectivity, and Immune Response. Virol. Sin. **36**, 774–783 (2021). https://doi.org/10.1007/s12250-021-00358-y

Chapter 16
Pseudotyped Viruses for the *Alphavirus* Chikungunya Virus

Jiajing Wu, Weijin Huang, and Youchun Wang

Abstract Members of the genus *Alphavirus* are mostly mosquito-borne pathogens that cause disease in their vertebrate hosts. Chikungunya virus (CHIKV), which is one member of the genus *Alphavirus* [1], has been a major health problem in endemic areas since its re-emergence in 2006. CHIKV is transmitted to mammalian hosts by the *Aedes* mosquito, causing persistent debilitating symptoms in many cases. At present, there is no specific treatment or vaccine. Experiments involving live CHIKV need to be performed in BSL-3 facilities, which limits vaccine and drug research. The emergence of pseudotyped virus technology offered the potential for the development of a safe and effective evaluation method. In this chapter, we review the construction and application of pseudotyped CHIKVs, the findings from which have enhanced our understanding of CHIKV. This will, in turn, enable the exploration of promising therapeutic strategies in animal models, with the ultimate aim of developing effective treatments and vaccines against CHIKV and other related viruses.

Keywords Chikungunya virus · Pseudotyped virus · Neutralization

J. Wu
Beijing Yunling Biotechnology Co., Ltd, Beijing, China

W. Huang
Division of HIV/AIDS and Sex-transmitted Virus Vaccines, National Institutes for Food and Drug Control (NIFDC) and WHO Collaborating Center for Standardization and Evaluation of Biologicals, Beijing, China

Y. Wang (✉)
Chinese Academy of Medical Sciences & Peking Union Medical College, Beijing, China

Institute of Medical Biology, Chinese Academy of Medicine Sciences & Peking Union Medical College, Kunming, China
e-mail: wangyc@nifdc.org.cn; wangyc@imbcams.com.cn

Y. Wang (ed.), *Pseudotyped Viruses*, Advances in Experimental Medicine and Biology 1407, https://doi.org/10.1007/978-981-99-0113-5_16

Abbreviations

BSL	Biosafety level
CHIKF	Chikungunya fever
CHIKV	Chikungunya virus
ECSA	East-Central-Southern African
HIV	Human immunodeficiency virus
MLV	Murine leukemia virus
MXRA8	Matrix remodeling-related protein 8
ORF	Open reading frame
PBNA	Pseudotyped virus-based neutralizing antibody assay
PRNT	Plaque reduction neutralization test
UTR	Untranslated region
VSV	Vesicular stomatitis virus

16.1 Biological Characteristics of Chikungunya Virus

Viruses belonging to the *Alphavirus* genus can infect humans and often cause symptoms such as joint pain and fever. According to the latest revision report published by the International Committee on Taxonomy of Viruses (ICTV) in 2021, the *Alphavirus* genus comprises 32 species, including Aura virus, Barmah Forest virus, Bebaru virus, Caaingua virus, Cabassou virus, Chikungunya virus (CHIKV), Sindbis virus, Eastern equine encephalitis virus, Western equine encephalitis virus, Mayaro virus, Semliki Forest virus, and Onyong-nyong virus.

CHIKV is an arthropod-borne *Alphavirus* that is transmitted to humans primarily via the bite of an infected mosquito. CHIKV is becoming widespread in Africa, South Asia, and Southeast Asia, with high infection rates [2, 3]. Infection of humans by CHIKV can cause Chikungunya fever (CHIKF), an acute febrile illness associated with severe, often debilitating, polyarthralgias, fever, rash, and even death [4–6]. CHIKV has emerged in more than 100 countries, with approximately one million people infected each year. In 2017, the World Health Organization (WHO) reported 10 potential infectious diseases for priority research, with CHIKV on the list.

16.1.1 Molecular Structure

CHIKV is an arbovirus of the *Alphavirus* genus, which belongs to the *Togaviridae* family [7]. It is a single-stranded positive-sense RNA virus with a full-length genome of 11.8 kb, including 5′ and 3′ untranslated regions (UTRs) and two open reading frames (ORFs) (Fig. 16.1) [8]. One ORF encodes four non-structural

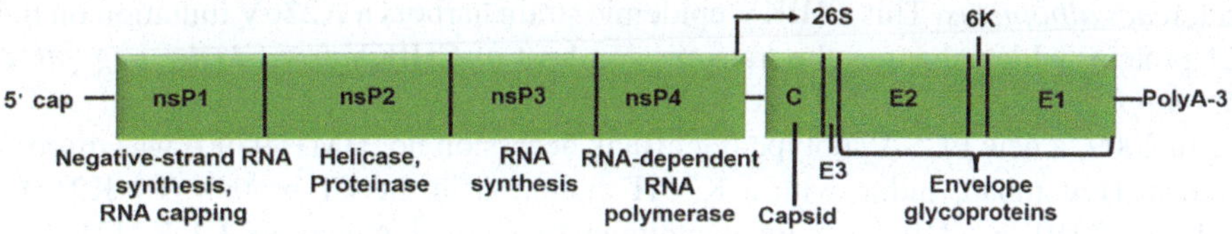

Fig. 16.1 Schematic diagram of the CHIKV genome structure

proteins, namely, nsP1, nsP2, nsP3, and nsP4; the other ORF encodes five structural proteins, namely, capsid proteins C, E3, E2, 6 K, and E1 [9].

Capsid protein C includes two functional domains: an N-terminal RNA binding domain and a C-terminal protease domain. The former binds to genomic RNA, while the latter is involved in the assembly of structural proteins [10, 11]. E1 contains a hydrophobic fusion loop, and under acidic conditions, the CHIKV membrane protein undergoes a conformational change, exposing the fusion loop and promoting nucleocapsid release [12]. Glycoprotein E2 binds to receptors on the host cell membrane to form an endosome and enter the cytoplasm [13, 14]. E3 prevents premature fusion of the heterodimer formed by E2 and E1 with the cell membrane [15]. Studies have shown that two amino acid residues, Gly91 and His230, play a key role in the membrane fusion of E1 [16]. 6 K is the signal peptide of glycoprotein E1 and consists of only 55–60 amino acids [15, 17]. Recently, through structural analysis, Chinese scholars found that the matrix remodeling-related protein 8 (MXRA8) molecule widely distributed on the surface of chondrocytes, muscle cells, and skeletal muscle cells is the receptor of CHIKV, which confirmed the mechanism of host cell invasion by CHIKV and provided a new target for the research and development of vaccines and broad-spectrum neutralizing antibodies [18–20].

16.1.2 Genotypes and Variants

According to genetic analysis, CHIKVs comprise one serotype and three genotypes, namely, the West African genotype, the East-Central-Southern African (ECSA) genotype, and the Asian genotype [21].

Sequence analysis of the isolated viruses revealed that the re-emerging strains of CHIKV originating from Reunion Island belonged to the ESCA genotype [22, 23], while the strains currently circulating in Southeast Asia include both Asian and ESCA genotypes [24–27]. The strains introduced into St. Maarten and the Americas belong to the Asian genotype [24, 25, 28]. Some evidence suggests that the CHIKV strain isolated in Brazil in 2014 has ESCA genes similar to those circulating in Angola [29].

Numerous studies have shown that the large-scale CHIKF outbreak in Reunion Island in 2005–2006 was caused by a CHIKV strain (ECSA genotype) transmitted via *Aedes albopictus*. This CHIKV epidemic strain harbors a A226V mutation on the E1 protein, which changes the transmission host of CHIKV from *Aedes aegypti* to *A. albopictus* [30–33].

In 2009, a new ECSA subtype (GenBank accession no. HM159390) was discovered in Hyderabad, India, with a K211E mutation in the E1 protein (E1-K211E) [34]. In 2010, a CHIKV strain containing two novel mutations E1-K211E and E2-V264A was found among all CHIKV isolates from New Delhi, India [35]. In 2011 and 2012, the abovementioned epidemic strains (E1-K211E and E2-V264A) were found in Tamil Nadu and Kolkata, respectively [34, 36].

16.1.3 Pathogenic Mechanisms and Biosafety Risk

CHIKV can be cultured on a variety of passage cell lines, such as C6/36, AP61, Vero, LLC- MK2, and BHK-21 cells [37]. CHIKV can also be isolated and cultured in vivo by intracranial inoculation of suckling rats aged 1–3 days.

CHIKV enters the host's skin 24–48 hours after a mosquito bite, resulting in viremia. Viremia manifests as a systemic proinflammatory response associated with leukopenia and mononucleosis. The virus circulates through the blood and attacks fibroblasts in muscles and joints, macrophages in lymphoid tissue, and endothelial cells in the liver [12, 38, 39].

CHIKV is considered to cause a high level of individual harm but a low level of group harm. Nucleic acid and serological testing of *A. albopictus*-inactivated serum and frozen specimens can be performed in biosafety level (BSL)-2 laboratories, but potential and confirmed CHIKV-positive samples have to be handled in BSL-3 facilities. It is important that researchers adhere to strict safety precautions to avoid unnecessary biosafety risks.

16.2 Construction of Pseudotyped CHIKV

Currently, there are three commonly used packaging backbone vectors for pseudotyped CHIKVs; these are human immunodeficiency virus (HIV-1)-based lentiviral vectors, vesicular stomatitis virus (VSV)-based vectors, and murine leukemia virus (MLV)-based vectors. In addition, when the abovementioned packaging backbone vector cannot construct a pseudotyped virus, a recombinant pseudotyped virus can be constructed by reverse genetics technology.

16.2.1 Construction of Pseudotyped CHIKV Using Different Vectors

16.2.1.1 Lentiviral Vectors

The HIV-1-based lentiviral system is the most commonly used packaging system for the construction of pseudotyped CHIKVs. As early as 2017, our group constructed the West African genotype pseudotyped CHIKV using the HIV-1 system [40]. The full-length CHIKV West African genotype (strain 37,997, GenBank accession no. AY726732) structural protein (C-E3-E2-6K-E1) was synthesized, and after codon optimization in mammalian cells, it was cloned into the eukaryotic expression vector pcDNA3.1(+), which constituted the CHIKV membrane protein expression plasmid pCMV3.1-CHIKV. Then, 293 T cells were transfected with the appropriate ratio (1:2) of membrane protein expression plasmid pCMV3.1-CHIKV and backbone plasmid pSG3Δenv.cmv.Fluc, and the supernatant was collected after 48 hours. The final prepared CHIKV pseudotyped virus titer was 1.0×10^7 $TCID_{50}$/ml. In 2019, Korean scholars constructed a pseudotyped CHIKV containing two reporter genes based on a lentivirus system [41]. In this study, a three-plasmid system was used to co-transfect 293 T cells with a full-length CHIKV ECSA genotype (strain KNIH/2009/77, GenBank accession no. MN158171.1), structural protein (C-E3-E2-6K-E1), expression plasmid pCMV2-FLAG-CHIKVst, a lentivirus-based dual vector (Luc2P-pLVX-IRES-ZsGreen1), and a packaging vector of lentivirus reporter (psPAX2). The supernatant was harvested to obtain the pseudotyped virus, CHIKVpseudo. There are two important findings from this study. First, when constructing the membrane protein expression plasmid, FLAG tag was added to aid the subsequent detection of protein expression. Second, the system contains two reporter genes, encoding luciferase and GFP. The tracer results of the two reporter genes can be compared simultaneously.

The packaging process of pseudotyped CHIKV based on lentiviral vectors is generally conducted as follows:

1. Seed 293 T cells one day before transfection.
2. Incubate the cells with DMEM (high glucose) containing 10% FBS for 12–18 h. The cells should be 80% confluent before transfection.
3. Co-transfect CHIKV structural protein expression plasmid and lentivirus-based packaging vector plasmid/plasmids using Lipofectamine 3000 or other transfection reagents. Dilute the plasmid with Opti-MEM if necessary.
4. Incubate at 37 °C for 4–6 h.
5. Remove the medium, add fresh DMEM with 10% FBS, and incubate at 37 °C overnight for another 42–44 h.
6. Harvest the culture supernatant. Filter using a 0.45 μM pore-size filter and store at ·70 °C until use.

All work involving pseudotyped CHIKV was performed in a BSL-2 facility.

16.2.1.2 VSV-Based Vectors

A pseudotyped CHIKV based on the VSV vector system has also been constructed [42]. Researchers constructed a West African genotype (strain 37,997, GenBank accession no. AY726732, E3-E2-6K-E1) CHIKV pseudotyped virus based on the VSV vector containing the GFP reporter gene and the firefly luciferase reporter gene, respectively, for the subsequent study of the virus entry mechanism.

The packaging process of pseudotyped CHIKV based on VSV vectors is generally conducted as follows:

1. Seed 293 T cells into a 10-cm dish one day before transfection.
2. Incubate the cells with 10 ml of DMEM (high glucose) containing 10% FBS. The cells should be 80% confluent before transfection.
3. Remove the medium, wash the cells once with the medium, and then add 5 ml of fresh medium.
4. Transfect 5 μg of CHIKV structural protein expression plasmid using Lipofectamine 3000 or other transfection reagents. Dilute the plasmid with Opti-MEM if necessary.
5. Incubate at 37 °C for 6 h.
6. Remove the medium, add fresh DMEM with 10% FBS, and incubate at 37 °C overnight for another 18 h.
7. Inoculate with a multiplicity of infection of 1.0 of VSVΔG/GFP*-G, VSVΔG/Luci*-G, or VSVΔG/SEAP*-G.
8. Adsorb the viruses at 37 °C for 6–8 h.
9. Remove the medium and wash with phosphate-buffered saline (PBS) or serum-free DMEM three times.
10. Add 10 ml of the medium and incubate at 37 °C overnight (17–18 h).
11. Harvest the culture supernatant. Filter using a 0.45 μM pore-size filter and store at ·70 °C until use.

All work involving pseudotyped CHIKV was performed in a BSL-2 facility.

16.2.1.3 MLV-Based Vectors

Like HIV-1, MLV is also a retrovirus that is commonly used as a pseudotyped viral packaging vector. French researchers constructed pseudotyped CHIKV using an MLV vector. Briefly, 293 T cells were co-transfected with the pTG5349 MLV packaging plasmid, the pTG13077 plasmid encoding an MLV-based vector containing a CMV-GFP internal transcriptional unit, a plasmid encoding the Gag-pol proteins of MLV, and a plasmid encoding the CHIKV La Réunion infectious clone (strain LR2006 OPY1, GenBank accession no. KT449801.1) viral envelope glycoproteins (C-E3-E2-6K-E1) [43]. After 24 hours of co-transfection with plasmids, the packaging effect of pseudotyped virus can be determined by observing the fluorescence signal intensity of GFP.

According to published studies and our previous research, all three genotypes of CHIKV can successfully be used to construct pseudotyped viruses. For pseudotyped CHIKV, the expression of E protein can be improved by removing capsid protein C, but whether or not the full-length structural protein is selected, the E2 and E1 protein genomes are essential for the construction of pseudotyped CHIKV [44].

16.2.2 CHIKV Infectious Clones and Virus-like Particles

Compared with other viruses, alphaviruses are more suitable for the construction of infectious clones and virus-like particles. In addition to pseudotyped viruses, several CHIKV infectious clones expressing different reporter genes have been developed to study viral replication and pathogenic mechanisms. The insertion position of the reporter gene in the *Alphavirus* genome is critical for the stability of infectious clones and the virulence of animals [45].

A common strategy for constructing *Alphavirus* infectious clones is to initiate reporter gene expression via an additional subgenomic promoter upstream of the true subgenomic promoter (5'-26S) or the E1 protein (3'-DP) downstream. Compared with 3'-DP, the 5'-26S method is more stable with minimal loss of viral virulence [45, 46]. A reproducible CHIKV infectious clone containing a near-infrared fluorescent protein (iRFP) reporter gene—CHIKV-iRFP—was developed, which was used to establish in vivo imaging of small animals. Briefly, the *iRFP* gene was fused to the full-length CHIKV genome (Asian lineage, GenBank accession no. KC488650) between nsP4 and the capsid (C), and another subgenomic (SG) T7 promoter was introduced downstream of the *iRFP* gene to facilitate the replication of recombinant viruses [47].

16.3 Application of Pseudotyped CHIKV

16.3.1 Neutralizing Assay Based on Pseudotyped CHIKV

The main application of pseudotyped CHIKV is to establish a pseudotyped virus-based neutralizing antibody assay (PBNA). The process of pseudotyped CHIKV infection of cells can be blocked by neutralizing antibodies in test samples. After a test sample interacts with the pseudotyped CHIKV, detecting the expression of the reporter gene carried by the pseudotyped virus can reflect the titer of the neutralizing antibody to be tested. Compared with plaque reduction neutralization test (PRNT), this method has the advantages of high safety, simple operation, easy access, and potential high throughput. In addition, PBNA allows for a reduction in the biosafety level of CHIKV neutralizing antibody detection (from BSL-3 to BSL-2), which was comparable to the detection results obtained with the live virus method.

16.3.1.1 Correlation of PBNA and PRNT

Chung et al. constructed a pseudotyped CHIKV based on a lentiviral vector and established a stable PBNA using the pseudotyped virus [41]. The activity of CHIKV E2 antibody was detected using this method, as well as using classical PRNT. The results showed that the use of PBNA yielded a mean Log IC_{50} value of 5.85 ± 0.09 pg/ml, while the use of PRNT yielded a Log IC_{50} value of 5.68 ± 0.06 pg/ml. The neutralization assay of CHIKVpseudo showed a CV value of 1.09%, similar to the PRNT CV value of 0.98%. In conclusion, PBNA using CHIKVpseudo is a safe, rapid, and reliable neutralization assay that can be used to evaluate the neutralizing activity of serum or monoclonal antibody against CHIKV.

16.3.1.2 Factors Closely Related to CHIKV PBNA

Undoubtedly, a high titer of pseudotyped virus reliably provides PBNA results. However, other factors can also affect the normal operation of PBNA.

(a) **Pseudotyped virus-infected cells:** In general, irrespective of the vector used to construct pseudotyped CHIKV, cells susceptible to pseudotyped virus should be selected as target cells for PBNA. Pseudotyped CHIKV has a wide range of cell tropism [48], and 293 T cells [40, 48] and HeLa cells [41] can be used as infective cells in PBNA. In addition, the source of cells should be clear and the number of passages should be controlled. For experiments with longer cell culture times (more than 48 hours), the evaporative effect in edge wells should be considered. When establishing this method, the cell seeding concentration range should be experimentally validated and optimized.
(b) **Negative and positive controls:** A perfect PBNA should have negative and positive controls to ensure the accuracy of the test results. If international or domestic standards cannot be obtained, the positive serum obtained after animal immunization can be used as a positive control.
(c) **Amount of CHIKV pseudotyped virus:** Adding too much virus can result in false negatives or can cause the inhibition rate to reverse the concentration gradient, which will affect the experimental results.
(d) **Incubation time for CHIKV pseudotyped virus:** The co-culture time of cells and pseudotyped CHIKV should be selected according to the specific characteristics of the pseudotyped virus. For example, the co-culture time for PBNA-based VSV vector is generally 24 h, while for the PBNA-based HIV backbone, the co-culture time is generally 48–72 h.
(e) **Cut-off value:** To determine the critical values for the negative and positive controls (cut-off value), the serum of the negative group with a similar level of background should be selected as the source of the negative sample for establishing the cut-off value. Generally, the test sample size should not be less than 100. The experiment should be repeated at least three times, and

appropriate methods should be used to determine the cut-off value, such as using the geometric mean ± 2SD or 3SD as the final cut-off value.

16.3.2 Establishment of an in Vivo Imaging Model of Small Animals

The pseudotyped virus can be used to establish a visual dynamic monitoring model in mice, to study the tissue distribution after virus infection, and to evaluate the in vivo efficacy of monoclonal antibodies, small molecule inhibitors, and vaccines. Using the accurate quantitative performance and imaging characteristics of the system, the infection site of pseudotyped virus in animals can be dynamically observed, the degree of infection can be judged according to the luminescence intensity and location, and the effect of monoclonal antibodies, drugs, and vaccines can be evaluated [49–51].

Our group infected mice with a HIV-1 vector-based pseudotyped CHIKV, and after optimizing variables, such as the mouse strain, infection route, and detection time, finally established a visual CHIKV pseudotyped virus infection model. Subsequently, the passive immunization effect of antiserum and the active immunization effect of the DNA vaccine were evaluated in vivo using the established mouse infection model [40].

16.3.3 Use of a Capture Antigen in CHIKV IgM Detection

The reference method for the detection of CHIKV-specific IgM is the IgM capture enzyme-linked immunosorbent assay (MAC-ELISA). In the MAC-ELISA method, CHIKV viral lysate is usually used as the antigen bound to the specific IgM to be detected. However, obtaining viral lysates is time-consuming and requires BSL-3 facilities. Chemical treatments such as beta-propiolactone must be used to inactivate the virus before use. However, beta-propiolactone treatment may alter the overall structure of enveloped viral proteins, which may modulate the affinity of antibodies against these proteins, thereby reducing the sensitivity of detection [52, 53].

To overcome these challenges, researchers used pseudotyped CHIKV as a substitute for CHIKV viral lysate [54]. The purified and lyophilized pseudotyped CHIKV was directly applied to paper-based analytical devices as the coating antigen, and the CHIKV IgM antibody detection results could be obtained within 8–10 min. The results showed that the sensitivity of this method to detection using patient serum was 70.6% and the specificity was ~98%. While these results are promising, they are only preliminary, and more extensive validation with a greater number of samples is needed before pseudotyped viruses can replace traditional antigens in serological tests.

16.3.4 Drug Screening

Until now, treatment for CHIKF has been mainly symptomatic because of the lack of an FDA-approved vaccine or effective antiviral drugs. One of the reasons for the delay in the development of anti-CHIKV drugs is the need for BSL-3 facilities in which to conduct the research, which limits the potential for high-throughput operations. Using pseudoviruses to screen small-molecule compounds that have already been approved allows for rapid drug repositioning. In addition, pseudotyped viruses can also be used to screen traditional Chinese medicines with antiviral activity, all of which have the advantages of high throughput and low cost [55, 56].

German scholars determined the anti-CHIKV activity of curcumin or *Boswellia serrata* gum resin extract (AKBA) using pseudotyped CHIKV based on a lentiviral vector [57]. Both compounds blocked the entry of pseudotyped CHIKV and inhibited pseudotyped CHIKV infection in vitro. The IC_{50} values were 6.75 μM for AKBA and 10.79 μM for curcumin.

16.3.5 The Mechanism of Viral Infection

An increasing number of reports have shown that pseudotyped viruses play an important role in the study of virus invasion mechanisms. Pseudotyped viruses have the advantages of being safe, easier to manipulate, easier to construct or discover mutations in coat genes, and easier to quantify the process of entry. In addition, the viral envelope protein itself can be studied using pseudotyped viruses, and the entry process of viral replication can be studied separately from other steps. Pseudotyped viruses are mainly used to study virus infection tropism, cell tropism, receptor recognition, and the virus entry mechanism in vitro [58, 59].

There is evidence that CHIKV enters host cells via a clathrin-dependent endocytic pathway [60]. Endosomal acidification inhibitors attenuate CHIKV infection. Therefore, it is believed that CHIKV particles bind to cell surface receptors and are internalized into the host cell endosome, and then endosomal acidification promotes fusion of the virus with the endosomal membrane through the CHIKV envelope (E) protein. Researchers used MLV vector-based pseudotyped CHIKV to infect 293 T and TE671 cells and found that virus particles containing the CHIKV E protein entered 293 T cells by endocytosis but were internalized into TE671 cell vesicles by macrophage phagocytosis [44]. Through vesicle acidification, the conformation of the E protein is altered, thereby activating its ability to promote membrane fusion. In addition, it was found that digestion of E protein by endosomal cathepsin B more efficiently promoted membrane fusion activity and enhanced infection.

16.4 Conclusion

The ongoing advances in the pseudotyped virus construction system have further confirmed it as an effective virological research strategy. Although some limitations of this system still need to be taken into account, for example, pseudotyped viruses cannot fully simulate the pathogenesis of live virus infection [61], and the innovative virus packaging system is needed for application.

Acknowledgments We thank Liwen Bianji (Edanz) (https://www.liwenbianji.cn) for editing the language of a draft of this book chapter.

References

1. Chen, R., et al.: ICTV Virus Taxonomy Profile: Togaviridae. J. Gen. Virol. **99**, 761–762 (2018). https://doi.org/10.1099/jgv.0.001072
2. Josseran, L., et al.: Chikungunya disease outbreak, Reunion Island. Emerg. Infect. Dis. **12**, 1994–1995 (2006). https://doi.org/10.3201/eid1212.060710
3. Kalantri, S.P., Joshi, R., Riley, L.W.: Chikungunya epidemic: an Indian perspective. Natl. Med. J. India. **19**, 315–322 (2006)
4. Guerrero-Arguero, I., et al.: Alphaviruses: host pathogenesis, immune response, and vaccine & treatment updates. J. Gen. Virol. **102**, 1 (2021). https://doi.org/10.1099/jgv.0.001644
5. Miner, J.J., et al.: Chikungunya viral arthritis in the United States: a mimic of seronegative rheumatoid arthritis. Arthritis Rheumatol. **67**, 1214–1220 (2015). https://doi.org/10.1002/art.39027
6. Schilte, C., et al.: Chikungunya virus-associated long-term arthralgia: a 36-month prospective longitudinal study. PLoS Negl. Trop. Dis. **7**, e2137 (2013). https://doi.org/10.1371/journal.pntd.0002137
7. Burt, F.J., Rolph, M.S., Rulli, N.E., Mahalingam, S., Heise, M.T.: Chikungunya: a re-emerging virus. Lancet. **379**, 662–671 (2012). https://doi.org/10.1016/S0140-6736(11)60281-X
8. Narwal, M., et al.: Crystal structure of chikungunya virus nsP2 cysteine protease reveals a putative flexible loop blocking its active site. Int. J. Biol. Macromol. **116**, 451–462 (2018). https://doi.org/10.1016/j.ijbiomac.2018.05.007
9. Khan, A.H., et al.: Complete nucleotide sequence of chikungunya virus and evidence for an internal polyadenylation site. J. Gen. Virol. **83**, 3075–3084 (2002). https://doi.org/10.1099/0022-1317-83-12-3075
10. Hong, E.M., Perera, R., Kuhn, R.J.: Alphavirus capsid protein helix I controls a checkpoint in nucleocapsid core assembly. J. Virol. **80**, 8848–8855 (2006). https://doi.org/10.1128/JVI.00619-06
11. Perera, R., Owen, K.E., Tellinghuisen, T.L., Gorbalenya, A.E., Kuhn, R.J.: Alphavirus nucleocapsid protein contains a putative coiled coil alpha-helix important for core assembly. J. Virol. **75**, 1–10 (2001). https://doi.org/10.1128/JVI.75.1.1-10.2001
12. Schwartz, O., Albert, M.L.: Biology and pathogenesis of chikungunya virus. Nat. Rev. Microbiol. **8**, 491–500 (2010). https://doi.org/10.1038/nrmicro2368
13. Ashbrook, A.W., et al.: Residue 82 of the chikungunya virus E2 attachment protein modulates viral dissemination and arthritis in mice. J. Virol. **88**, 12180–12192 (2014). https://doi.org/10.1128/JVI.01672-14
14. Smith, T.J., et al.: Putative receptor binding sites on alphaviruses as visualized by cryoelectron microscopy. Proc. Natl. Acad. Sci. U. S. A. **92**, 10648–10652 (1995)

15. Li, L., Jose, J., Xiang, Y., Kuhn, R.J., Rossmann, M.G.: Structural changes of envelope proteins during alphavirus fusion. Nature. **468**, 705–708 (2010). https://doi.org/10.1038/nature09546
16. Kuo, S.C., et al.: Cell-based analysis of chikungunya virus E1 protein in membrane fusion. J. Biomed. Sci. **19**, 44 (2012). https://doi.org/10.1186/1423-0127-19-44
17. Snyder, J.E., et al.: Functional characterization of the alphavirus TF protein. J. Virol. **87**, 8511–8523 (2013). https://doi.org/10.1128/JVI.00449-13
18. Basore, K., et al.: Cryo-EM structure of chikungunya virus in complex with the Mxra8 receptor. Cell. **177**, 1725–1737 e1716 (2019). https://doi.org/10.1016/j.cell.2019.04.006
19. Verma, J., Subbarao, N.: In silico identification of small molecule protein-protein interaction inhibitors: targeting hotspot regions at the interface of MXRA8 and CHIKV envelope protein. J. Biomol. Struct. Dyn. **1**, 19 (2022). https://doi.org/10.1080/07391102.2022.2048080
20. Yin, P., Kielian, M.: BHK-21 cell clones differ in chikungunya virus infection and MXRA8 receptor expression. Viruses. **13** (2021). https://doi.org/10.3390/v13060949
21. Schuffenecker, I., et al.: Genome microevolution of chikungunya viruses causing the Indian Ocean outbreak. PLoS Med. **3**, e263 (2006). https://doi.org/10.1371/journal.pmed.0030263
22. Arankalle, V.A., et al.: Genetic divergence of chikungunya viruses in India (1963-2006) with special reference to the 2005-2006 explosive epidemic. J. Gen. Virol. **88**, 1967–1976 (2007). https://doi.org/10.1099/vir.0.82714-0
23. Pialoux, G., Gauzere, B.A., Jaureguiberry, S., Strobel, M.: Chikungunya, an epidemic arbovirosis. Lancet Infect. Dis. **7**, 319–327 (2007). https://doi.org/10.1016/S1473-3099(07)70107-X
24. Lanciotti, R.S., Valadere, A.M.: Transcontinental movement of Asian genotype chikungunya virus. Emerg. Infect. Dis. **20**, 1400–1402 (2014). https://doi.org/10.3201/eid2008.140268
25. Sy, A.K., et al.: Molecular characterization of chikungunya virus, Philippines, 2011-2013. Emerg. Infect. Dis. **22**, 887–890 (2016). https://doi.org/10.3201/eid2205.151268
26. Powers, A.M., Brault, A.C., Tesh, R.B., Weaver, S.C.: Re-emergence of chikungunya and O'nyong-nyong viruses: evidence for distinct geographical lineages and distant evolutionary relationships. J. Gen. Virol. **81**, 471–479 (2000). https://doi.org/10.1099/0022-1317-81-2-471
27. Weaver, S.C., Forrester, N.L.: Chikungunya: evolutionary history and recent epidemic spread. Antivir. Res. **120**, 32–39 (2015). https://doi.org/10.1016/j.antiviral.2015.04.016
28. Leparc-Goffart, I., Nougairede, A., Cassadou, S., Prat, C., de Lamballerie, X.: Chikungunya in the Americas. Lancet. **383**, 514 (2014). https://doi.org/10.1016/S0140-6736(14)60185-9
29. Nunes, M.R., et al.: Emergence and potential for spread of chikungunya virus in Brazil. BMC Med. **13**, 102 (2015). https://doi.org/10.1186/s12916-015-0348-x
30. Tsetsarkin, K., et al.: Infectious clones of chikungunya virus (La Reunion isolate) for vector competence studies. Vector Borne Zoonotic Dis. **6**, 325–337 (2006). https://doi.org/10.1089/vbz.2006.6.325
31. Delatte, H., et al.: Aedes albopictus, vector of chikungunya and dengue viruses in Reunion Island: biology and control. Parasite. **15**, 3–13 (2008). https://doi.org/10.1051/parasite/2008151003
32. Vazeille, M., et al.: Two chikungunya isolates from the outbreak of La Reunion (Indian Ocean) exhibit different patterns of infection in the mosquito, Aedes albopictus. PLoS One. **2**, e1168 (2007). https://doi.org/10.1371/journal.pone.0001168
33. Bagny, L., Delatte, H., Quilici, S., Fontenille, D.: Progressive decrease in Aedes aegypti distribution in Reunion Island since the 1900s. J. Med. Entomol. **46**, 1541–1545 (2009). https://doi.org/10.1603/033.046.0644
34. Naresh Kumar, C.V., Sivaprasad, Y., Sai Gopal, D.V.: Genetic diversity of 2006-2009 chikungunya virus outbreaks in Andhra Pradesh, India, reveals complete absence of E1: A226V mutation. Acta Virol. **60**, 114–117 (2016)
35. Shrinet, J., et al.: Genetic characterization of chikungunya virus from New Delhi reveal emergence of a new molecular signature in Indian isolates. Virol. J. **9**, 100 (2012). https://doi.org/10.1186/1743-422X-9-100

36. Taraphdar, D., Chatterjee, S.: Molecular characterization of chikungunya virus circulating in urban and rural areas of West Bengal, India after its re-emergence in 2006. Trans. R. Soc. Trop. Med. Hyg. **109**, 197–202 (2015). https://doi.org/10.1093/trstmh/tru166
37. Sudeep, A.B., Vyas, P.B., Parashar, D., Shil, P.: Differential susceptibility & replication potential of Vero E6, BHK-21, RD, A-549, C6/36 cells & Aedes aegypti mosquitoes to three strains of chikungunya virus. Indian J. Med. Res. **149**, 771–777 (2019). https://doi.org/10.4103/ijmr.IJMR_453_17
38. Amin, P., et al.: Chikungunya: report from the task force on tropical diseases by the world Federation of Societies of intensive and critical care medicine. J. Crit. Care. (2018). https://doi.org/10.1016/j.jcrc.2018.04.004
39. Akahata, W., et al.: A virus-like particle vaccine for epidemic chikungunya virus protects nonhuman primates against infection. Nat. Med. **16**, 334–338 (2010). https://doi.org/10.1038/nm.2105
40. Wu, J., Zhao, C., Liu, Q., Huang, W., Wang, Y.: Development and application of a bioluminescent imaging mouse model for chikungunya virus based on pseudovirus system. Vaccine. **35**, 6387–6394 (2017). https://doi.org/10.1016/j.vaccine.2017.10.007
41. Chung, W.C., Hwang, K.Y., Kang, S.J., Kim, J.O., Song, M.J.: Development of a neutralization assay based on the pseudotyped chikungunya virus of a Korean isolate. J. Microbiol. **58**, 46–53 (2020). https://doi.org/10.1007/s12275-020-9384-0
42. Tong, W., Yin, X.X., Lee, B.J., Li, Y.G.: Preparation of vesicular stomatitis virus pseudotype with chikungunya virus envelope protein. Acta Virol. **59**, 189–193 (2015). https://doi.org/10.4149/av_2015_02_189
43. Theillet, G., et al.: Comparative study of chikungunya virus-like particles and Pseudotyped-particles used for serological detection of specific immunoglobulin M. Virology. **529**, 195–204 (2019). https://doi.org/10.1016/j.virol.2019.01.027
44. Izumida, M., Hayashi, H., Tanaka, A., Kubo, Y.: Cathepsin B protease facilitates chikungunya virus envelope protein-mediated infection via endocytosis or macropinocytosis. Viruses. **12**, 1 (2020). https://doi.org/10.3390/v12070722
45. Kummerer, B.M., Grywna, K., Glasker, S., Wieseler, J., Drosten, C.: Construction of an infectious chikungunya virus cDNA clone and stable insertion of mCherry reporter genes at two different sites. J. Gen. Virol. **93**, 1991–1995 (2012). https://doi.org/10.1099/vir.0.043752-0
46. Sun, C., Gardner, C.L., Watson, A.M., Ryman, K.D., Klimstra, W.B.: Stable, high-level expression of reporter proteins from improved alphavirus expression vectors to track replication and dissemination during encephalitic and arthritogenic disease. J. Virol. **88**, 2035–2046 (2014). https://doi.org/10.1128/JVI.02990-13
47. Zhang, H.L., et al.: Visualization of chikungunya virus infection in vitro and in vivo. Emerg Microbes Infect. **8**, 1574–1583 (2019). https://doi.org/10.1080/22221751.2019.1682948
48. Hu, D., et al.: Chikungunya virus glycoproteins pseudotype with lentiviral vectors and reveal a broad spectrum of cellular tropism. PLoS One. **9**, e110893 (2014). https://doi.org/10.1371/journal.pone.0110893
49. Tian, Y., et al.: Development of in vitro and in vivo neutralization assays based on the pseudotyped H7N9 virus. Sci. Rep. **8**, 8484 (2018). https://doi.org/10.1038/s41598-018-26822-6
50. Kong, Y., Cirillo, J.D.: Fluorescence imaging of mycobacterial infection in live mice using fluorescent protein-expressing strains. Methods Mol. Biol. **1790**, 75–85 (2018). https://doi.org/10.1007/978-1-4939-7860-1_6
51. Dhadve, A., Thakur, B., Ray, P.: Construction of dual modality optical reporter gene constructs for bioluminescent and fluorescent imaging. Methods Mol. Biol. **1790**, 13–27 (2018). https://doi.org/10.1007/978-1-4939-7860-1_2
52. Fan, C., et al.: Beta-propiolactone inactivation of coxsackievirus A16 induces structural alteration and surface modification of viral capsids. J. Virol. **91**, 1 (2017). https://doi.org/10.1128/JVI.00038-17

53. Bonnafous, P., et al.: Treatment of influenza virus with beta-propiolactone alters viral membrane fusion. Biochim. Biophys. Acta. **1838**, 355–363 (2014). https://doi.org/10.1016/j.bbamem.2013.09.021
54. Theillet, G., et al.: Detection of chikungunya virus-specific IgM on laser-cut paper-based device using pseudo-particles as capture antigen. J. Med. Virol. **91**, 899–910 (2019). https://doi.org/10.1002/jmv.25420
55. Madrid, P.B., et al.: A systematic screen of FDA-approved drugs for inhibitors of biological threat agents. PLoS One. **8**, e60579 (2013). https://doi.org/10.1371/journal.pone.0060579
56. Zhang, X., et al.: Characterization of the inhibitory effect of an extract of Prunella vulgaris on Ebola virus glycoprotein (GP)-mediated virus entry and infection. Antivir. Res. **127**, 20–31 (2016). https://doi.org/10.1016/j.antiviral.2016.01.001
57. von Rhein, C., et al.: Curcumin and Boswellia serrata gum resin extract inhibit chikungunya and vesicular stomatitis virus infections in vitro. Antivir. Res. **125**, 51–57 (2016). https://doi.org/10.1016/j.antiviral.2015.11.007
58. Sanders, D.A.: No false start for novel pseudotyped vectors. Curr. Opin. Biotechnol. **13**, 437–442 (2002). https://doi.org/10.1016/s0958-1669(02)00374-9
59. Steffen, I., Simmons, G.: Pseudotyping viral vectors with emerging virus envelope proteins. Curr. Gene Ther. **16**, 47–55 (2016)
60. Bernard, E., et al.: Endocytosis of chikungunya virus into mammalian cells: role of clathrin and early endosomal compartments. PLoS One. **5**, e11479 (2010). https://doi.org/10.1371/journal.pone.0011479
61. Li, Q., Liu, Q., Huang, W., Li, X., Wang, Y.: Current status on the development of pseudoviruses for enveloped viruses. Rev. Med. Virol. **28**, 1 (2018). https://doi.org/10.1002/rmv.1963

Chapter 17
Pseudotyped Virus for *Flaviviridae*

Leiliang Zhang, Xiao Wang, Annan Ming, and Wenjie Tan

Abstract Members of *Flaviviridae* are enveloped single positive-stranded RNA viruses including hepacivirus, pestivirus, pegivirus, and mosquito-transmitted flavivirus, which are important pathogens of infectious diseases and pose serious threats to human health. Pseudotyped virus is an artificially constructed virus-like particle, which could infect host cells similar to a live virus but cannot produce infectious progeny virus. Therefore, pseudotyped virus has the advantages of a wide host range, high transfection efficiency, low biosafety risk, and accurate and objective quantification. It has been widely used in biological characteristics, drug screening, detection methods, and vaccine evaluation of *Flaviviridae* viruses like hepatitis C virus, Japanese encephalitis virus, dengue virus, and Zika virus.

Keywords Flavivirus · Pseudovirus · HCV · JEV · DENV · ZIKV

Abbreviations

AvFc	Avaren FC
bNAb	Broadly neutralizing antibodies
C	Capsid

L. Zhang (✉)
Shandong Provincial Hospital Affiliated to Shandong First Medical University, Jinan, Shandong, China

Department of Pathogen Biology, School of Clinical and Basic Medical Sciences, Shandong First Medical University & Shandong Academy of Medical Sciences, Jinan, Shandong, China

X. Wang · A. Ming
Department of Pathogen Biology, School of Clinical and Basic Medical Sciences, Shandong First Medical University & Shandong Academy of Medical Sciences, Jinan, Shandong, China

W. Tan (✉)
NHC Key Laboratory of Biosafety, National Institute for Viral Disease Control and Prevention, Chinese Center for Disease Control and Prevention, Beijing, China
e-mail: tanwj@ivdc.chinacdc.cn

Y. Wang (ed.), *Pseudotyped Viruses*, Advances in Experimental Medicine and Biology 1407, https://doi.org/10.1007/978-981-99-0113-5_17

CCA	Cholangiocarcinoma
DENV	Dengue virus
E	Envelope
Env	Envelope
GSCs	Glioma stem cells
HCC	Hepatocellular carcinoma
HCV	Hepatitis C virus
HCVpp	HCV pseudovirus particle
HIV	Human immunodeficiency virus
HMG	High mannose glycan
HMW	High-molecular-weight
HPSE	Heparanase
JEV	Japanese encephalitis virus
LCFA-CoA	Long-chain fatty acyl-coenzyme A
MLV	Murine leukemia virus
NSAIDs	Non-steroidal anti-inflammatory drugs
ORF	Open reading frame
prM	Pre-membrane
PRNT	Reduction neutralization test
SR-BI	Scavenger receptor class B type I
TBEV	Tick-borne encephalitis virus
VSV	Vesicular stomatitis virus
WNV	West Nile virus
YFV	Yellow fever virus
ZIKV	Zika virus

17.1 Introduction

Flaviviridae includes four virus genera, namely, flavivirus, pestivirus, pegivirus, and hepacivirus. Hepatitis C virus (HCV) is a positive-stranded RNA virus belonging to the hepacivirus genus of *Flaviviridae* [1]. It has high genetic variability as 8 genotypes and more than 100 subtypes have been identified so far. The polyprotein encoded by the HCV 9.6 kb genome could be processed into three structural proteins, namely, core protein, envelope protein E1 and E2, and seven non-structural proteins (NS proteins), including P7, NS2, NS3, NS4A, NS4B, NS5A, and NS5B [2]. Core protein affects many host cellular processes such as host cell transcription, lipid metabolism, and apoptosis, which has potential carcinogenicity. Envelope protein E2 is responsible for the specific binding of host receptors CD81 and scavenger receptor class B type I (SR-BI), while viral residues 290–312 in the central hydrophobic region of E1 protein affect the infectivity of HCV and control the virus to achieve pH-dependent membrane fusion [3, 4]. In addition, the interaction of NS proteins plays critical roles in the assembly of infectious virus particles. NS2 coordinates with P7, NS3, NS5A, and E1/E2 complexes to promote nucleocapsid assembly. NS5B is an RNA-dependent RNA polymerase, which functions in viral RNA replication [5].

HCV is transmitted through contact with contaminated blood, while there is a limited risk of vertical transmission or sexual transmission [6]. HCV infection causes severe liver diseases including chronic hepatitis, liver cirrhosis, and hepatocellular carcinoma (HCC) [7]. Patients with chronic HCV infection will experience fibrosis, cirrhosis, and eventually decompensate liver function or severe hepatocellular carcinoma. In addition, HCV infection, as a systemic disease, is related to a variety of extrahepatic organ dysfunction and can cause a variety of extrahepatic diseases, including atherosclerosis, glucose and lipid metabolism disorders, changes in iron metabolism pathways, and lymphoproliferative diseases beyond the traditional liver manifestations of liver cirrhosis and hepatocellular carcinoma [8]. Recent epidemiological studies suggest that HCV infection is also closely related to cholangiocarcinoma (CCA). HCV infection in hepatic progenitor cells or bile duct cells may be the direct cause of CCA [9].

Flavivirus is widely distributed in the world with many species, including Japanese encephalitis virus (JEV), dengue virus (DENV), Zika virus (ZIKV), West Nile virus (WNV), yellow fever virus (YFV), and tick-borne encephalitis virus (TBEV). Morphological structure, genomic characteristics, and biological characteristics of flavivirus are highly similar. This kind of single positive-stranded RNA virus possesses a spherical particle with icosahedral symmetrical nucleocapsid. The viral genome with its length of 11 kb encodes three structural proteins: capsid (C) protein, pre-membrane (prM) protein, envelope (E) protein, and seven non-structural proteins (NS1, NS2A, NS2B, NS3, NS4A, NS4B, NS5) [10]. E protein is the main glycoprotein on the virus envelope, which has abilities of binding to cell surface receptors and mediating membrane fusion. It also has specific antigen epitopes and neutralizing antigen epitopes, which can stimulate the production of neutralizing antibodies and play an important role in virus pathogenesis and immunity [11].

Flavivirus uses arthropods as vectors and causes zoonotic natural foci diseases through insect bites, showing clinical symptoms such as fever, rash, skin mucosal hemorrhage, encephalitis, neonatal malformation, and so on. Its distribution and transmission have seasonal and regional characteristics. However, in recent years, the epidemic focus of JEV, YFV, DENV, and other viruses has gradually expanded from tropical regions such as Africa and South America to the global reach, and its prevention and control have become a serious public health problem worldwide [12]. However, there is a lack of specific treatment for flavivirus infection in general.

Pseudotyped virus is a kind of virus-like particle based on one virus whose nucleic acid is coated by the envelope protein encoded by another virus to form recombinant virus particles. Its genome retains the characteristics of the backbone virus, but it loses the ability of self-replication and can only replicate in the host cell for one round [13]. This psedotyped virus has been widely used in various studies of *Flaviviridae*. This review will summarize the construction strategies of *Flaviviridae* pseudotyped virus and the application of the pseudotyped virus in many fields.

17.2 Construction Strategies of Pseudotyped *Flaviviridae*

Pseudotyped virus of the enveloped is usually constructed by 2-plasmid, 3-plasmid, or 4-plasmid transfection system, and the envelope protein is embedded in the framework of a backbone vector. The three most commonly used viral vectors are human immunodeficiency virus (HIV), vesicular stomatitis virus (VSV), and murine leukemia virus (MLV), whose packaging systems have respective advantages [14]. The backbone plasmid contains the gene sequences required for virus transcription, packaging, and integration other than the envelope protein. Through co-transfection into the packaging cells, the envelope protein is expressed on the surface of the packaging cells, and the virus particles expressed by the skeleton plasmid are coated by the envelope protein when budding, thus forming pseudotyped viruses. The vector gene sequence usually carries reporter genes expressing luciferase or fluorescent protein for quantitative analysis of virus infection. The yield of pseudotyped virus is affected by many factors, such as the expression level of the envelope protein, the selection of the packaging system, transfection conditions, and so on [13].

17.2.1 Construction Strategies of HCV Pseudotyped Virus

For a long time, due to the lack of an effective cell culture system to support the massive reproduction of HCV, the research on HCV virus-host interaction at the cellular and molecular levels has been hindered. To overcome this obstacle, two groups of scientific research teams jointly proposed the HCV pseudovirus particle (HCVpp) model in 2003—the unmodified HCV glycoproteins E1 and E2 were assembled on the core particles of retrovirus and lentivirus to construct an infectious pseudotyped virus [15]. The existing construction strategy of HCVpp is to transfect three plasmids encoding retroviral genome expressing a reporter gene, gag-pol protein of HIV or MLV, and HCV E1 and E2 glycoprotein in 293 T cells [16].

With the development of various experimental techniques, HCVpp models have been continuously optimized. The interaction of CD81 with envelope glycoprotein E2 promotes HCV entry [17]. Researchers found that the knockdown of CD81 in 293 T cells can significantly enhance the infectivity of most clinically screened isolates without affecting their particle antigenicity [18]. It was previously known that HCV NS2 protein could participate in membrane rearrangement, and P7 oligomer, as an ion channel, can regulate the pH of newly formed virions [19]. On this basis, Soares HR et al. proposed that co-expression of P7NS2 ORF could enhance the infectivity of the pseudotyped virus in the H77 strain [19].

17.2.2 Construction Strategies of JEV Pseudotyped Virus

JEV could invade the central nervous system with many critical symptoms and high mortality. Lee HJ et al. constructed pHCMV-E without prM and pHCMV PrM/E expressing PrM with MLV as a vector to explore the biological function of the prM protein of JEV [20]. Tani H et al. used VSV ΔG/Luc-*g vector, replacing its G protein sequence with luciferase gene and co-transfecting it with JEV E protein [21]. To improve the packaging efficiency of lentivirus vectors, Liu H et al. introduced cleavable signal peptide sequences into HIV-1 pseudotyped viruses and constructed three pseudotyped viruses with different signal intensities [22]. MLV carrier and VSV ΔG/Luc-*g vector are widely used in the follow-up studies of the JEV [23–27].

17.2.3 Construction Strategies of DENV Pseudotyped Virus

Dengue fever is an arbovirus disease with the most widespread distribution and incidence worldwide. Hu HP et al. developed the lentivirus vector pNL4–3.Luc.R-E- [28]. The prM/E expression plasmid pCB-D2 (or enhanced pCB-D2VSV) of DENV2 was co-transfected with the vector, and its entry into cells was pH-dependent. When exploring the role of the anchoring sequence of DENV capsid protein, Rana J et al. used the WNV replicon (WNV Rep) carrying the enhanced green fluorescent protein reporter gene as the backbone vector to construct pseudotyped virus particles together with the plasmid encoding DENV2 C-Ca-PrM–E [29].

17.2.4 Construction Strategies of ZIKV Pseudotyped Virus

ZIKV was first isolated in 1947 from a rhesus monkey in the Zika forest of Uganda. It is transmitted in a cycle of human-mosquito-human. pHIV-luciferase, pCMV-ΔR8.2, and envelope (Env) protein plasmids were co-transfected into HEK293T cells to produce pseudotyped virus [30]. In addition to using HIV as a vector, Rana J et al. co-transfected WNV replicon carrying enhanced green fluorescent protein reporter gene with S protein/prM-E protein and infected Vero cells [31]. Kretschmer M et al. constructed four Z-HIV*luc* pseudotyped viruses with envelope proteins (pMEZ1-Z4) and infected Vero-B4 cells [32]. In the optimization of the packaging scheme, it was found that the highest infection rate could be obtained by using 8 μg pME-Z1 plasmid and 37 μg HIV-1 vector pNL Luc AM plasmid. According to the analysis of luciferase activity, Z1-HIV*luc* showed the highest RLU value and was the most effective subtype of infection. Then, a three-plasmid system Z1-LENTI*luc* was constructed with lentivirus as vector (the packaging plasmid psPAX2 expressed gag and gag-pol but did not express Nef, and the transfer plasmid pLenti-luciferase-P2A-Neo encoded the HIV-1 genome and carried the reporter gene). Ruiz-Jimenez F et al. transfected the envelope protein with lentiviral vector HIV-1 (pNL4.3.Luc.R-G-) or retroviral vector MLV (phCMV-5349) and added 24 μl polyethyleneimine to the culture medium to improve the titer [33].

17.3 Application of Pseudotyped *Flaviviridae*

As an enveloped virus, when *Flaviviridae* invades host cells, its surface protein recognizes target cell receptors and starts the process of adsorption and penetration. Pseudotyped viruses carrying related reporter genes have been widely used to study the interaction between flavivirus surface proteins and host cells, such as virus receptor recognition, virus cytotropism and infectivity, and virus invasion process. See Table 17.1 for the application status of some pseudotyped viruses in recent years.

After HCVpp infects cells, the reporter gene in the transgenic plasmid (such as EGFP, Luc, or β- the expression of gal, etc.) can be directly and conveniently observed or determined by fluorescence technology. HCVpp has been developed as a method to screen HCV host entry factors. At the same time, it has also been applied to explore the function of HCV envelope glycoprotein, identify the neutralization of antibodies, and develop new anti-HCV molecules.

Table 17.1 Construction strategies and applications of pseudotyped viruses for flaviviruses

Virus	Backbone/envelope	Application	References
JEV	pHCMV/E, prM-E	Explore the function of PRM protein, neutralizing antibody detection, and lenti-virus vector optimization	20, 22, 25, 27
JEV	VSVΔG/Luc-*G/E	Recognition of virus receptors, factors affecting virus replication, and ways of virus invading host cells	21, 23, 24, 26
DENV	pNL4–3.Luc.R-E–/prM/E	Construction and function determination of pseudovirus	28
DENV	WNV-rep/C-ca-PrM–E	Explore the function of capsid protein anchoring sequence	29
ZIKV	pCMV-ΔR8.2/env	Screening of antiviral drugs	30
ZIKV	pNL-Luc-AM/pME Z1–4, psPAX2, pLenti-luc-P2A-neo/pME-Z1	Antitumor therapy	32
ZIKV	WNV-rep/env	Explore the function of capsid protein anchoring sequence	31
ZIKV	pNL4.3.Luc.R-G–/env, phCMV-5349/env	Recognize specific receptors	33
YFV	WNVVII-GFP/C- PrM-E	Neutralizing antibody detection	63, 64

17.3.1 Study the Interaction between Virus and Host Cell

At present, HCVpp has been widely used to identify HCV-specific host entry factors or to screen new host entry factors on a large scale [16]. CD81 was identified to participate in the infection process of HCV and proposed a reasonable model: cholesterol-mediated CD81 closed conformation determines the entry of HCV [17]. Claudin-12 was identified as a novel HCV entry factor through HCVpp system [34]. In addition to HCV host entry factors, heparanase (HPSE) enhanced HCV release by promoting CD63 production and exosome secretion without affecting the entry of HCVpp into cells. That is, HPSE was not involved in virus entry or viral RNA replication [35].

HCV envelope proteins E1 and E2 drive virus-host membrane fusion by forming heterodimers. The envelope glycoprotein in HCVpp is produced by plasmids that only express E1 and E2, so it is easier to carry out stable site-specific mutagenesis, and it is also convenient for researchers to study a large number of mutants at the same time [16]. Many researchers used the pseudotyped virus model to replace live HCV to study the function of the envelope glycoprotein. Pfaff-Kilgore JM et al. used different E1 and E2 library mutants to construct HCV pseudotyped virus models and identified 97 envelope glycoprotein residues related to viral infection in HCV, providing new ideas for the development of antiviral drugs [36].

Moreover, HCV envelope glycoprotein is closely related to cell fusion. The envelope virus will change its conformation when invading the cell, to complete the fusion with the host cell membrane. The construction of the HCV pseudotyped virus model enables the study of HCV glycoprotein fusion on the particle surface rather than on the cell surface. Researchers used cells or liposomes as receptor membranes and generated HCVpp as donor membranes to replace the cell surface expression of HCV glycoproteins, which could optimize the membrane fusion assay of HCV entering host cells [37].

17.3.2 Neutralizing Antibody Detection and Vaccine Effect Evaluation

HCVpp is widely used to study the neutralization of HCV by immune serum or HCV-specific monoclonal antibodies [38]. Through the HCVpp model researchers proved that a newly proposed DNA vaccination strategy with high immunogenicity could induce neutralizing antibodies in mice [39]. Dent M et al. found that the fusion protein Avaren FC (AvFc), which is composed of Avaren lectin and the crystallizable FC fragment of human immunoglobulin G1, has anti-HCV activity. The underlying mechanism of HCV may be through binding with high mannose glycan (HMG) on the dimer of E1/E2 envelope protein, blocking their association with cellular receptors and virus entry [40]. Center RJ et al. used sequential reduction and oxidation strategies to refold the two high-yield monomer E2 species D123 and disulfide bond minimization version (D123A7) into disulfide-linked high-molecular-weight (HMW)-like substances. They proved that it could neutralize homologous and heterologous HCV genotypes, block the interaction between E2 and CD81, and provide a new direction for the development of higher quality and lower cost HCV vaccines [41]. Sabrina J et al. found that the cross genotype Ar3 specific neutralizing antibody could neutralize HCVpp and protect patients from chronic HCV infection for a long time [42]. Khera T et al. tested the neutralization effect of vaccine-induced antibodies by H77 (GT1a) HCVpp determination [43]. The results showed that viruses lacking HVR1 and selected glycosylation sites would expose cross-neutralizing antibody epitopes and CD81 binding sites. The recombinant E2 protein with these mutations could induce strong cross-binding antibodies but could not cross neutralize [43].

HCV vaccine development requires accurate and repeatable measurement of the neutralization span of vaccine-induced antibodies [44]. Salas JH et al. selected 15 HCVpp for neutralization assay to test the diversity of their genes and antigens. They found that the difference in neutralization sensitivity amid HCVpp was not related to the genetic distance between E1E2 clones. To make a meaningful comparison of neutralization effects of vaccine antibodies from different laboratories, defining the neutralization range of HCV antibodies requires the use of viruses that span multiple neutralization sensitivities, rather than only based on genetic diversity [44].

Due to the abnormal diversity of circulating virus strains, an effective preventive HCV vaccine may need to induce broadly neutralizing antibodies (bNAb) recognizing one or more of the most conserved epitopes on the heterodimer of HCV E1E2 [45]. Therefore, understanding the epitope specificity of neutralizing antibodies is also crucial for the design of vaccines. Ahsan A et al. studied the amino acids 412–423, 523–532, and 432–443 of envelope protein E2, the three conserved linear epitopes of HCV, with the help of HCVpp, and found that they contributed to the spontaneous clearance of HCV [46]. By constructing the HCVpp model, Brasher NA et al. proposed that polyclonal plasma neutralized HCV mainly by targeting known bNAb epitopes, and the antibody reaction of non-neutralizing domains is independent of HCV neutralization [47]. However, for the preparation of neutralizing antibodies with highly variable epitopes, Mosa AI et al. found that the immune serum from mice vaccinated with HCV bivalent vaccine was more effective in neutralizing HCVpp than any or both monovalent vaccines. These results suggested that neutralizing antibodies against hypervariable pathogens could be widely induced by designing vaccines against conserved residues in hypervariable epitopes [48].

At present, vaccination is the key measure to prevent epidemic encephalitis B and dengue fever, while the research and development of the ZIKV vaccine are still in the pre-clinical or clinical research stage. Neutralizing antibodies are the key tools to evaluate the effectiveness of vaccines, so effective neutralizing antibody detection methods are very important for the quality evaluation of vaccine production.

The E protein and prM protein expressed by pseudotyped virus have immunogenicity, and their conformation is highly similar to that of natural virus protein.

They can be used as neutralizing antigens to screen neutralizing antibodies and then evaluate their effect and active sites. Flavivirus pseudotyped virus that replaces live viruses has been widely used in clinical cases and clinical trials of several viral vaccines. JEV and YFV often construct pseudotyped viruses to evaluate the titer of neutralizing antibodies [25, 27, 63, 64] (Table 17.1). The pseudotyped virus-containing reporter gene is usually mixed with serum, and the neutralizing antibody in it can block the pseudotyped virus-infected cells. The degree of decline of the detected fluorescent signal is positively correlated with the antibody titer.

Compared with the traditional neutralizing antibody detection methods, using a pseudotyped virus system to evaluate the efficacy of vaccines has the advantages of objectivity, accuracy, high speed, and safe. The plaque reduction neutralization test (PRNT) is a special serological test, and it is also the gold standard for detecting and quantifying JEV neutralizing antibodies [25]. However, PRNT requires the use of a live virus, which has the disadvantages of high requirements for laboratory biosafety conditions (BSL-3), requiring a large number of professional operators, time-consuming, and strong subjectivity of results. In 2014, Lee HJ et al. used MLV/Env (coated with E protein and reporter gene β-galactosidase) pseudotyped virus and a simple, rapid, and safe Vero cell neutralization detection system was established [25]. Taking 50% virus reduction as the standard for measuring antibody titer, JEV pseudotyped virus determination was closely related to the results of PRNT, and the use time was shortened by more than twice. The observation of the results is objective and simple (X-gal staining), which saves manpower and material resources, and has strong repeatability and reproducibility.

17.3.3 Screening of Antiviral Drugs

Although direct-acting antiviral agents (DAA) could eradicate HCV, its 2193 genes with epigenetic and transcriptional modifications will still be preserved [49]. To this end, researchers used HCVpp to explore whether new anti-HCV molecules had an impact on the entry of the virus. In a recent study, HCVpp was applied to detect the anti-HCV activity of schisandronic acid derivatives in vitro [50].

As a molecule transformed from Herpesvirus saimiri and overexpressed in human T cells, the potential antiviral effect of IL-26 has attracted much attention. Beaumont É et al. demonstrated that the antiviral effect of IL-26 was achieved by binding viral RNA and inhibiting its replication, rather than interfering with the entry of HCV [51]. Similarly, Zheng X et al. found that hawthorn extracts #106 and #110 inhibited HCV infection in a dose-dependent manner. The constructed HCVpp analysis showed that compounds #106 and #110 could specifically inhibit the replication of HCV RNA but did not inhibit the virus entry or translation process [52]. In addition, Magri A et al. developed a new amido urea derivative to suppress HCV. The HCVpp model proved that the drug did not affect the entry of HCV. By testing the replication-defective replicon, they concluded that the antiviral effect of spermine amido urea derivative 8A was achieved by inhibiting viral translation [53].

With the help of HCVpp, Zhang H et al. found that the flavonoid triazole hybrid 10 m and 10r inhibited the production of HCV by acting at entry step [54]. Li X et al. showed that long-chain fatty acyl-coenzyme A (LCFA-CoA) achieved an antiviral effect by targeting long-chain saturated fatty acids and coenzyme groups. In addition, their study found that LCFA-CoA specifically targeted the steps of virus attachment and binding and also inhibited the intercellular transmission of the virus [55]. By HCVpp model, Hung TC et al. demonstrated that the plant alkaloid berberine played an antiviral role by interfering with HCV envelope glycoproteins E1 and E2 [56]. Similarly, the two compounds harzianoic acids A and B isolated from the sponge-associated fungus *Trichoderma harzianum* proposed by Li B et al. could target HCV envelope protein E1/E2 and CD81 of host cells to inhibit HCV entry [57]. Shahid M et al. demonstrated through the HCV pseudotyped virus model that the engineered microvirin variant LUMS11 with the same domain can effectively inhibit the entry of HCV into host cells by interfering with HCV E1/E2 mediated virus invasion [58].

In addition to identifying some new anti-HCV molecules, Tong Y et al. found that photocatalytic titanium dioxide can attack the viral genome to inactivate HCVpp and effectively inhibit HCV infection [59]. To sum up, there are many kinds of new anti-HCV molecules, and their mechanisms for blocking the virus are also different.

Pseudotyped viruses infect cells in the same process as natural viruses, so they have been widely applied to the screening and evaluation of antiviral drugs. AXL protein is a member of the TAM receptor tyrosine kinase family, and it is also the host factor required by SARS-CoV-2 [60] and ZIKV [61] to enter cells. Pan T et al. found that non-steroidal anti-inflammatory drugs (NSAIDs) could downregulate the expression of AXL on cell surface to inhibit the entry and replication of ZIKV, which was an effective strategy to control ZIKV infection [30]. First, through a high-throughput screening of drugs with Env/HIV-1 pseudotyped virus, ten non-steroidal anti-inflammatory drugs including aspirin, ibuprofen, and naproxen were identified, which can effectively inhibit the entry of pseudotyped virus into cells. The drug was pre-cultured with a pseudotyped virus, and no inhibition of infection was observed. The drug treatment of target cells showed a robust inhibition of viral infection, indicating that the drug interfered with the expression of some factors on the surface of host cells. Subsequently, researchers found that the phosphorylation level of AXL protein decreased significantly at this time. Nowadays, there is a general lack of specific drugs for flavivirus infection. In the future, it is expected to screen antiviral compounds targeting enveloped viruses through known cell receptors or cofactors, with the help of the molecular mechanism of the action of related drugs and the involved signal pathways, to develop more potential treatment strategies.

17.3.4 Research on Antitumor Therapy

The drug resistance of glioblastoma multiforme comes from its cell subgroup glioma stem cells (GSCs). ZIKV has strong neurotropism, which could break through the four barriers of blood-brain, blood-eye, blood-testis, and blood-fetus, leading to the significant death of neural stem cells, inhibiting the development of the fetal brain, and causing congenital microcephaly and Guillain-Barre syndrome. Based on this feature, ZIKV can show oncolytic activity to GSCs, resulting in the loss of self-renewal and proliferation of glioblastoma multiforme [62]. Kretschmer M et al. constructed pseudotyped virus with *p*ME-Z1 to *p*ME-Z4 plasmid and HIV-1 vector pNL-Luc-AM, tested the input efficiency of Z1–Z4 HIV*luc* with Vero-B4 cells, selected the subtype Z1-HIV*luc* with the highest infection efficiency, and infected glioma derived cell lines U87 and 86HG39 with it [32]. Then, the three-plasmid systems including pME-Z1, HIV-1 packaging plasmid psPAX2, and lentivirus vector pLenti-luciferase-P2A-Neo were expanded. The corresponding Z1-LENTI/*luc* pseudotyped virus also infected U87 and 86HG39 cells. ZIKV *pr*ME coated lentivirus pseudotyped virus has certain selectivity to glioma-derived cells and is a potential targeted antitumor tool.

17.4 Summary and Prospect

Compared with natural viruses, the sequences of pseudotyped viruses expressing envelope proteins or capsid proteins are deleted or modified and are usually replaced by reporter genes, so that they maintain the characteristics of entry to host cells, but they cannot produce infectious virus particles. Therefore, they will not pollute the environment and do not need to be operated in BSL-3 or above laboratories. Pseudotyped viruses can also carry specific reporter genes and use chemiluminescence to achieve the objective and accurate quantification of viruses. In recent years, pseudotyped viruses are increasingly used in the process of the virus entering host cells, neutralizing antibody epitopes, neutralizing antibody titer detection, antiviral drug screening, virus gene therapy, visual animal model establishment, and other research. Nowadays, there is still a lack of specific drugs for *Flaviviridae* infection, and vaccines for some virus species have not been widely introduced in clinical applications. Pseudotyped viruses will become a good genetic tool to promote the in-depth study of the interaction mechanism between *Flaviviridae* and the host, as well as the development of related prevention and control methods.

To obtain pseudotyped viruses with a high titer, it is usually necessary to optimize and screen the plasmid gene structure, codon sequence, transfection condition, virus titer, cell concentration, host cell line, and so on. In the HIV packaging system, pseudotyped virus titer can be improved by optimizing the proportion and absolute amount of host cells, transfection reagents, and plasmid DNA. In the VSV packaging system, pseudotyped virus titer is often improved by optimizing harvest time. In the future, it is expected to develop enzymes or reagents that can optimize the titer of flavivirus and find more virus subtypes suitable for pseudotyped virus testing.

At present, *Flaviviridae* infection is still an important public safety issue. This review collected the literatures on *Flaviviridae* pseudotyped virus published in recent years and summarized the optimization strategies and related applications of

Flaviviridae pseudotyped virus model construction. Development of anti-*Flaviviridae* drugs and effective vaccines are needed. We hope that highly effective antiviral drugs targeting entry can be developed in the future, which could provide novel ideas and new directions for the prevention and treatment of *Flaviviridae* infection.

Acknowledgements This work was supported by grants from National Natural Science Foundation of China [81871663 and 82072270] and Academic promotion programme of Shandong First Medical University [2019LJ001].

References

1. Yato, K., et al.: Induction of neutralizing antibodies against hepatitis C virus by a subviral particle-based DNA vaccine. Antivir. Res. **199**, 105266 (2022). https://doi.org/10.1016/j.antiviral.2022.105266
2. Agnetti, J., Desterke, C., Gassama-Diagne, A.: Impact of HCV infection on hepatocyte polarity and plasticity. Pathogens (Basel, Switzerland). **11**, 1 (2022). https://doi.org/10.3390/pathogens11030337
3. Dearborn, A.D., Marcotrigiano, J.: Hepatitis C virus structure: defined by what it is not. Cold Spring Harb. Perspect. Med. **10** (2020). https://doi.org/10.1101/cshperspect.a036822
4. Banda, D.H., et al.: A central hydrophobic E1 region controls the pH range of hepatitis C virus membrane fusion and susceptibility to fusion inhibitors. J. Hepatol. **70**, 1082–1092 (2019). https://doi.org/10.1016/j.jhep.2019.01.033
5. Zheng, F., Li, N., Xu, Y., Zhou, Y., Li, Y.P.: Adaptive mutations promote hepatitis C virus assembly by accelerating core translocation to the endoplasmic reticulum. J. Biol. Chem. **296**, 100018 (2021). https://doi.org/10.1074/jbc.RA120.016010
6. Gruszewska, E., Grytczuk, A., Chrostek, L.: Glycosylation in viral hepatitis. Biochim. Biophys. Acta, Gen. Subj. **1865**, 129997 (2021). https://doi.org/10.1016/j.bbagen.2021.129997
7. Roger, S., Ducancelle, A., Le Guillou-Guillemette, H., Gaudy, C., Lunel, F.: HCV virology and diagnosis. Clin. Res. Hepatol. Gastroenterol. **45**, 101626 (2021). https://doi.org/10.1016/j.clinre.2021.101626
8. Chaudhari, R., Fouda, S., Sainu, A., Pappachan, J.M.: Metabolic complications of hepatitis C virus infection. World J. Gastroenterol. **27**, 1267–1282 (2021). https://doi.org/10.3748/wjg.v27.i13.1267
9. Navas, M.C., et al.: Hepatitis C virus infection and cholangiocarcinoma: an insight into epidemiologic evidences and hypothetical mechanisms of oncogenesis. Am. J. Pathol. **189**, 1122–1132 (2019). https://doi.org/10.1016/j.ajpath.2019.01.018
10. Chambers, T.J., Hahn, C.S., Galler, R., Rice, C.M.: Flavivirus genome organization, expression, and replication. Annu. Rev. Microbiol. **44**, 649–688 (1990)
11. Chu, J.-H.J., Chiang, C.-C.S., Ng, M.-L.: Immunization of flavivirus West Nile recombinant envelope domain III protein induced specific immune response and protection against West Nile virus infection. J. Immunol. **178**, 2699–2705 (2007)
12. Gubler, D.J.: Human arbovirus infections worldwide. Ann. N. Y. Acad. Sci. **951**, 13–24 (2001). https://doi.org/10.1111/j.1749-6632.2001.tb02681.x
13. Li, Q., Liu, Q., Huang, W., Li, X., Wang, Y.: Current status on the development of pseudoviruses for enveloped viruses. Rev. Med. Virol. **28** (2018). https://doi.org/10.1002/rmv.1963

14. Xiang, Q., Li, L., Wu, J., Tian, M., Fu, Y.: Application of pseudovirus system in the development of vaccine, antiviral-drugs, and neutralizing antibodies. Microbiol. Res. **258**, 126993 (2022). https://doi.org/10.1016/j.micres.2022.126993
15. Bartosch, B., Dubuisson, J., Cosset, F.L.: Infectious hepatitis C virus pseudo-particles containing functional E1-E2 envelope protein complexes. J. Exp. Med. **197**, 633–642 (2003). https://doi.org/10.1084/jem.20021756
16. Riva, L., Dubuisson, J.: Similarities and differences between HCV Pseudoparticle (HCVpp) and cell culture HCV (HCVcc) in the study of HCV. Methods in molecular biology (Clifton, N.J.). **1911**, 33–45 (2019). https://doi.org/10.1007/978-1-4939-8976-8_2
17. Palor, M., et al.: Cholesterol sensing by CD81 is important for hepatitis C virus entry. J. Biol. Chem. **295**, 16931–16948 (2020). https://doi.org/10.1074/jbc.RA120.014761
18. Kalemera, M.D., et al.: Optimized cell systems for the investigation of hepatitis C virus E1E2 glycoproteins. J. Gen. Virol. **102** (2021). https://doi.org/10.1099/jgv.0.001512
19. Soares, H.R., et al.: Enhancing hepatitis C virus pseudoparticles infectivity through p7NS2 cellular expression. J. Virol. Methods. **274**, 113714 (2019). https://doi.org/10.1016/j.jviromet.2019.113714
20. Lee, H.J., et al.: The prM-independent packaging of pseudotyped Japanese encephalitis virus. Virol. J. **6**, 115 (2009). https://doi.org/10.1186/1743-422X-6-115
21. Tani, H., et al.: Involvement of ceramide in the propagation of Japanese encephalitis virus. J. Virol. **84**, 2798–2807 (2010). https://doi.org/10.1128/JVI.02499-09
22. Liu, H., et al.: Introducing a cleavable signal peptide enhances the packaging efficiency of lentiviral vectors pseudotyped with Japanese encephalitis virus envelope proteins. Virus Res. **229**, 9–16 (2017). https://doi.org/10.1016/j.virusres.2016.12.007
23. Kambara, H., et al.: Involvement of cyclophilin B in the replication of Japanese encephalitis virus. Virology. **412**, 211–219 (2011). https://doi.org/10.1016/j.virol.2011.01.011
24. Zhu, Y.-Z., et al.: Japanese encephalitis virus enters rat neuroblastoma cells via a pH-dependent, dynamin and caveola-mediated endocytosis pathway. J. Virol. **86**, 13407–13422 (2012). https://doi.org/10.1128/JVI.00903-12
25. Lee, H.J., et al.: Comparison of JEV neutralization assay using pseudotyped JEV with the conventional plaque-reduction neutralization test. J. Microbiol. **52**, 435–440 (2014). https://doi.org/10.1007/s12275-014-3529-y
26. Xu, Q., et al.: E3 ubiquitin ligase Nedd4 promotes Japanese encephalitis virus replication by suppressing autophagy in human neuroblastoma cells. Sci. Rep. **7**, 45375 (2017). https://doi.org/10.1038/srep45375
27. Lee, H.J., et al.: Retention of neutralizing antibodies to Japanese encephalitis vaccine in age groups above fifteen years in Korea. Int. J. Infect. Dis. **100**, 53–58 (2020). https://doi.org/10.1016/j.ijid.2020.08.037
28. Hu, H.P., Hsieh, S.C., King, C.C., Wang, W.K.: Characterization of retrovirus-based reporter viruses pseudotyped with the precursor membrane and envelope glycoproteins of four serotypes of dengue viruses. Virology. **368**, 376–387 (2007). https://doi.org/10.1016/j.virol.2007.06.026
29. Rana, J., Slon Campos, J.L., Poggianella, M., Burrone, O.R.: Dengue virus capsid anchor modulates the efficiency of polyprotein processing and assembly of viral particles. J. Gen. Virol. **100**, 1663–1673 (2019). https://doi.org/10.1099/jgv.0.001346
30. Pan, T., et al.: Nonsteroidal anti-inflammatory drugs potently inhibit the replication of zika viruses by inducing the degradation of AXL. J. Virol. **92** (2018). https://doi.org/10.1128/JVI.01018-18
31. Rana, J., et al.: Role of capsid anchor in the morphogenesis of zika virus. J. Virol. **92** (2018). https://doi.org/10.1128/JVI.01174-18
32. Kretschmer, M., et al.: Zikavirus prME envelope Pseudotyped human immunodeficiency virus Type-1 as a novel tool for glioblastoma-directed virotherapy. Cancer. **12** (2020). https://doi.org/10.3390/cancers12041000
33. Ruiz-Jimenez, F., et al.: Challenges on the development of a pseudotyping assay for zika glycoproteins. J. Med. Microbiol. **70** (2021). https://doi.org/10.1099/jmm.0.001413

34. Huang, J., Yin, P., Zhang, L.: COPII cargo claudin-12 promotes hepatitis C virus entry. J. Viral Hepat. **26**, 308–312 (2019). https://doi.org/10.1111/jvh.13026
35. Gallard, C., et al.: Heparanase-1 is upregulated by hepatitis C virus and favors its replication. J. Hepatol. (2022). https://doi.org/10.1016/j.jhep.2022.01.008
36. Pfaff-Kilgore, J.M., et al.: Sites of vulnerability in HCV E1E2 identified by comprehensive functional screening. Cell Rep. **39**, 110859 (2022). https://doi.org/10.1016/j.celrep.2022.110859
37. Denolly, S., Cosset, F.L., Freitas, N.: Membrane fusion assays for studying entry hepatitis C virus into cells. Methods in molecular biology (Clifton, N.J.). **1911**, 219–234 (2019). https://doi.org/10.1007/978-1-4939-8976-8_15
38. Bailey, J.R., Urbanowicz, R.A., Ball, J.K., Law, M., Foung, S.K.H.: Standardized method for the study of antibody neutralization of HCV Pseudoparticles (HCVpp). Methods in molecular biology (Clifton, N.J.). **1911**, 441–450 (2019). https://doi.org/10.1007/978-1-4939-8976-8_30
39. Masavuli, M.G., et al.: A hepatitis C virus DNA vaccine encoding a secreted, oligomerized form of envelope proteins is highly immunogenic and elicits neutralizing antibodies in vaccinated mice. Front. Immunol. **10**, 1145 (2019). https://doi.org/10.3389/fimmu.2019.01145
40. Dent, M., et al.: Safety and efficacy of Avaren-fc Lectibody targeting HCV high-mannose glycans in a human liver chimeric mouse model. Cell. Mol. Gastroenterol. Hepatol. **11**, 185–198 (2021). https://doi.org/10.1016/j.jcmgh.2020.08.009
41. Center, R.J., et al.: Enhancing the antigenicity and immunogenicity of monomeric forms of hepatitis C virus E2 for use as a preventive vaccine. J. Biol. Chem. **295**, 7179–7192 (2020). https://doi.org/10.1074/jbc.RA120.013015
42. Merat, S.J., et al.: Cross-genotype AR3-specific neutralizing antibodies confer long-term protection in injecting drug users after HCV clearance. J. Hepatol. **71**, 14–24 (2019). https://doi.org/10.1016/j.jhep.2019.02.013
43. Khera, T., et al.: Functional and immunogenic characterization of diverse HCV glycoprotein E2 variants. J. Hepatol. **70**, 593–602 (2019). https://doi.org/10.1016/j.jhep.2018.11.003
44. Salas, J.H., et al.: An antigenically diverse, representative panel of envelope glycoproteins for hepatitis C virus vaccine development. Gastroenterology. **162**, 562–574 (2022). https://doi.org/10.1053/j.gastro.2021.10.005
45. Bonsignori, M., Marcotrigiano, J.: HCV neutralization goes elite. Immunity. **55**, 195–197 (2022). https://doi.org/10.1016/j.immuni.2022.01.010
46. Ahsan, A., et al.: Characterization of linear epitope specificity of antibodies potentially contributing to spontaneous clearance of hepatitis C virus. PLoS One. **16**, e0256816 (2021). https://doi.org/10.1371/journal.pone.0256816
47. Brasher, N.A., et al.: B cell immunodominance in primary hepatitis C virus infection. J. Hepatol. **72**, 670–679 (2020). https://doi.org/10.1016/j.jhep.2019.11.011
48. Mosa, A.I., et al.: A bivalent HCV peptide vaccine elicits pan-genotypic neutralizing antibodies in mice. Vaccine. **38**, 6864–6867 (2020). https://doi.org/10.1016/j.vaccine.2020.08.066
49. Hamdane, N., et al.: HCV-induced epigenetic changes associated with liver cancer risk persist after sustained virologic response. Gastroenterology. **156**, 2313–2329.e2317 (2019). https://doi.org/10.1053/j.gastro.2019.02.038
50. Zhang, K.X., et al.: Synthesis and in vitro anti-HCV and antitumor evaluation of Schisandronic acid derivatives. Medicinal chemistry (Shariqah (United Arab Emirates)). **17**, 974–982 (2021). https://doi.org/10.2174/1573406416999200818150053
51. Beaumont, É., et al.: IL-26 inhibits hepatitis C virus replication in hepatocytes. J. Hepatol. **76**, 822–831 (2022). https://doi.org/10.1016/j.jhep.2021.12.011
52. Zheng, X., et al.: Identification of natural compounds extracted from crude drugs as novel inhibitors of hepatitis C virus. Biochem. Biophys. Res. Commun. **567**, 1–8 (2021). https://doi.org/10.1016/j.bbrc.2021.06.022
53. Magri, A., Mokrane, O., Lauder, K., Patel, A.H., Castagnolo, D.: Synthesis, biological evaluation and mode of action studies of novel amidinourea inhibitors of hepatitis C virus (HCV). Bioorg. Med. Chem. Lett. **29**, 724–728 (2019). https://doi.org/10.1016/j.bmcl.2019.01.008

54. Zhang, H., et al.: Flavonoid-triazolyl hybrids as potential anti-hepatitis C virus agents: synthesis and biological evaluation. Eur. J. Med. Chem. **218**, 113395 (2021). https://doi.org/10.1016/j.ejmech.2021.113395
55. Li, X., et al.: Long-chain fatty acyl-coenzyme a suppresses hepatitis C virus infection by targeting virion-bound lipoproteins. Antivir. Res. **177**, 104734 (2020). https://doi.org/10.1016/j.antiviral.2020.104734
56. Hung, T.C., et al.: Berberine inhibits hepatitis C virus entry by targeting the viral E2 glycoprotein. Phytomedicine : international journal of phytotherapy and phytopharmacology. **53**, 62–69 (2019). https://doi.org/10.1016/j.phymed.2018.09.025
57. Li, B., et al.: Harzianoic acids a and B, new natural scaffolds with inhibitory effects against hepatitis C virus. Bioorg. Med. Chem. **27**, 560–567 (2019). https://doi.org/10.1016/j.bmc.2018.12.038
58. Shahid, M., et al.: An engineered Microvirin variant with identical structural domains potently inhibits human immunodeficiency virus and hepatitis C virus cellular entry. Viruses. **12** (2020). https://doi.org/10.3390/v12020199
59. Tong, Y., et al.: Photo-catalyzed TiO(2) inactivates pathogenic viruses by attacking viral genome. Chemical engineering journal (Lausanne, Switzerland : 1996). **414**, 128788 (2021). https://doi.org/10.1016/j.cej.2021.128788
60. Wang, S., et al.: AXL is a candidate receptor for SARS-CoV-2 that promotes infection of pulmonary and bronchial epithelial cells. Cell Res. **31**, 126–140 (2021). https://doi.org/10.1038/s41422-020-00460-y
61. Nowakowski, T.J., et al.: Expression analysis highlights AXL as a candidate zika virus entry receptor in neural stem cells. Cell Stem Cell. **18**, 591–596 (2016). https://doi.org/10.1016/j.stem.2016.03.012
62. Zhu, Z., et al.: Zika virus has oncolytic activity against glioblastoma stem cells. J. Exp. Med. **214**, 3145 (2017). https://doi.org/10.1084/jem.20171093
63. Bellier, B., et al.: DNA vaccines expressing retrovirus-like particles are efficient immunogens to induce neutralizing antibodies. Vaccine. **27**, 5772–5780 (2009). https://doi.org/10.1016/j.vaccine.2009.07.059
64. Mercier-Delarue, S., et al.: Screening test for neutralizing antibodies against yellow fever virus, based on a flavivirus pseudotype. PLoS One. **12**, e0177882 (2017). https://doi.org/10.1371/journal.pone.0177882

Chapter 18
Replicating-Competent VSV-Vectored Pseudotyped Viruses

Fei Yuan and Aihua Zheng

Abstract Vesicular stomatitis virus (VSV) is prototype virus in the family of *Rhabdoviridae*. Reverse genetic platform has enabled the genetic manipulation of VSV as a powerful live viral vector. Replicating-competent VSV is constructed by replacing the original VSV glycoprotein gene with heterologous envelope genes. The resulting recombinant viruses are able to replicate in permissive cells and incorporate the foreign envelope proteins on the surface of the viral particle without changing the bullet-shape morphology. Correspondingly, the cell tropism of replicating-competent VSV is determined by the foreign envelope proteins. Replicating-competent VSVs have been successfully used for selecting critical viral receptors or host factors, screening mutants that escape therapeutic antibodies, and developing VSV-based live viral vaccines.

Keywords Replicating-competent · Receptor screening · Escape mutation screening · Vaccine development

Abbreviations

ACE2	Angiotensin-converting enzyme 2
BSL	Biosafety level
BVDV	Bovine viral diarrhea virus
CCR5	C-C Motif Chemokine Receptor 5
CMV	Cytomegalovirus
CPE	Cytopathic effect
CXCR4	C-X-C Motif Chemokine Receptor 4
EBOV	Ebola virus
ELISA	Enzyme-linked immunosorbent assay

F. Yuan · A. Zheng (✉)
State Key Laboratory of Integrated Management of Pest Insects and Rodents, Institute of Zoology, Chinese Academy of Sciences, Beijing, China
e-mail: zhengaihua@ioz.ac.cn

Y. Wang (ed.), *Pseudotyped Viruses*, Advances in Experimental Medicine and Biology 1407, https://doi.org/10.1007/978-981-99-0113-5_18

Env	Envelope
FDA	Food and Drug Administration
G	Glycoprotein
GFP	Green fluorescent protein
GP	Glycoprotein
HBsAg	Hepatitis B surface antigen
HCMV	Human cytomegalovirus
HIV	Human immunodeficiency virus
HOPS	Homotypic fusion and vacuole protein sorting
I.p.	Intraperitoneal
L	Large protein
LASV	Lassa virus
M	Matrix protein
mAbs	Monoclonal antibodies
MARV	Marburg virus
N	Nucleocapsid
nAbs	Neutralizing antibodies
NPC1	Niemann-Pick type C1
P	Phosphoprotein
PFU	Plaque-forming units
RBD	Receptor-binding domain
RSV	Respiratory syncytial virus
S	Spike
SHIV	Simian-human immunodeficiency virus
VSV	Vesicular stomatitis virus
ZEBOV	Zaire ebolavirus

18.1 Construction of Replicating-Competent VSV Viruses

Vesicular stomatitis virus (VSV) is a single, negative-stranded RNA virus that belongs to the *Rhabdoviridae* family. The 11Kb VSV genome is relatively simple, comprising five genes in the order N-P-M-G-L. These genes are transcribed to generate five subgenomic mRNAs encoding five structural proteins: the nucleocapsid (N) protein, the phosphoprotein (P), the matrix (M) protein, the glycoprotein (G), and large protein (L). The P and L proteins form the RNA-dependent RNA polymerase, which is responsible for genome transcription and replication. As the distance between the gene and the promoter increases, the transcription attenuation occurs following the gene order (N > P > M > G > L) [1].

18.1.1 History of Replicating-Competent VSV Viruses

In 1995, two groups reported recovery of infectious VSV particles from plasmid DNA using reverse genetic system [2, 3]. This has enabled further genetic engineering of VSV vector. VSV is able to tolerate large inserts up to 4.5Kb size and express multiple foreign genes from the extra transcription units [4]. As an expression vector, recombinant VSVs have expressed many soluble and transmembrane proteins at high levels [4–6]. Most viral glycoproteins could incorporate into the surface of the recombinant virions without any sequence change [4, 7, 8]. However, some envelope proteins have to modify the cytoplasmic domain for them to incorporate into the VSV particle, e.g., HIV Env protein [9, 10]. Incorporation of heterologous proteins does not alter the bullet morphology of recombinant VSV particles as shown by electron microscopy pictures [4, 6, 7].

In many early studies, additional heterologous envelope genes were introduced into the full length of VSV genome. The resulting recombinant viruses thus express and exhibit at least two envelope proteins on their surface, including VSV glycoprotein. As we all know, the envelope proteins determine the cellular and tissue tropism of the enveloped viruses. VSV glycoprotein shows a broad host range and it mediates the infection of VSV in nearly all kinds of cell lines. The existence of VSV G protein would interfere the investigation of entry features of the target viruses, thus greatly limiting the extensive application of VSV vector. To solve the problem, VSV△G vector was generated, with the G coding sequence being deleted from the genome. The foreign envelope genes were usually constructed in the place between M and L genes of VSV, where the VSV G located [11–13]. The resulting recombinant virus could be rescued by complementing VSV G protein in trans and propagated in baby hamster kidney cells (BHK-21) expressing G. In order to study the entry mediated by the envelope proteins of interest, the recombinant virus was subjected on non-complementing BHK-21 cells for one round infection. This step caused 1000-fold of viral titer decrease, and there were still few traces of G proteins on the VSV's surface. By this means, researchers have discovered that the VSV△G-CC4, which lacked VSV G but express HIV receptor CD4 and coreceptor CXCR4 instead, could specifically infect, propagate, and kill a human T cell line that expresses HIV envelope proteins on the surface [12]. On the other hand, another study described a recombinant VSV△G expressing GFP and the hybrid HIV envelope protein (extracellular and transmembrane domains of gp160 and cytoplasmic domain of VSV G) [13]. This virus, named as VSV△G-gp160G-GFP, specifically infected Hela cells expressing HIV receptor CD4 and coreceptor CXCR4. The infection could be blocked by antibodies targeting CD4 and CXCR4, as well as CXCR4 ligand. In addition, this study describes propagation of VSV△G-gp160G-GFP in Hela-CD4 cells, although the titer was approximately 10–100-fold lower than the titer of viruses propagated in BHK-21 cells.

In 2004, a Canada group reported rescue of the replicating-competent VSVs expressing glycoproteins of Marburg virus, Ebola virus, and Lassa virus with or without VSV G in the genome [14]. To the best of our knowledge, it is the first time to use non-BHK-21 cells for amplification of the recombinant VSV. Instead, the rescued viruses were blind passaged on VeroE6 cells, which were not only susceptible to these viruses but also able to produce high titers of viral stocks. Notably, the growth was attenuated and the maximum viral titers were reduced when VSV G was replaced by the foreign transmembrane proteins. This indicates that the package and budding efficiency of VSV∆G vector is not as optimal when G is present in the viruses. The viruses, which expressed the envelope genes of EBOV, LASV, or MARV in place of VSV G, could not replicate in Jurkat cells, indicating altered cell tropism.

Taking together, VSV lacking G could be used as a powerful surrogate virus tool to study the entry pathways of a variety of enveloped viruses and developed as a promising vaccine vehicle. The replication ability of recombinant VSV∆G in proper cell lines provides new insights into the future applications of VSV∆G vector.

18.1.2 Rescue Method of Replicating-Competent VSV Viruses

Reverse genetics system was developed to recover the negative-sense RNA viruses from cDNA. Briefly, cells are transfected with a cDNA plasmid encoding the positive-sense VSV RNA together with three helper or supporting plasmids encoding nucleocapsid (N), phosphoprotein (P), and polymerase (L). All these plasmids are driven by the T7 promoter. With the presence of bacteriophage T7 RNA polymerase, the mRNAs of three viral proteins and the full-length antigenome of VSV are generated. Once the proteins are being translated, the antigenome of VSV is encapsidated by the VSV N protein and form the nucleocapsid with the polymerase complex composed of P and L proteins. The antigenomic VSV will then be used as a template to yield the negative-sense, genomic RNA, thus initiating the downstream mRNA synthesis as well as RNA replication.

The VSV full-length plasmid contains all five genes of the positive-sense VSV antigenome (N, P, M, G, and L) in order, which are flanked by T7 promoter, leader sequence, trailer sequence, HDV (hepatitis delta virus) ribozyme, and T7 terminator (Fig. 18.1). This plasmid is used for recovery of the wild-type VSV virus [3]. The use of VSV antigenome could avoid potential hybridization of genomic RNA with mRNAs from the helper plasmids. Since the VSV plasmid and three helper plasmids are under the control of T7 promoter, a source of T7 RNA polymerase is required in

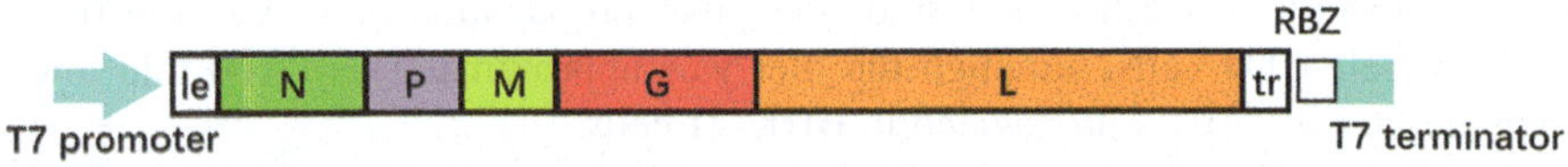

Fig. 18.1 Antigenome structure of VSV vector. le, leader sequence; N, nucleocapsid gene; P, phosphoprotein gene; M, matrix protein gene; G, glycoprotein gene; L, polymerase gene; tr, trailer sequence; RBZ, hepatitis delta virus ribozyme

this reverse genetics system. Initially, a recombinant vaccinia virus that expresses T7 RNA polymerase (vTF7–3) was used to infect the BHK-21 cells before transfection. In the following step, to remove the vaccinia viruses, the supernatants from the transfected cells have to be filtered on the 0.22 μm filters or supplemented with vaccinia virus inhibitor such as araC (1-β-D-arabinofuranosylcytosine) [2, 3]. Later, recombinant VSV is successfully recovered from a vaccinia virus-free system using BSR-T7/5 cells, which are derived from BHK-21 cells and constitutively express T7 RNA polymerase [15]. Recently, most researchers express T7 polymerase through co-transfection an expression plasmid under the control of CMV (cytomegalovirus) promoter [16].

To generate replicating-competent VSV-vectored pseudotyped viruses, the G coding sequence has to be deleted from VSV full-length plasmid and the foreign gene has to be inserted into the plasmid. Although theoretically, the foreign genes could be inserted into any position of the VSV genome, they are mostly constructed in the place of G. In order to facilitate the gene insertion, two restriction sites (MluI/NotI) are introduced between matrix (M) and polymerase (L) gene. Thus, when the foreign gene is constructed into the VSV vector through the two sites, the VSV glycoprotein gene is simultaneously deleted. Five supporting plasmids include one encoding T7 polymerase and four encoding N, P, G, L of VSV respectively,as shown in Fig. 18.2. Supplementing G protein at the transfection step will help the recovery of the recombinant viruses that lacking G in the genome. Once the virus particles have been generated, the subsequent passage in the permissive cells would no longer need the presence of G protein (Fig. 18.2).

The detailed rescue method is as follows.

1. In a certified BSL 2 lab, seed 8 x 10^5 BHK-21 cells in a 6-cm dish.
2. The second day, BHK-21 cells should grow to approximately 70–80% confluence at the time of transfection.
3. The VSV backbone plasmid and five supporting plasmids were co-transfected into BHK-21 cells using Lipofectamine 3000 (Invitrogen) according to the manufacturer's instruction or calcium phosphate transfection method. Here, we introduce the Lipofectamine transfection method.

   (a) In tube 1, mix VSV plasmid and T7 polymerase, N, P, G, L expression plasmids at a ratio of 10:50:8:4:1:1 with opti-MEM (Invitrogen). Mix well and incubate at room temperature for 5 min.
   (b) In tube 2, mix Lipofectamine with opti-MEM. Mix well and incubate at room temperature for 5 min.
   (c) Mix tube 1 and 2. Incubate at room temperature for 15 min.
   (d) Remove media from the 6-well dish. Add the DNA/Lipofectamine mixtures dropwise to the cell monolayer in the dish.
   (e) Put the dish in a 37 °C, 5% CO_2 incubator, and incubate for 48 hours.

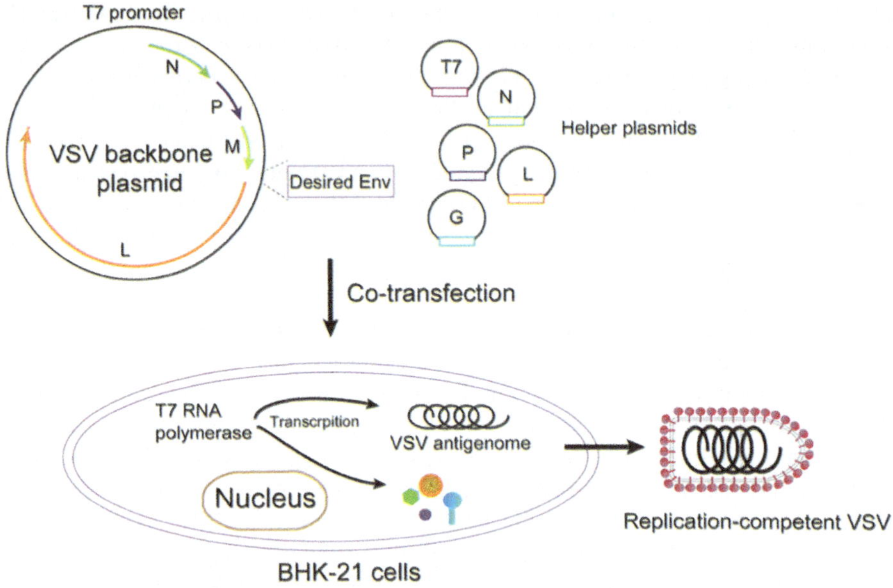

Fig. 18.2 Diagram of replicating-competent VSV recovery system

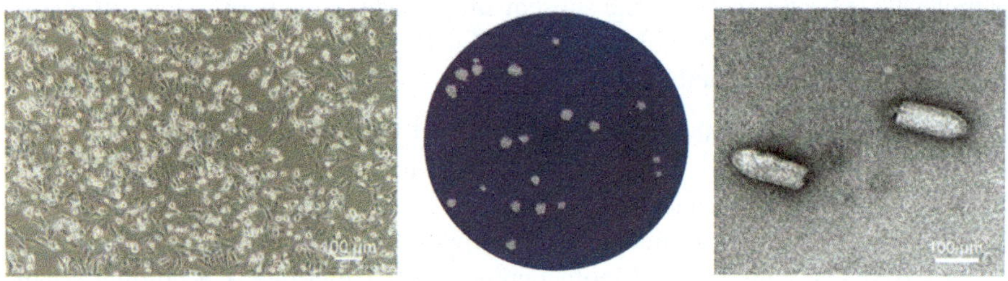

Fig. 18.3 Typical VSV-induced cytopathic effect, plaques, and bullet-shaped viral particles after negative staining under electron microscope

4. Forty-eight hours after transfection, collect the supernatants and blind passage onto fresh permissive cells (80–90% confluent at the time of infection).
5. Cells are incubated in a 37 °C, 5% CO_2 incubator for 48–72 hours. Successful recovery of replication-competent VSV is confirmed by VSV-induced CPE, which are seen as rounded and detached cells (Fig. 18.2).
6. Passage the rescued virus on permissive cells at MOI of 0.01 for several times to reach the desired virus titer. Virus titer could be determined by plaque forming assay, TCID 50 assay, or immunofluorescence assay. The VSVs could be visualized by electron microscopy after negative staining (Fig. 18.3).

18.2 Application of Replicating-Competent VSV

18.2.1 Screening of Viral Host Factors/Receptors

Viral receptors are the first cellular barrier for virus infection. Commonly, the enveloped viruses bind to the cell surface through the attachment factors and then internalize either by direct fusion with the membrane or via endocytosis. For some viruses, single receptor is able to mediate both binding and fusion, e.g., ACE2 for SARS-CoV [17]. For other viruses, multiple receptors are required for viral entry, e.g., CD4 and CCR5/CXCR4 for HIV-1 [18]. Despite the different entry pathways, searching for functional entry factors is very important for understanding the virus-host interaction and providing the most effective therapeutic target.

Although the traditional biochemical approaches such as co-immunoprecipitation and virus overlay protein blot assay have successfully identified several virus receptors, more and more researchers start to apply genetic screen strategy for receptor hunting [19]. There are two types of genetic screens: gain-of-function screen and loss-of-function screen. These screens can be performed on genome-wide scale in a high-throughput manner, which enables more effective screen.

The first application of recombinant VSV in receptor screening successfully identified Niemann-Pick C1 (NPC1) as the cellular receptor for Ebola virus [20]. Ebola virus is the member of *Filoviridae* family and causes a highly fatal haemorrhagic fever after infection. It is classified as category A bioterrorism agents, and management of it requires biosafety level (BSL) 4 containment. Therefore, a replicating-competent VSV harboring the Ebola virus glycoprotein (GP), designated as rVSV-GP-EboV, was used as a selection agent to investigate Ebola virus entry [21]. The recombinant VSV has the following advantages. First, it incorporates the foreign envelope proteins on its surface instead of the naïve glycoprotein, thus acquiring the entry features of the viruses of interest. Second, the cytolytic effects of VSV ensure killing all the infected cells. More importantly, since VSV can be handled in BSL 2 facilities, it serves as a surrogate to study the entry mechanisms of highly pathogenic viral pathogens.

A haploid cell line, named HAP1, was used to derive the mutated cell library. Unlike its predecessor KBM-7 cells, HAP1 is susceptible to a wider range of viruses including Ebola virus [20, 22]. To generate random mutagenesis, HAP1 cells were transduced by a retroviral gene-trap vector [23], which contains a splice acceptor site. The integration of the vector into the genome leads to truncated mRNA transcripts, thus resulting in ablation of gene expression. The library was then exposed to rVSV-GP-EboV, and cells resistant to killing were selected and expanded. This was followed by mapping the sites of mutation by parallel deep sequencing and determining the genes with enriched mutations in resistant population compared to unselected population (Fig. 18.4). The screen discovered six members of HOPS (homotypic fusion and vacuole protein-sorting) complex and NPC1, as critical host factors for filovirus entry.

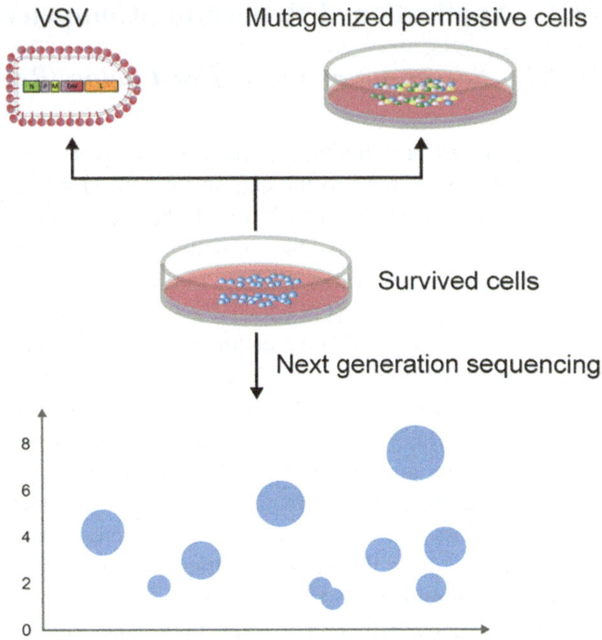

Fig. 18.4 Schematic diagram of receptor screening process using replicating-competent VSV as selection agent

HOPS complex is known to mediate the fusion of endosome to lysosome. It comprises six subunits: VPS11, VPS16, VPS18, VPS33A, VPS39, and VPS41. Cells with knockout mutations in VPS11 and VPS33A are resistant to rVSV-GP-EboV infection, however maintain the susceptibility to other unrelated viruses.

The biggest hit of the screen is NPC1, which encodes an endo/lysosomal cholesterol transporter. Loss of NPC1 expression in a variety of cell lines and primary fibroblasts from Niemann-Pick disease patients confers resistance to infection by Ebola virus and Marburg virus, and the susceptibility was restored by ectopic expression of NPC1. The viruses were arrested within vesicular compartments of the NPC1-deficient cells, implicating the crucial role of NPC1 in viral membrane fusion and virus release to the cytosol. In addition, heterozygous NPC1 knockout mice showed resistance to lethal challenge of Ebola virus and Marburg virus.

Combination of replicating-competent VSV and genome-wide knock out screen is a reliable and effective strategy to uncover essential host factors/receptors for virus entry and infection. Till now, the strategy has been applied for six viruses, four of which are BSL 4 pathogens (Table 18.1). This highlights the advantages of recombinant VSV as a valuable selection tool for the study of the entry mechanisms of viruses that are either difficult to culture in vitro or requiring high-level containment.

Table 18.1 Application of replicating-competent VSV in screening receptors/host factors

Pathogen	Family	BSL level	Host factors/ receptors	Strategy	Selection agent	Ref
Ebola virus (EBOV)	*Filoviridae*	4	NPC1, HOPS complex	Haploid screen	rVSV-GP-EboV	[20]
Lassa virus (LASV)	*Arenaviridae*	4	LAMP1, ST3GAL4, DAG1	Haploid screen	rVSV-GP-LASV	[22]
Andes virus (ANDV)	*Bunyaviridae*	4	SREBP-2, SCAP, MBTPS1P, MBTPS2P	Haploid screen	rVSV-ANDV	[24, 25]
Lujo virus (LUJV)	*Arenaviridae*	4	NRP2, CD63	Haploid screen	VSV-LUJV	[26]
Sin Nombre virus (SNV) Andes virus (ANDV)	*Bunyaviridae*	4	Protocadherin-1	Haploid screen	rVSV-ANDV Gn/Gc	[27]
Lymphocytic choriomeningitis virus (LCMV)	*Arenaviridae*	2/3	Heparan sulfate	CRISPR-Cas9 knockout screen	VSV-GP (LCMV GP variants)	[28]
SARS-CoV-2	*Coronaviridae*	3	ACE2, SMARCA4	CRISPR-Cas9 knockout screen	rcVSV-SARS-CoV-2-S	[29]

18.2.2 Screening of Mutations that Escape Therapeutic mAbs

Antibody therapies are considered as a promising approach to prevent and treat infectious viral diseases. Over 40 therapeutic mAbs are currently approved by the Food and Drug Administration (FDA) or in clinical trials, targeting viral pathogens such as HIV, Ebola virus, Influenza virus, HCMV, and RSV. Most recently, the FDA has authorized several antibody cocktails for pre- and pro-prophylaxis against COVID-19. However, the emergence of the escape mutants due to rapid mutation rate poses great threat to the efficacy of these therapeutic antibodies.

Replicating-competent VSV is a powerful tool to screen the escape mutants under the selective pressure of a single antibody or antibody cocktails. Unlike yeast screening strategy, which solely depends on the binding property between antibodies and viral envelope proteins or receptor, replicative VSV screening is based on the neutralizing activities of antibodies. When the antibody is unable to completely inhibit virus infection at certain concentration, the escape mutations occur, accumulate, and then become dominant. Therefore, to develop ideal antibodies for therapeutic use, we need to screen mAbs not prone to generate escape mutations.

Soon after the COVID-19 outbreak in early 2020, Regeneron Pharmaceuticals Inc. generated a collection of human neutralizing antibodies (nAbs) against SARS-CoV-2 spike protein from humanized mice and convalescent patients [30]. The top four most potent nAbs (REGN10987, REGN10989, REGN10933, and REGN10934) which recognize the receptor-binding domain (RBD) of SARS-CoV-2 and specifically block RBD interaction with ACE2 were chosen for escape mutants screening [31]. These nAbs also showed high neutralizing potency against pseudotype viruses expressing 16 circulating S variants. To screen for escape mutants, the replicating-competent VSV harboring full-length SARS-CoV-2 spike in place of original glycoprotein (VSV-SARS-CoV-2-S) was used. VSV-SARS-CoV-2-S (1.5×10^6 pfu) was pre-incubated with fivefold serial diluted individual antibody or antibody combinations at room temperature for 30 min before applied on Vero E6 monolayer in 12-well plate. The selection of each passage lasted for 4 days at 37 °C in order to better mimic the infection of the authentic viruses. Then the supernatants in the well containing the highest concentration of nAbs with VSV induced cytopathic effect (CPE, >20%) were harvested. For the second round screen, the passage 1 (P1) supernatants were added to fresh Vero E6 cells in the presence of the same dilutions of nAbs and incubated for another 4 days. Cells in the wells with the highest antibody concentration showing detectable CPE from both passages were used to extract RNA for deep sequencing (Fig. 18.5).

After the first passage, multiple mutations emerged under the pressure of 4 tested single antibodies, a few of which became dominant (>90%) by the second passage. Accordingly, the survived mutants were resistant to these Abs at a concentration of 10 μg/ml or 50 μg/ml. On the other hand, two antibody combinations (REGN10989 + REGN10987, REGN10987 + REGN10933) maintained the neutralization potency against VSV-SARS-CoV-2-S by the second passage. Meanwhile, the selection of the two cocktails did not result in any mutation in the S encoding sequences. In contrast, one combination (REGN10989 + REGN10934) rapidly selected the escape mutants, which lose the ability to neutralize VSV-SARS-CoV-2-S, similar to the selection by individual antibodies. Glu at 484 is key residue for the neutralizing activities of these two antibodies, and single mutation of E484 was enough to resist neutralization. The results indicate that if the two antibodies recognize distinct or partially overlapping regions of the RBD, the cocktails could prevent rapid generation of escape mutants. If the binding epitopes of the two antibodies were completely competing, the virus could readily escape neutralization.

18.2.3 Vaccine Development

Viral vectored vaccines have shown great promise since they generate robust protective immunity. The first viral vector used to develop vaccine is vaccinia virus. It was designed to express hepatitis B surface antigen (HBsAg) [32] and showed protective efficacy in non-human primary animals [33]. Since then, more and more viral vectors were generated as vaccine carrier. The commonly used vaccine vectors include adenovirus, poxvirus, rhabdovirus, alphavirus, and many vaccines based on these vectors are currently in human clinical trials [34]. To date, VSV vectored Ebola Zaire Vaccine (ERVEBO) is the first FDA-approved viral

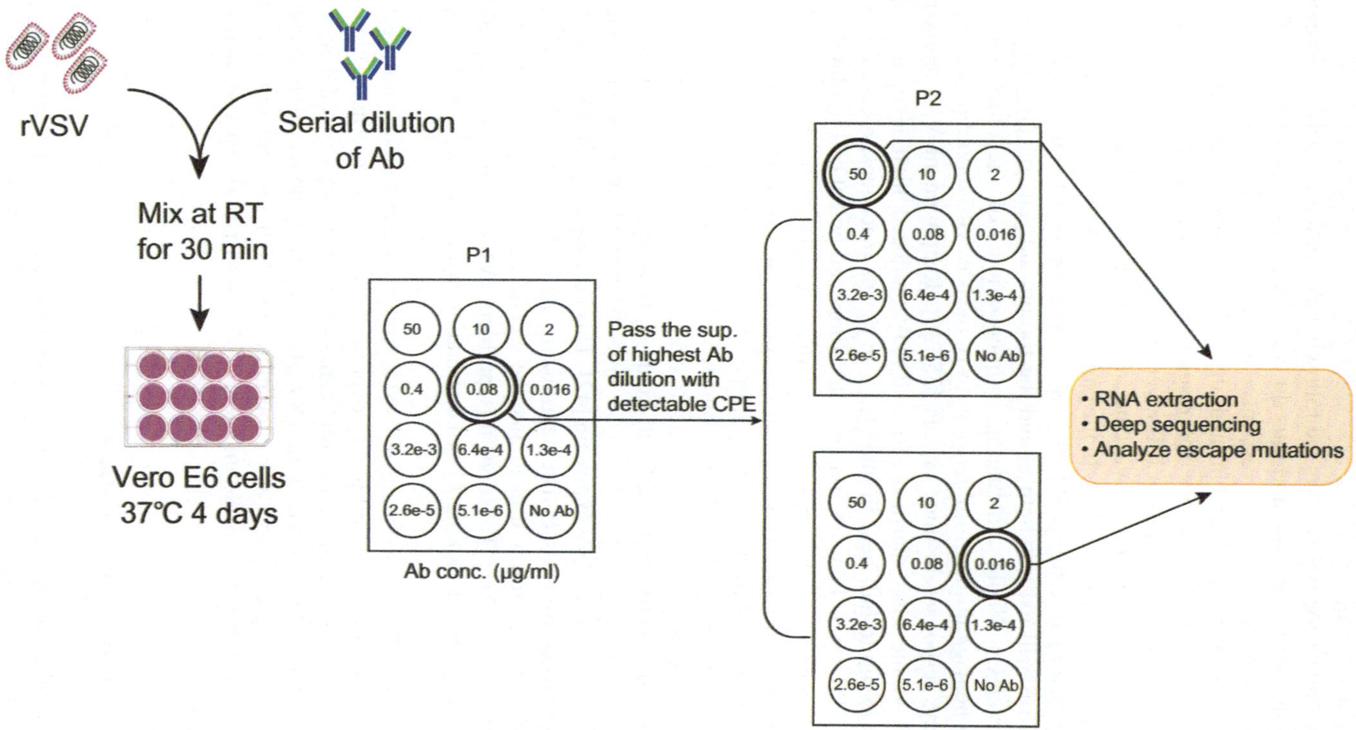

Fig. 18.5 The scheme of screening escape mutants from neutralizing antibodies using replicating-competent VSV

vectored vaccine for human use. Therefore, VSV is considered to be one of the most prospective live vectors.

As a vaccine vector, VSV offers the following advantages. 1. The seroprevalence of VSV in human is very low and thus do not have preexisting immunity. 2. VSV induces strong humoral immune responses as well as cellular immunity in animal models. 3. It can be administered intranasally to induce mucosal immunity, which is suitable for developing vaccines for respiratory pathogens. 4. VSV will never integrate into the host genome, so it is a very safe viral vector.

Based on VSV vector, a lot of vaccine candidates have been developed and were highly effective in animal models. In the early stage, VSV-based vaccine candidates were developed by inserting additional envelope genes in the genome of VSV. The resulting virus will express both G and the foreign antigens on the surface of the virus particles. Single intranasal immunization of recombinant VSVs expressing hemagglutinin of influenza virus or measles virus could generate protective immune responses in mice or rats [35, 36]. Recombinant VSV carrying E2 protein of bovine viral diarrhea virus (BVDV) induced high level neutralizing antibodies for up to half year in mice [37]. In rhesus macaques, recombinant VSV harboring HIV gag and env as immunogens stimulated potent cell immunity, which effectively controlled HIV replication fourteen months after challenge [38]. In addition, VSV expressing HIV env protein could induce specific antibodies with broad-spectrum neutralization activity [39].

The drawback of this type of recombinant VSV vaccines is the existence of VSV glycoprotein, the key determinant of neurovirulence [40, 41]. Since VSV G protein mediate the virus entry into nearly all kinds of cells, the recombinant VSVs thus exhibit very wide cell tropism. In VSV infected animals and humans, the most antibodies were against VSV G protein [42]. This become an obstacle if prime-boost vaccination procedure is used. The neutralizing antibodies against VSV G induced by the first immunization will neutralize the second dose, thus greatly affecting the efficacy of the vaccines. To solve this problem, VSV vector carrying different serotype G proteins or lacking G or combination with other type of immunogens were used.

VSV lacking G has been utilized as a replication-competent viral vector for vaccine development. The abilities to express desired immunogens at high levels and grow to high titers make the replication-competent VSVs excellent vaccine candidates. Many vaccines based on VSVΔG vector have demonstrated safety and immunogenicity in pre-clinical as well as clinical trials. In the following section, we will mainly focus on the development of VSV-based EBOV vaccine and SARS-CoV-2 vaccine.

18.2.3.1 VSV-Based EBOV Vaccine

The replication-competent VSV that harbors the GP gene of Zaire EBOV in replacement of VSV G gene was designated as VSVΔG/ZEBOVGP and first reported in 2004 [14]. Two doses of intraperitoneal injection of VSVΔG/ZEBOVGP did not cause any clinical symptoms and offered complete protection against lethal dose mouse-adapted Zaire ebolavirus challenge in mice. In contrast, gamma-inactivated VSV∆G/ZEBOVGP virus failed to provide protection, suggesting that replication ability of VSV∆G/ZEBOVGP is important in inducing protective immunity. The efficacy of VSV∆G/ZEBOVGP was subsequently evaluated in non-human primates [43]. Single intramuscular injection of 10^7 plaque-forming units (PFU) VSV∆G/ZEBOVGP fully protected cynomolgus macaques from lethal ZEBOV challenge (10^3 PFU) 28 days after vaccination. Despite transient viremia after vaccination, VSV∆G/ZEBOVGP did not cause any diseases in the monkeys. Likewise, VSV∆G/ZEBOVGP could confer protection against aerosol challenge of ZEBOV in cynomolgus macaques [44].

Further studies investigated the efficacy of VSV∆G/ZEBOVGP using different immunization dosage and routes in mouse models [45]. One dose of as low as 2 PFU VSV∆G/ZEBOVGP injection induced protective immune responses similar to 2 x 10^4 PFU. The mice vaccinated via intranasal, intraperitoneal, or intramuscular routes were fully protected from lethal virus challenge up to 10^6 LD_{50} of mouse adapted ZEBOV. Even two-dose oral immunization showed protective efficacy. Similarly, in cynomolgus macaque model, intranasal, oral, or intramuscular immunization of $2x10^7$ PFU VSV∆G/ZEBOVGP protected the animals from systemic ZEBOV challenge [46]. And mucosal routes stimulated more potent immune responses than intramuscular injection.

The time for VSV∆G/ZEBOVGP to develop protective immunity was also investigated. In one study, the minimum time for the mice to develop immunity was seven days after i.p. vaccination of 2 x 10^4 PFU VSV∆G/ZEBOVGP [45]. In another study, i.p. injection of 2 x 10^5 PFU VSV∆G/ZEBOVGP 24 hours prior challenge also conferred 100% protection [47]. The shortest time for Syrian hamsters to elicit protective immunity was three days after i.p. immunization of 10^5 PFU VSV∆G/ZEBOVGP [48]. On the other hand, the mice were completely protected up to 9 months after vaccination [45, 49]. Likewise, challenge of guinea pigs with lethal dose of guinea pig-adapted EBOV provided 100% protection as long as 18 months after vaccination [49]. These studies indicate that VSV∆G/ZEBOVGP not only induces rapid immunity but confers long-term protection.

Other than successful pre-exposure protection in animal models, VSV∆G/ZEBOVGP also showed great potential in post-exposure treatment against Ebola virus infection. In mice, 30 min or 24 hours post exposure of 10^3 LD_{50} mouse-adapted ZEBOV, i.p. injection of 2 x 10^5 PFU VSV∆G/ZEBOVGP resulted in 100% survival of the animals, with mild clinical symptoms [47]. In guinea pigs, immunization of VSV∆G/ZEBOVGP 1 or 24 hours after lethal ZEBOV challenge resulted in 83% and 50% survival, respectively, with weight loss to different extent [47]. All hamsters were fully protected when treated 1 day after ZEBOV infection [48]. Notably in rhesus macaques, only 50% of the animals survived if treated 20–30 min after ZEBOV challenge [47].

Although wild-type VSV infection in humans only causes mild symptoms and mostly self-limiting, safety is still a significant concern for VSV-based vaccine usage in immune-compromised individuals. To address this concern, the safety of

VSVΔG/ZEBOVGP was evaluated in mice and non-human primates with compromised immune systems. Immunization of NOD-SCID mice with 2 x 10^5 PFU VSVΔG/ZEBOVGP did not cause any sign of illness, such as ruffling, hunching, or lack of mobility [45]. Similarly, simian-human immunodeficiency virus (SHIV) infected rhesus macaques, which had impaired $CD4^+$ T cells, $CD8^+$ T cells, and memory $CD4^+$ T cells, showed no signs of vaccine-associated illness after immunization of 10^7 PFU VSVΔG/ZEBOVGP [50]. Transient viremia was observed on day 2 after immunization. Notably, VSVΔG/ZEBOVGP provided protection in four of six SHIV-infected monkeys against lethal EBOV challenge, suggesting VSVΔG/ZEBOVGP was still efficacious in immune-compromised populations.

Soon after the VSV-based Ebola vaccine showed the protective efficacy in multiple animal models, the first double-blind phase 1 clinical trial has been initiated in two locations of Maryland, USA. A total of 78 participants received single or two doses of 3 x 10^6, 2 x 10^7, or 10^8 PFU Ebola virus vaccine candidate (now called rVSV-ZEBOV) or placebo [51]. The second dose was administered at day 28 using the same amount of vaccines. No serious adverse events were reported in this study. The common adverse events included injection site pain, fatigue, myalgia, headache, fever, and chills, which were mild to moderate and self-limiting. The seroconversion occurred in 100% vaccinated participants by day 28 as determined by ELISA. Single dose of rVSV-ZEBOV induced neutralizing antibody responses in all the vaccinees by day 28, in a dose-dependent manner. The second dose significantly increased the geometric mean titers of neutralizing antibodies from day 28 to day 56. However, there were no significant differences in the neutralizing antibodies titers between one-dose and two-dose strategy at day 180. In another phase 1 trial, a total of 158 participants from Europe and Africa were injected intramuscularly once with rVSV-ZEBOV at doses ranging from 3 x 10^5 to 2 x 10^7 PFU or placebo to assess the safety and immunogenicity of the vaccine [52]. Although mild-to-moderate reactogenicity was frequent, no serious adverse effects were reported after the vaccination. Transient vaccine viremia was detected in most of the participants. However, the VSV RNA was not detectable in saliva and urine samples. All participants generated high levels of antibodies against ZEBOV GP assessed by ELISA at 28 days after vaccination and the antibody titers persisted until 6 months. Neutralizing antibody responses were elicited in 85% vaccinees and higher doses of vaccines correlated with higher neutralizing antibody titers. Four phase 1 trials were later launched in Canada, USA, Switzerland, and Gabon [53–56]. The safety and immunogenicity of the vaccine was also evaluated in adolescents (13–17 years) and children (6–12 years) in Gabon [53]. All these trials proved that rVSV-ZEBOV vaccine was well tolerated and could stimulate long-term GP-specific binding antibody as well as neutralizing antibody responses after single immunization. The phase 1b results from the USA supported using 2 x 10^7 PFU dose for vaccination in the further trials [53, 55].

A cluster-randomized phase 3 trial was first launched in Guinea then extended to Sierra Leone in 2015 to evaluate the protective efficacy of rVSV-ZEBOV vaccine [57, 58]. This trial, Ebola ça Suffit! (meaning "Ebola this is enough"), applied a ring vaccination strategy, which was used in the final stage of smallpox eradication. After case confirmation of Ebola virus disease, clusters (ring) of all contacts and contacts of contacts received either immediate or delayed (21 days later) vaccination with single dose of 2×10^7 PFU rVSV-ZEBOV. The incidence of Ebola virus disease was compared in the two trial groups. No case of Ebola virus disease occurred 10 days or more after vaccination among 3775 individuals who received immediate vaccination versus 23 cases among 4507 individuals who received delayed vaccination and never vaccinated in both groups. These results indicated that rVSV-ZEBOV offered estimated 100% protection efficacy against Ebola virus disease in high-risk regions. An individually randomized controlled phase II/III trial, which enrolled healthcare and frontline workers in several most affected districts in Sierra Leone, showed similar results [59].

From multiple clinical trials conducted in different locations, rVSV-ZEBOV has proven itself to be a safe and effective vaccine for human to prevent Ebola virus disease. As a result, rVSV-ZEBOV (under the name Ervebo by Merck) has become the first FDA-approved live viral-vectored vaccine in humans at the end of 2019 [60].

18.2.3.2 VSV-Based SARS-CoV-2 Vaccine

To facilitate the study of entry and neutralization of SARS-CoV-2 in biosafety level 2 laboratory, Case et al. firstly reported generation of replication-competent VSV expressing eGFP as a marker and the spike (S) protein of SARS-CoV-2 [61]. A mutation from cysteine to stop mutation at position 1253 emerged after serial passages, resulting in the deletion of 21 amino acids in the cytoplasmic tail of S protein. This virus, named as VSV-SARS-CoV-2-$S_{\Delta 21}$, propagated more efficiently in Vero cells and was genetically stable within 12 passages. The same or longer deletion (24 amino acids) at the S cytoplasmic tail was described by two other groups [62, 63], suggesting that it may play an important role for viral adaptation to grow in Vero cells.

The efficacy of VSV-vectored SARS-CoV-2 vaccines was firstly evaluated in rodent animals. Intraperitoneal immunization of BALB/c mice or intranasal immunization of human ACE2 transgenic C57BL/6 J mice with 10^6 PFU of VSV-eGFP-SARS-CoV-2 induced high levels of specific antibodies targeting receptor-binding domain of S as well as neutralizing antibodies against SARS-CoV-2 [61]. Two doses of VSV-eGFP-SARS-CoV-2 significantly decreased the viral RNAs in most organs and limited virus-induced lung inflammation after SARS-CoV-2 challenge, indicating that the vaccine offered effective protection [61]. Protection was also observed in mice received the immune sera from VSV-eGFP-SARS-CoV-2 vaccinated mice [61]. An Israel group constructed a similar VSV-based SARS-CoV-2 vaccine without eGFP marker and evaluated vaccine safety and efficacy in hamster model [63]. No weight loss or signs of disease were observed even when the hamsters received intramuscular dose as high as 10^8 PFU, proving the safety of the vaccine. Single-dose vaccination induced high level of neutralizing antibodies and protected hamsters against SARS-CoV-2 challenge, as manifested by less weight loss, reduced lung damage and viral loads. Later, Li et al. compared neutralizing antibody levels elicited by different immunization routes in mouse and non-human primate models [64]. Surprisingly, intramuscular injection of rVSV-SARS-CoV-2 was more immunogenic in mice; however, nAb titers induced by one intranasal shot were much higher than those by the intramuscular injection in macaques. Given that NHPs have a close phylogenetic relationship with humans, the intranasal route may be more suitable for human vaccination. To further enhance the vaccine immunogenicity, a chimeric vaccine was developed by replacing the RBD domain of SARS-CoV spike with that from the SARS-CoV-2 [64]. This vaccine elicited significantly increased nAbs in mice and aged macaques compared with rVSV-SARS-CoV-2 and fully protected mice against SARS-CoV-2 challenge.

During the SARS-CoV-2 pandemic, many variants of concern (VOC) emerged, such as alpha, beta, delta, and omicron. It was reported that the intranasal vaccination of rVSV-SARS2 vaccine protected hamsters against the challenge of alpha and beta variants at 10 days post vaccination [65]. Furthermore, recombinant VSVs carrying the spike of beta variant elicited more potent neutralizing antibody responses than that of original strain in mice and hamsters, indicating that mutated spike could be better immunogen for COVID-19 vaccine [66].

Until now, two VSV-based COVID-19 vaccine candidates have been evaluated in clinical trials. In early 2020, Merck discontinued the development of V590, their COVID-19 vaccine candidate based on recombinant VSV platform, after phase 1 trial (ClinicalTrials.gov-NCT04569786). According to Merck's announcement, V590 generated weaker immune responses than those following natural infection and reported for other COVID-19 vaccines. Another rVSV-SARS-CoV-2-S vaccine, named BriLife developed by Israel Institute for Biological Research, is now in phase 2 trial (ClinicalTrials.gov-NCT04608305). BriLife acquired several mutations during development, including E484D, Q493R, N501Y in the receptor-binding domain, and R685G at the furin cleavage site. The BriLife vaccinees received two intramuscular dose at 28-day interval, and the serum samples collected 14 days after second dose showed neutralizing activities against all the SARS-CoV-2 variants tested, including alpha, beta, gamma, delta, and omicron [67].

References

1. Iverson, L.E., Rose, J.K.: Localized attenuation and discontinuous synthesis during vesicular stomatitis virus transcription. Cell. **23**, 477–484 (1981). https://doi.org/10.1016/0092-8674(81)90143-4
2. Whelan, S.P., Ball, L.A., Barr, J.N., Wertz, G.T.: Efficient recovery of infectious vesicular stomatitis virus entirely from cDNA clones. Proc.Natl.Acad.Sci.USA. **92**, 8388–8392 (1995)
3. Lawson, N.D., Stillman, E.A., Whitt, M.A., Rose, J.K.: Recombinant vesicular stomatitis viruses from DNA. Proc.Natl.Acad.Sci.USA. **92**, 4477–4481 (1995). https://doi.org/10.1073/pnas.92.10.4477

4. Schnell, M.J., Buonocore, L., Kretzschmar, E., Johnson, E., Rose, J.K.: Foreign glycoproteins expressed from recombinant vesicular stomatitis viruses are incorporated efficiently into virus particles. Proc.Natl.Acad.Sci.USA. **93**, 11359–11365 (1996). https://doi.org/10.1073/pnas.93.21.11359
5. Schnell, M.J., Buonocore, L., Whitt, M.A., Rose, J.K.: The minimal conserved transcription stop-start signal promotes stable expression of a foreign gene in vesicular stomatitis virus. J. Virol. **70**, 2318–2323 (1996). https://doi.org/10.1128/JVI.70.4.2318-2323.1996
6. Haglund, K., Forman, J., Kraussllch, H.G., Rose, J.K.: Expression of human immunodeficiency virus type 1 gag protein precursor and envelope proteins from a vesicular stomatitis virus recombinant: high-level production of virus-like particles containing HIV envelope. Virologie. **268**, 112–121 (2000). https://doi.org/10.1006/viro.1999.0120
7. Kretzschmar, E., Buonocore, L., Schnell, M.J., Rose, J.K.: High-efficiency incorporation of functional influenza virus glycoproteins into recombinant vesicular stomatitis viruses. J.Virol. **71**, 5982–5989 (1997). https://doi.org/10.1128/JVI.71.8.5982-5989.1997
8. Kahn, J.S., Schnell, M.J., Buonocore, L., Rose, J.K.: Recombinant vesicular stomatitis virus expressing respiratory syncytial virus (RSV) glycoproteins: RSV fusion protein can mediate infection and cell fusion. Virologie. **254**, 81–91 (1999). https://doi.org/10.1006/viro.1998.9535
9. Johnson, J.E., Schnell, M.J., Buonocore, L., Rose, J.K.: Specific targeting to CD4+ cells of recombinant vesicular stomatitis viruses encoding human immunodeficiency virus envelope proteins. J.Virol. **71**, 5060–5068 (1997). https://doi.org/10.1128/JVI.71.7.5060-5068.1997
10. Johnson, J.E., Rodgers, W., Rose, J.K.: A plasma membrane localization signal in the HIV-1 envelope cytoplasmic domain prevents localization at sites of vesicular stomatitis virus budding and incorporation into VSV virions. Virologie. **251**, 244–252 (1998). https://doi.org/10.1006/viro.1998.9429
11. Roberts, A., Buonocore, L., Price, R., Forman, J., Rose, J.K.: Attenuated vesicular stomatitis viruses as vaccine vectors. J.Virol. **73**, 3723–3732 (1999). https://doi.org/10.1128/JVI.73.5.3723-3732.1999
12. Schnell, M.J., Johnson, J.E., Buonocore, L., Rose, J.K.: Construction of a novel virus that targets HIV-1-infected cells and controls HIV-1 infection. Cell. **90**, 849–857 (1997). https://doi.org/10.1016/s0092-8674(00)80350-5
13. Boritz, E., Gerlach, J., Johnson, J.E., Rose, J.K.: Replication-competent rhabdoviruses with human immunodeficiency virus type 1 coats and green fluorescent protein: entry by a pH-independent pathway. J.Virol. **73**, 6937–6945 (1999). https://doi.org/10.1128/JVI.73.8.6937-6945.1999
14. Garbutt, M., et al.: Properties of replication-competent vesicular stomatitis virus vectors expressing glycoproteins of filoviruses and arenaviruses. J.Virol. **78**, 5458–5465 (2004). https://doi.org/10.1128/jvi.78.10.5458-5465.2004
15. Harty, R.N., Brown, M.E., Hayes, F.P., Wright, N.T., Schnell, M.J.: Vaccinia virus-free recovery of vesicular stomatitis virus. J. Mol. Microbiol. Biotechnol. **3**, 513–517 (2001)
16. Farzani, T.A., Chov, A., Herschhorn, A.: A protocol for displaying viral envelope glycoproteins on the surface of vesicular stomatitis viruses. STAR Protoc. **1**, 100209 (2020). https://doi.org/10.1016/j.xpro.2020.100209
17. Li, W., et al.: Angiotensin-converting enzyme 2 is a functional receptor for the SARS coronavirus. Nature. **426**, 450–454 (2003). https://doi.org/10.1038/nature02145
18. Mercer, J., Lee, J.E., Saphire, E.O., Freeman, S.A.: SnapShot: enveloped virus entry. Cell. **182**, 786-786 e781 (2020). https://doi.org/10.1016/j.cell.2020.06.033
19. Pillay, S., Carette, J.E.: Hunting viral receptors using haploid cells. Annual review of virology. **2**, 219–239 (2015). https://doi.org/10.1146/annurev-virology-100114-055119
20. Carette, J.E., et al.: Ebola virus entry requires the cholesterol transporter Niemann-pick C1. Nature. **477**, 340–343 (2011). https://doi.org/10.1038/nature10348
21. Wong, A.C., Sandesara, R.G., Mulherkar, N., Whelan, S.P., Chandran, K.: A forward genetic strategy reveals destabilizing mutations in the ebolavirus glycoprotein that alter its protease

dependence during cell entry. J.Virol. **84**, 163–175 (2010). https://doi.org/10.1128/JVI.01832-09
22. Jae, L.T., et al.: Virus entry. Lassa virus entry requires a trigger-induced receptor switch. Science. **344**, 1506–1510 (2014). https://doi.org/10.1126/science.1252480
23. Carette, J.E., et al.: Haploid genetic screens in human cells identify host factors used by pathogens. Science. **326**, 1231–1235 (2009). https://doi.org/10.1126/science.1178955
24. Petersen, J., et al.: The major cellular sterol regulatory pathway is required for Andes virus infection. PLoS Pathog. **10**, e1003911 (2014). https://doi.org/10.1371/journal.ppat.1003911
25. Kleinfelter, L.M., et al.: Haploid genetic screen reveals a profound and direct dependence on cholesterol for hantavirus membrane fusion. mBio. **6**, e00801 (2015). https://doi.org/10.1128/mBio.00801-15
26. Raaben, M., et al.: NRP2 and CD63 are host factors for lujo virus cell entry. Cell Host Microbe. **22**, 688–696. e685 (2017). https://doi.org/10.1016/j.chom.2017.10.002
27. Jangra, R.K., et al.: Protocadherin-1 is essential for cell entry by New World hantaviruses. Nature. **563**, 559–563 (2018). https://doi.org/10.1038/s41586-018-0702-1
28. Volland, A., et al.: Heparan sulfate proteoglycans serve as alternative receptors for low affinity LCMV variants. PLoS Pathog. **17**, e1009996 (2021). https://doi.org/10.1371/journal.ppat.1009996
29. Wei, J., et al.: Genome-wide CRISPR screens reveal host factors critical for SARS-CoV-2 infection. Cell. **184**, 76–91 (2021). https://doi.org/10.1016/j.cell.2020.10.028
30. Hansen, J., et al.: Studies in humanized mice and convalescent humans yield a SARS-CoV-2 antibody cocktail. Science. **369**, 1010–1014 (2020). https://doi.org/10.1126/science.abd0827
31. Baum, A., et al.: Antibody cocktail to SARS-CoV-2 spike protein prevents rapid mutational escape seen with individual antibodies. Science. **369**, 1014–1018 (2020). https://doi.org/10.1126/science.abd0831
32. Smith, G.L., Mackett, M., Moss, B.: Infectious vaccinia virus recombinants that express hepatitis B virus surface antigen. Nature. **302**, 490–495 (1983). https://doi.org/10.1038/302490a0
33. Moss, B., Smith, G.L., Gerin, J.L., Purcell, R.H.: Live recombinant vaccinia virus protects chimpanzees against hepatitis B. Nature. **311**, 67–69 (1984). https://doi.org/10.1038/311067a0
34. Vrba, S.M., Kirk, N.M., Brisse, M.E., Liang, Y., Ly, H.: Development and applications of viral vectored vaccines to combat zoonotic and emerging public health threats. Vaccines (Basel). **8** (2020). https://doi.org/10.3390/vaccines8040680
35. Roberts, A., et al.: Vaccination with a recombinant vesicular stomatitis virus expressing an influenza virus hemagglutinin provides complete protection from influenza virus challenge. J. Virol. **72**, 4704–4711 (1998). https://doi.org/10.1128/JVI.72.6.4704-4711.1998
36. Schlereth, B., et al.: Successful mucosal immunization of cotton rats in the presence of measles virus-specific antibodies depends on degree of attenuation of vaccine vector and virus dose. J. Gen.Virol. **84**, 2145–2151 (2003). https://doi.org/10.1099/vir.0.19050-0
37. Grigera, P.R., et al.: Presence of bovine viral diarrhea virus (BVDV) E2 glycoprotein in VSV recombinant particles and induction of neutralizing BVDV antibodies in mice. Virus Res. **69**, 3–15 (2000). https://doi.org/10.1016/s0168-1702(00)00164-7
38. Rose, N.F., et al.: An effective AIDS vaccine based on live attenuated vesicular stomatitis virus recombinants. Cell. **106**, 539–549 (2001). https://doi.org/10.1016/s0092-8674(01)00482-2
39. Rose, N.F., Roberts, A., Buonocore, L., Rose, J.K.: Glycoprotein exchange vectors based on vesicular stomatitis virus allow effective boosting and generation of neutralizing antibodies to a primary isolate of human immunodeficiency virus type 1. J.Virol. **74**, 10903–10910 (2000). https://doi.org/10.1128/jvi.74.23.10903-10910.2000
40. Cooper, D., et al.: Attenuation of recombinant vesicular stomatitis virus-human immunodeficiency virus type 1 vaccine vectors by gene translocations and g gene truncation reduces neurovirulence and enhances immunogenicity in mice. J.Virol. **82**, 207–219 (2008). https://doi.org/10.1128/JVI.01515-07

41. Clarke, D.K., et al.: Synergistic attenuation of vesicular stomatitis virus by combination of specific G gene truncations and N gene translocations. J.Virol. **81**, 2056–2064 (2007). https://doi.org/10.1128/JVI.01911-06
42. Kelley, J.M., Emerson, S.U., Wagner, R.R.: The glycoprotein of vesicular stomatitis virus is the antigen that gives rise to and reacts with neutralizing antibody. J.Virol. **10**, 1231–1235 (1972). https://doi.org/10.1128/JVI.10.6.1231-1235.1972
43. Jones, S.M., et al.: Live attenuated recombinant vaccine protects nonhuman primates against Ebola and Marburg viruses. Nat.Med. **11**, 786–790 (2005). https://doi.org/10.1038/nm1258
44. Geisbert, T.W., et al.: Vesicular stomatitis virus-based vaccines protect nonhuman primates against aerosol challenge with Ebola and Marburg viruses. Vaccine. **26**, 6894–6900 (2008). https://doi.org/10.1016/j.vaccine.2008.09.082
45. Jones, S.M., et al.: Assessment of a vesicular stomatitis virus-based vaccine by use of the mouse model of Ebola virus hemorrhagic fever. J. Infect. Dis. **196**(Suppl 2), S404–S412 (2007). https://doi.org/10.1086/520591
46. Qiu, X., et al.: Mucosal immunization of cynomolgus macaques with the VSVDeltaG/ZEBOVGP vaccine stimulates strong ebola GP-specific immune responses. PLoS One. **4**, e5547 (2009). https://doi.org/10.1371/journal.pone.0005547
47. Feldmann, H., et al.: Effective post-exposure treatment of Ebola infection. PLoS Pathog. **3**, e2 (2007). https://doi.org/10.1371/journal.ppat.0030002
48. Tsuda, Y., et al.: Protective efficacy of a bivalent recombinant vesicular stomatitis virus vaccine in the Syrian hamster model of lethal Ebola virus infection. J. Infect. Dis. **204**(Suppl 3), S1090–S1097 (2011). https://doi.org/10.1093/infdis/jir379
49. Wong, G., et al.: Immunization with vesicular stomatitis virus vaccine expressing the Ebola glycoprotein provides sustained long-term protection in rodents. Vaccine. **32**, 5722–5729 (2014). https://doi.org/10.1016/j.vaccine.2014.08.028
50. Geisbert, T.W., et al.: Vesicular stomatitis virus-based ebola vaccine is well-tolerated and protects immunocompromised nonhuman primates. PLoS Pathog. **4**, e1000225 (2008). https://doi.org/10.1371/journal.ppat.1000225
51. Regules, J.A., et al.: A recombinant vesicular stomatitis virus Ebola vaccine. N.Engl.J.Med. **376**, 330–341 (2017). https://doi.org/10.1056/NEJMoa1414216
52. Agnandji, S.T., et al.: Phase 1 trials of rVSV Ebola vaccine in Africa and Europe. N.Engl. J. Med. **374**, 1647–1660 (2016). https://doi.org/10.1056/NEJMoa1502924
53. Agnandji, S.T., et al.: Safety and immunogenicity of rVSVDeltaG-ZEBOV-GP Ebola vaccine in adults and children in Lambarene, Gabon: a phase I randomised trial. PLoS Med. **14**, e1002402 (2017). https://doi.org/10.1371/journal.pmed.1002402
54. ElSherif, M.S., et al.: Assessing the safety and immunogenicity of recombinant vesicular stomatitis virus Ebola vaccine in healthy adults: a randomized clinical trial. CMAJ. **189**, E819–E827 (2017). https://doi.org/10.1503/cmaj.170074
55. Heppner Jr., D.G., et al.: Safety and immunogenicity of the rVSVG-ZEBOV-GP Ebola virus vaccine candidate in healthy adults: a phase 1b randomised, multicentre, double-blind, placebo-controlled, dose-response study. Lancet Infect. Dis. **17**, 854–866 (2017). https://doi.org/10.1016/S1473-3099(17)30313-4
56. Huttner, A., et al.: The effect of dose on the safety and immunogenicity of the VSV Ebola candidate vaccine: a randomised double-blind, placebo-controlled phase 1/2 trial. Lancet Infect. Dis. **15**, 1156–1166 (2015). https://doi.org/10.1016/S1473-3099(15)00154-1
57. Henao-Restrepo, A.M., et al.: Efficacy and effectiveness of an rVSV-vectored vaccine in preventing Ebola virus disease: final results from the Guinea ring vaccination, open-label, cluster-randomised trial (Ebola ca Suffit!). Lancet. **389**, 505–518 (2017). https://doi.org/10.1016/S0140-6736(16)32621-6
58. Henao-Restrepo, A.M., et al.: Efficacy and effectiveness of an rVSV-vectored vaccine expressing Ebola surface glycoprotein: interim results from the Guinea ring vaccination cluster-randomised trial. Lancet. **386**, 857–866 (2015). https://doi.org/10.1016/S0140-6736(15)61117-5

59. Widdowson, M.A., et al.: Implementing an Ebola vaccine study - Sierra Leone. MMWR Suppl. **65**, 98–106 (2016). https://doi.org/10.15585/mmwr.su6503a14
60. Ollmann Saphire, E.: A Vaccine against Ebola Virus. Cell. **181**, 6 (2020). https://doi.org/10.1016/j.cell.2020.03.011
61. Case, J.B., et al.: Replication-competent vesicular stomatitis virus vaccine vector protects against SARS-CoV-2-mediated pathogenesis in mice. Cell Host Microbe. **28**, 465–474 e464 (2020). https://doi.org/10.1016/j.chom.2020.07.018
62. Dieterle, M.E., et al.: A replication-competent vesicular stomatitis virus for studies of SARS-CoV-2 spike-mediated cell entry and its inhibition. Cell Host Microbe. **28**, 486–496 e486 (2020). https://doi.org/10.1016/j.chom.2020.06.020
63. Yahalom-Ronen, Y., et al.: A single dose of recombinant VSV-ΔG-spike vaccine provides protection against SARS-CoV-2 challenge. Nat. Commun. **11**, 6402 (2020). https://doi.org/10.1038/s41467-020-20228-7
64. Li, H., et al.: Enhanced protective immunity against SARS-CoV-2 elicited by a VSV vector expressing a chimeric spike protein. Signal Transduct. Target. Ther. **6**, 389 (2021). https://doi.org/10.1038/s41392-021-00797-9
65. O'Donnell, K.L., et al.: Optimization of single-dose VSV-based COVID-19 vaccination in hamsters. Front. Immunol. **12**, 788235 (2021). https://doi.org/10.3389/fimmu.2021.788235
66. Ding, L.S., et al.: Growth, antigenicity, and immunogenicity of SARS-CoV-2 spike variants revealed by a live rVSV-SARS-CoV-2 virus. Front. Med (Lausanne). **8**, 793437 (2021). https://doi.org/10.3389/fmed.2021.793437
67. Yahalom-Ronen, Y., et al.: Neutralization of SARS-CoV-2 variants by rVSV-DeltaG-spike-elicited human sera. Vaccines (Basel). **10** (2022). https://doi.org/10.3390/vaccines10020291

The manufacturer's authorised representative in the EU is Springer Nature Customer Service Centre GmbH, Europaplatz 3, 69115 Heidelberg, Germany. If you have any concerns regarding our products, please contact ProductSafety@springernature.com

Printed and bound by CPI Group (UK) Ltd, Croydon, CR0 4YY
30/09/2026
02225244-0009